Raspberry Pi®

A Wiley Brand

Raspberry Pi®

4th Edition

by Sean McManus and Mike Cook

A Wiley Brand

Raspberry Pi® For Dummies®

Published by: **John Wiley & Sons, Inc.**, 111 River Street, Hoboken, NJ 07030-5774, www.wiley.com

Copyright © 2021 by John Wiley & Sons, Inc., Hoboken, New Jersey

Published simultaneously in Canada

For general information on our other products and services, please contact our Customer Care Department within the U.S. at 877-762-2974, outside the U.S. at 317-572-3993, or fax 317-572-4002. For technical support, please visit www.wiley.com/techsupport.

Wiley publishes in a variety of print and electronic formats and by print-on-demand. Some material included with standard print versions of this book may not be included in e-books or in print-on-demand. If this book refers to media such as a CD or DVD that is not included in the version you purchased, you may download this material at http://booksupport.wiley.com. For more information about Wiley products, visit www.wiley.com.

Library of Congress Control Number: 2021941416

ISBN 978-1-119-79682-4 (pbk); ISBN 978-1-119-79686-2 (ebk); ISBN 978-1-119-79687-9 (ebk)

SKY10028353_072221

Contents at a Glance

Table of Contents

Introduction

Raspberry Pi computers are at the forefront of the maker movement, where people make their own inventions using a mixture of traditional craft skills and modern coding and electronics knowledge. They've also given more and more people access to a computer that provides a gateway into programming, electronics, and the world of Linux — the technically powerful (and free) rival to Windows and Mac OS. As a supercheap computer, the Raspberry Pi is also being pressed into service in media centers and as a family computer for games, music, photo editing, and word processing.

You might be a geek who relishes learning new technologies, or you might be someone who wants a new family computer to use with the children. In either case, *Raspberry Pi For Dummies*, 4th Edition, helps you get started with your Raspberry Pi and teaches you about some of the many fun and inspiring things you can do with it.

About This Book

Raspberry Pi For Dummies, 4th Edition, provides a concise and clear introduction to the terminology, technology, and techniques that you need to get the most from your Pi. With this book as your guide, you'll learn how to

>> Set up your Raspberry Pi.

>> Discover and install great free software you can use on your Raspberry Pi.

>> Use the desktop environment to run programs, manage files, surf the web, and view photos.

>> Use the Linux command line to manage your Raspberry Pi and its files.

>> Use the Raspberry Pi as a productivity tool.

>> Edit photos.

>> Play music and video.

>> Create animations and arcade games with the child-friendly Scratch programming language.

» Write your own games and other programs using the Python programming language.

» Compose music by programming with Sonic Pi.

» Get started with electronics, from an introduction to soldering to the design and creation of electronic projects controlled by the Raspberry Pi.

Incidentally, within this book, you may note that some web addresses break across two lines of text. If you're reading this book in print and want to visit one of these web pages, simply key in the web address exactly as it's noted in the text, pretending as though the line break doesn't exist. If you're reading this as an ebook, you've got it easy — just click or tap the web address to be taken directly to the web page.

Foolish Assumptions

Raspberry Pi For Dummies, 4th Edition, is written for beginners, by which we mean people who have never used a similar computer. However, we do have to make a few assumptions in writing this book, because we wouldn't have enough space for all its cool projects if we had to start by explaining what a mouse is! Here are our assumptions:

» **You are familiar with other computers, such as Windows or Apple computers.** In particular, we assume that you're familiar with using windows, icons, and the keyboard and mouse, and that you know the basics of using your computer for things like browsing the Internet, installing software, or copying files.

» **The Raspberry Pi is not your only computer.** At times, you'll need to have access to another computer — for example, to create your SD or microSD card for the Pi. (See Chapter 2.) When it comes to networking, we assume you already have a router set up with an Internet connection and a spare port that you can plug the Raspberry Pi into if you're using a wired connection.

» **The Raspberry Pi is your first Linux-based computer.** If you're a Linux ninja, this book still gives you a solid reference on the Raspberry Pi and the version of Linux it uses, but no prior Linux knowledge is required.

» **You share our excitement.** The Raspberry Pi can open up a world of possibilities to you!

Other than these assumptions, we hope this book is approachable for everyone. The Raspberry Pi is being adopted in classrooms and youth groups, and this book is a useful resource for teachers and students. The Raspberry Pi is also finding its way into many homes, where people of all ages (from children to adults) are using it for education and entertainment.

Icons Used in This Book

If you've read other *For Dummies* books, you know that they use icons in the margin to call attention to particularly important or useful ideas in the text. In this book, we use four such icons:

TIP

The Tip icon highlights expert shortcuts or simple ideas that can make life easier for you.

REMEMBER

Although we'd like to think that reading this book is an unforgettable experience, we've highlighted some points that you might want to particularly commit to memory. They're either important takeaways, or they're fundamental to the project you're working on.

WARNING

As you would do on the road, slow down when you see a Warning icon. It highlights an area where things could go wrong.

TECHNICAL
STUFF

Arguably, the whole book talks about technical stuff, but this icon highlights something that's *particularly* technical. We've tried to avoid unnecessary jargon and complexity, but some background information can give you a better understanding of what you're doing, and sometimes we do need to get quite techy, given the sophistication of the projects you're doing. Paragraphs highlighted with this icon might be worth rereading, to make sure you understand, or you might decide that you don't need to know that much detail. It's up to you!

Beyond the Book

In addition to what you're reading right now, this book comes with a free access-anywhere Cheat Sheet with tips on installing software and using Scratch. To get this Cheat Sheet, simply go to www.dummies.com and type **Raspberry Pi Dummies Cheat Sheet** in the Search box.

Also be sure to check out this book's companion website (www.dummies.com/go/raspberrypifd4e), where you can download the code listings that appear throughout this book.

Both of us maintain our own personal websites too, which contain some additional information on the Raspberry Pi. Mike's is at www.thebox.myzen.co.uk/Raspberry/Punnet.html, and Sean's is at www.sean.co.uk.

Where to Go from Here

It's up to you how you read this book. It's been organized to take you on a journey from acquiring and setting up your Raspberry Pi to learning the software that comes with it, and from writing your own programs to finally creating your own electronics projects. Some chapters build on knowledge gained in earlier chapters, especially the sections on Scratch and Python — and all of Part 5.

We understand, though, that some projects or topics might interest you more than others, and you might need help in some areas right now. When a chapter assumes knowledge from elsewhere, we include cross-references to help you quickly find what you might have missed. We also include some signposts to future chapters, so you can skip ahead to a later chapter if it provides the quickest answer for you.

If you haven't set up your Pi yet, start with Part 1. If you have your Pi up and running, Part 2 shows you how to use the software on it. Part 3 covers productivity, creativity, and entertainment software. To flex your programming muscles, perhaps for the first time, read Part 4. You can learn Scratch, Python, or Sonic Pi here, and feel free to start with any one of those languages. The Python chapters provide a good foundation for Part 5, where you can start building your own electronics projects.

1

Setting Up Your Raspberry Pi

Get to know the Raspberry Pi and what other equipment you will need to be able to use it.

Download the Linux operating system and prepare a microSD card for use on your Raspberry Pi.

Connect your Raspberry Pi to the power, keyboard, mouse, and screen.

Install and test the Raspberry Pi Camera Module.

Change the settings on your Raspberry Pi.

Chapter 1

Introducing the Raspberry Pi

The Raspberry Pi is perhaps the most inspiring computer available today. Although most of the computing devices being used (including phones, tablets, and game consoles) are designed to stop people from tinkering with them, the Raspberry Pi is exactly the opposite. It invites you to prod it, play with it, and create with it. It comes with the tools you need to start creating your own software (or *programming*), and you can connect your own electronic inventions to it. Some models are cheap enough that breaking them won't break the bank, so you can experiment with confidence.

Lots of people are fired up about the Raspberry Pi's potential, and they're discovering exciting new ways to use it. Dave Akerman (www.daveakerman.com) and friends attached one to a weather balloon and sent it nearly 40 kilometers high to take pictures of the Earth from near space using a webcam. (You can read about Dave's ballooning project in Chapter 20.)

Professor Simon Cox and his team at the University of Southampton connected 64 Raspberry Pi boards to build an experimental supercomputer, held together by Lego bricks. In the supercomputer (see Figure 1-1), the Raspberry Pis work together to solve a single problem. The project has been able to cut the cost of a supercomputer from millions of dollars to thousands or even hundreds of dollars, making supercomputing much more accessible to schools and students. Others have also experimented with combining the processing power of multiple Pis. There's even an off-the-shelf kit you can use to combine four Raspberry Pi Zeros with a full-size Raspberry Pi (the Cluster HAT from Pimoroni) so that you can experiment with running programs across multiple Pis at the same time.

FIGURE 1-1: Two of the Raspberry Pi boards used in the University of Southampton's supercomputer, with the rest of the supercomputer in the background.

Courtesy of Simon Cox and Glenn Harris, University of Southampton.

The Pi is also being used to make fitness gadgets, gaming devices, electric skateboards, and much more, as you discover in Chapter 20.

Although those projects are grabbing headlines, another story is less visible but more important: the thousands of people of all ages who are taking their first steps in computer science, thanks to the Raspberry Pi.

Both of the authors of this book used computers in the 1980s, when the notion of a home computer first became a reality. Back then, computers were less friendly than they are today. When you switched them on, you were faced with a flashing cursor and had to type something in to get it to do anything. As a result, though, a whole generation grew up knowing at least a little bit about how to give the computer commands, and how to create programs for it. As computers started to use mice and windows, people didn't need those skills any more, and they lost touch with them.

Eben Upton, designer of the Raspberry Pi, noticed the slide in skill levels when he was working at Cambridge University's computer laboratory in 2006. Students applying to study computer science started to have less experience with programming than students of the past did. Upton and his university colleagues hatched the idea of creating a computer that would come supplied with all the tools needed to program it — and would sell for a target price of $25 (about £20). It had to be able to do other interesting things, too, so that people were drawn to use it, and it had to be robust enough to survive being pushed in and out of school bags hundreds of times.

That idea started a six-year journey that led to the Raspberry Pi you probably have on your desk you as you read this book. It was released in February 2012, and sold half a million units by the end of the quarter. By July 2017, there were more than 14 million Raspberry Pis in homes, schools, and workplaces, 10 million of them made in the UK. More than 30 million Raspberry Pi computers have now been sold. It is, by a large margin, the best-selling British computer of all time.

Introducing the Raspberry Pi Range

Over the years, the Raspberry Pi has evolved, increasing its memory, improving its performance, and adding features. So which one should you get? Here's an overview designed to help you decide.

Raspberry Pi 4 Model B

This model is a circuit board with components and sockets stuck on it, as shown in Figure 1-2. In an age when most computing devices are sleek and shiny boxes, the spiky Pi, with tiny codes printed in white all over it, seems alien. That's a big part of its appeal, though: Many of the cases you can buy for the Raspberry Pi are transparent because people love the look of it.

The Raspberry Pi 4 is the latest Raspberry Pi board. It features the following:

>> Up to 8GB of memory

>> Four USB ports (two USB 2 ports and two higher-speed USB 3 ports)

>> Built-in Wi-Fi and Bluetooth and a Gigabit Ethernet port for a wired Internet or network connection

>> A headphones-style audio-out socket

>> 40 general-purpose input/output (GPIO) pins, which you can use to connect your own electronics projects or specially designed add-ons (see Chapter 21)

FIGURE 1-2:
The Raspberry Pi
4 Model B
(center), Model A+
(top right),
.and Pi Zero W
(top left).

>> Support for two monitors at resolutions of up to 4K

>> Compatibility with the Raspberry Pi Camera Module

>> Power over Ethernet (PoE) support when used with the Raspberry Pi PoE HAT, which enables you to use your Ethernet cable for both networking and powering your Pi

Like previous Pi models, the Raspberry Pi 4 is about the size of a deck of cards. As with any current Raspberry Pi, it uses a microSD card for storage. Its price is around $35 for 2GB of memory or $75 for 8GB of memory.

The Raspberry Pi Desktop Kit is also available, which includes the accessories you'll need, except for the monitor.

The Raspberry Pi 4 is our recommendation for the most powerful budget-friendly Raspberry Pi. You may be able to use it with your own keyboard and mouse to save money. The GPIO pins are great for electronics projects.

TECHNICAL
STUFF

It's called the Model B, incidentally, as a tribute to the BBC Microcomputer that was popular in the UK in the 1980s. It's sobering to think that the BBC Micro cost about ten times the price of a Raspberry Pi, which, thanks to 40 years of progress in computer science, has more than 15,600 times more memory.

Raspberry Pi 400

The Raspberry Pi 400 (see Figure 1-3) takes even more inspiration from the classic computers of the '80s by building the Raspberry Pi 4 computer into a computer keyboard. It makes the whole setup much more compact, because you don't have the separate Pi unit on the table, with a cable going to the keyboard.

FIGURE 1-3:
The Raspberry Pi 400 hides the computer inside the keyboard.

There are performance improvements, too. The Raspberry Pi 400 is faster than the Raspberry Pi 4, and it's designed with passive cooling built in.

The Raspberry Pi 400 is a white keyboard, with all the sockets on the back of it. It features the following:

>> 4GB of memory.

>> Three external USB ports (one USB 2 port and two higher-speed USB 3 ports). This is fewer than the four ports you get on a Raspberry Pi 4. The fourth port is used to connect the keyboard inside the case.

>> Built-in Wi-Fi and Bluetooth and a Gigabit Ethernet port for a wired Internet or network connection.

- » 40 GPIO pins, but these are on the back of the case, not on the top surface. You'll need to use an extension cable or board to use the pins easily and to use add-on boards (see Chapter 21). Although add-on boards can be connected directly, few will work well because their top surface will face away from you.

- » Support for two monitors at resolutions of up to 4K.

- » No compatibility with the Raspberry Pi Camera Module. You can use a USB camera, as you can on any Raspberry Pi computer.

There is no audio out socket, so you'll need to pass audio through your monitor.

The Raspberry Pi 400 costs $70. The Raspberry Pi 400 Personal Computer Kit adds the accessories you'll need, except for the monitor. The Raspberry Pi 400 is a fantastic value, but it's more expensive than the bare board. We recommend the Raspberry Pi 400 if your budget will bear it and you plan to use the Raspberry Pi as a desktop computer. For electronics projects, we find the bare board easier to use.

TIP

The official Raspberry Pi keyboard and the Raspberry Pi 400 look the same. If you have both on your desk, put a sticker on one of them; otherwise, you'll waste time trying to use the wrong one!

Raspberry Pi 3 Model A+

The Model A+ is a cut-down bare-board Raspberry Pi. It's useful for projects that need lower power consumption — typically battery-based projects. It is suitable for robots and projects in remote locations, where a wired electricity supply isn't viable and batteries must be used instead.

It features the following:

- » 512MB of memory

- » One USB 2 port

- » Built-in Wi-Fi and Bluetooth

- » A headphones-style audio-out socket

- » 40 GPIO pins

- » Compatibility with the Raspberry Pi Camera Module

This model has a price of $20. The Model A+ is slightly shorter on the long side than the Raspberry Pi 3, measuring 2½ inches by 2 inches.

Raspberry Pi Zero

The Raspberry Pi Foundation astounded everyone when it gave the Raspberry Pi Zero computer away with the print edition of its magazine *The MagPi*. We'd seen cover-mounted CDs and even tapes long ago, but never a computer before.

There are three models: Raspberry Pi Zero, Raspberry Pi Zero W (adding wireless networking), and Raspberry Pi Zero WH (adding wireless networking and GPIO pins).

The Raspberry Pi Zero family features the following:

>> A lightweight, smaller board measuring just 2½ inches by 1 inch.

>> A single-core 1 GHz processor. This is less powerful than the bigger boards. The Model B and A+ are quad-core, which means there are four processing units inside the chip that can all work at the same time. The quad-core processors run at a higher frequency, too. Here, you get a single core running at a lower frequency.

>> 512MB of memory.

>> One Micro USB port.

>> Built-in Wi-Fi and Bluetooth, only on the Raspberry Pi Zero W and Zero WH.

>> 40 GPIO pins, only on the Raspberry Pi Zero WH. On other models, you can solder your own pins.

>> Compatibility with the Raspberry Pi Camera Module, only on the Raspberry Pi Zero W and Zero WH.

You'll also need a converter for the Mini HDMI socket, and for the Micro USB socket, so you should expect to spend a bit more than the price of the Pi (and have a bit more complexity in your setup). Billed as the $5 computer, the Raspberry Pi Zero has at times been difficult to get hold of, which is perhaps not surprising given the phenomenal demand for it.

The Raspberry Pi Zero is great for compact electronics projects that don't need the performance of a Model B or Model A+.

Older models

Of course, the older Raspberry Pis are still out there. Recent models usually remain in production while there is demand, and you can buy secondhand versions online from websites such as eBay. Generally speaking, the newer the model, the faster its performance. Memory upgrades have made a difference, as well as the use of more powerful processors, as the Pi has evolved. There are plenty of uses for the Pi that don't need especially fast performance, though, so you might find that an older Pi is perfect for your project. If you want to support the Raspberry Pi Foundation while buying cheaper, secondhand boards, you can donate to the foundation online.

The older models are described in this list:

>> **Raspberry Pi 1 Model B with 256MB memory:** Although it's called Model B, this was the first Raspberry Pi to be released, in February 2012. The Raspberry Pi Model B features an Ethernet connection for the Internet and two USB ports. It uses an SD card for storage.

>> **Raspberry Pi 1 Model B with 512MB memory:** Released in October 2012, the Raspberry Pi Model B had twice the memory capacity. This improved the speed of some software, especially applications that used images heavily.

>> **Raspberry Pi 1 Model A:** The Model A, released in February 2013, is a stripped-down version of the Model B. It has just one USB port and doesn't have an Ethernet port for connecting to the Internet. It has 256MB of memory.

>> **Raspberry Pi 1 Model B+:** The Model B+, released in July 2014, has been described by the Raspberry Pi Foundation as "the final evolution of the original Raspberry Pi." It runs all the same software as the previous versions of the Raspberry Pi, but it has four USB ports, more GPIO pins for connecting electronics projects to the Pi, and lower power consumption and better audio than the Model B. In common with the Model B, it has 512MB of memory. Although all previous versions use SD cards for data storage, the Model B+ introduced the smaller microSD cards, which are now standard on the Raspberry Pi.

>> **Raspberry Pi 2 Model B:** Launched in February 2015, this model doubled the memory on the Model B+ to 1GB. It increased performance, compared to the Model B+, while retaining its physical features. Over the years the Pi's performance has been improved through new software releases as well as updates to the hardware. The Pi 2 represents an immediately noticeable speed-up, compared to the Model B+.

>> **Raspberry Pi 3 Model B:** Launched in February 2016, this model has a new 64-bit processor, which means it can handle data in bigger chunks than the previous 32-bit processor. The Raspberry Pi 3 Model B is 50 percent to 60 percent faster than the Raspberry Pi 2 Model B when working in 32-bit mode.

>> **Raspberry Pi 3 Model B+:** Launched in March 2018, this model has a faster processor and improved networking speeds. It introduced support for PoE, which enables the Raspberry Pi to be powered through the Ethernet cable. You'll need to add the Raspberry Pi PoE HAT accessory.

REMEMBER

If you're using anything earlier than the Model B+, you'll need full-size SD cards (not microSD) for storage, and you'll only have 26 GPIO pins to play with. Current add-ons are unlikely to be compatible with the early boards, so check their requirements before you buy.

Many of the projects in this book will work on older Raspberry Pi models (indeed, they first appeared in previous editions of this book when those models were the latest thing). But for best performance, we recommend using a current model, if possible.

WHAT'S THE RASPBERRY PI COMPUTE MODULE?

You'll also see the Raspberry Pi Compute Module in the online stores alongside the Raspberry Pi, but this is something quite different.

The Compute Module, or C for short, is designed for industrial use and intended to be built into a product you're manufacturing. The modules tend to follow the release of the main Raspberry Pi models. There are also light versions available that correspond to the Model A of the Raspberry Pi. They're built on a SODIMM board, which is what is sometimes used for PC memory modules. You're supposed to design your own board to plug the Compute Module into, but a development kit is available with a C module and an example motherboard containing all the normal plug-in connectors (see www.raspberrypi.org/products/compute-module-development-kit-2). Note, however, that this is an expensive way to buy what is otherwise a normal Raspberry Pi. Currently, the Raspberry Pi Compute Module 4 is the latest one, but the dev kit uses the Compute Module 3.

We only mention the Compute Module here in case you wonder what it is: It's not covered further in this book, and it's almost certainly not what you want to buy for your first Raspberry Pi.

RASPBERRY PI PICO: A MICROCONTROLLER, NOT A COMPUTER

The Raspberry Pi Pico is a radical new departure for the Raspberry Pi Foundation. Whereas previous devices were general-purpose computers, Raspberry Pi Pico is a microcontroller. A microcontroller is usually built into a device that does one job, such as a heating system or a microwave oven.

You can use the Raspberry Pi Pico for your electronics projects. You program it by connecting it to a computer. It's similar to the Arduino, which you might have heard of, but the Pico uses the Raspberry Pi Foundation's own custom chip.

The big advantage of a microcontroller is that there is no operating system to get in the way of things, so you can get precise control over the signals coming from its pins. This is important for things like audio generation and motor/servo control.

The Raspberry Pi Pico can be programmed using either MicroPython or C, which are both programming languages. (A programming language is a way of giving instructions to a computer or computing device – Part 4 introduces you to some programming languages). MicroPython is a version of Python optimized for running on microcontrollers. There are a few differences in some instructions, but MicroPython mostly looks the same as Python. You can program a Pico using Thonny, a Python programming tool available in Raspberry Pi OS. You get the option of saving your code into the Pico's memory or your computer. Any code saved into a file called `main.py` will run automatically when power is applied to the Pico, independently of whether you have a computer attached.

Programming a Pico in C, however, is not for the fainthearted. It requires a long process to prepare the C code for compiling or the use of a complex piece of software. We expect it to get easier, but at the moment we would recommend MicroPython instead.

The Raspberry Pi Pico is extremely cheap: It costs just $4, and it doesn't need an additional microSD card for storage.

You can find more information on the Raspberry Pi Pico in Chapter 17, but our focus in this book is on the Raspberry Pi computers and not the microcontroller. When we say "the Raspberry Pi," we're referring to the computers.

Figuring Out What You Can Do with a Raspberry Pi

The Raspberry Pi is a fully featured computer, and you can do almost anything with it that you can do with a desktop computer.

Instead of running Windows or macOS, the Raspberry Pi uses an operating system called Linux. It's a leading example of open source, a completely different philosophy to the commercial software industry. Rather than being created within the heavily guarded walls of a company, with its design treated as a trade secret, Linux is built by companies and expert volunteers working together. Anyone is free to inspect and modify the source code (a bit like the recipe) that makes it work. You don't have to pay to use Linux, and you're allowed to share it with other people, too.

You probably won't be able to run the software you have on your other computers on your Raspberry Pi. It won't run Windows or Mac software, and not all Linux software works on the Raspberry Pi. But a lot of Linux software that is compatible with the Raspberry Pi is available and is free of charge.

The Raspberry Pi has a graphical windows desktop to start and manage programs (see Chapter 4) as well as a shell for accepting text commands (see Chapter 5). You can use it for browsing the Internet (see Chapter 4), for word processing and spreadsheets (see Chapter 6), or for editing photos (see Chapter 7). You can use it for playing back music or video (see Chapter 8) or for playing games (see Chapter 19). You can use the built-in software to write your own music, too (see Chapter 14). It's the perfect tool for homework, but it's also a useful computer for writing letters, managing your accounts, and paying bills online.

The Raspberry Pi is at its best, however, when it's being used to learn how computers work, and how you can create your own programs or electronics projects using them. It comes with Scratch (see Chapter 9), a visual programming language that enables people of all ages to create their own animations and games while learning some of the core concepts of computer programming along the way.

It also comes with Python (see Chapter 11), a professional programming language used by YouTube, Google, and Industrial Light & Magic (the special effects gurus for the *Star Wars* films), among many others.

It has GPIO pins on it that you can use to connect up your own circuits to the Raspberry Pi, so you can use your Raspberry Pi to control other devices and to receive and interpret signals from them. In Part 5, we show you how to build some electronic projects controlled by the Raspberry Pi. In Chapter 21, we show you some add-ons you can connect to the GPIO pins.

Getting Your Hands on a Raspberry Pi

One of the great things about the Raspberry Pi is that it's established a community of businesses that have created products for it, or have shared in its success by selling it. You can now buy the Raspberry Pi from a wide range of electronics companies for hobbyists. Global retailers include Pimoroni (www.pimoroni.com), The Pi Hut (https://thepihut.com), and Adafruit (www.adafruit.com). It's also available from the Raspberry Pi's distributors, RS Components (www.rs-components.com) and Element14 (www.element14.com).

You might also be able to buy it from your local computer or electronics store, although you'll probably find it's only available as part of a kit there. Shops often bundle the Raspberry Pi with other items you need to use it. It can be convenient to get everything at once, but it might not represent the cheapest way to get started.

Determining What Else You Need

The creators of Raspberry Pi have stripped costs to the bone to enable you to own a fully featured computer for less than $35, so you'll need to scavenge or buy a few other bits and pieces in order to use your Pi. We say *scavenge* because the things you need are exactly the kind of things many people have lying around their house or garage already, or can easily pick up from friends or neighbors. In particular, if you're using a Raspberry Pi as your second computer, you probably have most of the peripherals you need.

WARNING

Not all devices are compatible. In particular, incompatible USB hubs, keyboards, and mice can cause problems that are hard to diagnose. USB hubs that feed power back into your Raspberry Pi through the Pi's USB port (known as *backpowering*) could potentially cause damage to the Raspberry Pi if they feed in too much power.

A list of compatible and incompatible devices is maintained at https://elinux.org/RPi_VerifiedPeripherals, and you can check online reviews to see whether others have experienced difficulties using a particular device with the Raspberry Pi.

TIP

If you're buying new devices, you can minimize the risk by buying recommended devices from Raspberry Pi retailers.

In any case, you should set a little bit of money aside to spend on accessories. The Raspberry Pi is inexpensive, but buying a keyboard, mouse, USB hub, and cables can easily double or triple your costs, and you may have to resort to that if what you have on hand turns out not to be compatible.

The following sections offer a roundup of what else you may need.

Essentials

There are a few things that are essential to get your Raspberry Pi up and running:

» **Monitor:** The Raspberry Pi has a high-definition video feed and uses an HDMI (high-definition multimedia interface) or Micro HDMI connection for it. If your monitor has an HDMI socket, you can connect the Raspberry Pi directly to it. If your monitor does not support HDMI, it probably has a DVI socket, and you can get a simple and cheap converter that enables you to connect an HDMI cable to it. Older VGA (video graphics array) monitors require a device to convert the HDMI signal into a VGA one. If you're thinking of buying a converter, check online first to see whether it works with the Raspberry Pi. A lot of cheap cables are just cables, when what you need is a device that converts the signal from HDMI format to VGA, not one that just fits into the sockets on the screen and your Raspberry Pi. These converters can be quite expensive, so Gert van Loo has designed a device that uses the Raspberry Pi's GPIO pins to connect to a VGA monitor. He's published the design specs so that anyone can build one, and sell it if they want to, too. Take a look at eBay if you need one, and you might well find what you need. For more information, check out https://github.com/fenlogic/vga666. (If your monitor is connected using a blue plug and the connector has three rows of five pins in it, it's probably a VGA monitor.)

» **TV:** You can connect your Raspberry Pi to a high-definition TV using the HDMI socket and should experience a crisp picture. If you have an old television in the garage, you can also press it into service for your Raspberry Pi. The Pi can send a composite video signal, so it can use a TV as its display. When we tried this, it worked but the text lacked definition, which made it difficult to read. You'll need to get a cable with the right connector to fit your Pi: The original Model A and Model B have a dedicated RCA video socket, but current models use the headphone socket for RCA video output, too.

» **USB keyboard and mouse:** The Raspberry Pi only supports wired USB keyboards and mice. If you're still using ones with PS/2 connectors (round rather than flat), you may be able to use a PS/2 to USB adapter. Official Raspberry Pi keyboards and mice are available with an attractive white and red design. You can use Bluetooth devices, but you'll need to use a wired keyboard and mouse to set them up.

TIP

When the Raspberry Pi behaves unpredictably, it can be because the keyboard is drawing too much power, so avoid keyboards with too many flashing lights and features.

» **SD card or microSD card:** The Raspberry Pi doesn't have a hard drive built into it, so it uses a microSD card (current models) or SD card (older models, earlier than the Model B+) as its main storage. You probably have some SD cards that you use for your digital camera, although you might need to get a higher-capacity one. We recommend a 16GB card as a minimum for Raspberry Pi OS, but you can use a 4GB card if you use a media center operating system (OS) like LibreELEC (see Chapter 8 for a guide to LibreELEC). SD and microSD cards have different class numbers that indicate how fast you can copy information to and from them. You will be fine with a Class 6 or higher. If you buy an official Raspberry Pi kit, it includes a microSD card with Raspberry Pi OS already installed on it.

Note: In this book, when we say microSD card, we also mean SD card if that's what you're using. If we're talking about something that's different for SD cards, we tell you.

» **SD or microSD card writer:** Many PCs today have a slot for SD or microSD cards, so you can easily copy photos from your camera to your computer. If yours doesn't, you might want to consider getting an SD or microSD card writer to connect to your computer. You can use it to copy software to an SD card for use with your Raspberry Pi, but you won't be able to use it to copy files from your Raspberry Pi to a Windows computer. You can also use the card writer to create a backup copy of your Raspberry Pi's files and software. (You can read about making back-ups in Chapter 4.)

» **Power supply:** To power your Raspberry Pi, you need to use a 5V power supply. The Raspberry Pi 4 and Raspberry Pi 400 use a USB-C connector, and earlier models use a USB-C Micro USB connector. Although you may have mobile phone and tablet chargers that fit, many of them can't deliver enough current (up to 2,500 milliamperes for a Raspberry Pi 3 Model A+, and up to 3,000 milliamperes for Raspberry Pi 4), which can make the Raspberry Pi perform unreliably. It's worth checking to see whether you have a 5V charger that may do the job (it should say on it how much current it provides), but for best results, we recommend buying a compatible charger from the same company that you buy your Raspberry Pi from. There is an official Raspberry Pi 4 power supply available, which has plug styles for the United States, Canada, United Kingdom, Australia, New Zealand, Europe, India, and China.

Don't try to power the Pi by connecting its power port to the USB port on your PC with a cable, because your computer probably can't provide enough power for your Pi. You can also power the Pi through the GPIO pins, but you could damage the Raspberry Pi if there is a spike in current or the wrong voltage is applied. If you want to provide power through the GPIO pins, a safer approach is to use a hardware-attached-on-top (HAT) device designed to sit on the GPIO pins and provide the consistent power you need while protecting the Pi underneath. For portable applications, you can power the Raspberry Pi using a battery pack designed for mobile phone charging. The Raspberry Pi Foundation advises that you should only use batteries to power your Raspberry Pi if you know what you're doing, because there's a risk of damaging your Raspberry Pi. There is an official Raspberry Pi PoE HAT if you want to power your Pi through an Ethernet cable.

For more details on the power requirements of various Raspberry Pi models, consult the FAQ at www.raspberrypi.org/documentation/faqs.

>> **Cables:** You'll need cables to connect it all up, too. In particular, you need an HDMI cable (if you're using an HDMI or DVI monitor), an HDMI-to-DVI adapter (if you're using a DVI monitor), an RCA cable (if you're connecting to an older TV), an audio cable (if you're connecting the audio jack to your stereo), and an Ethernet cable (for networking on models with an Ethernet port). The Raspberry Pi 4 and 400 use Micro HDMI connections, so you'll need a cable that connects Micro HDMI to (normal) HDMI for your monitor, or an adapter. Note that the Raspberry Pi 2 and later (including Raspberry Pi 4) send the RCA video signal through a 3.5mm jack (headphone socket). Earlier models had a dedicated RCA socket. You need a different cable, depending on which version of the Pi's design you have, if you plan to use RCA. If you have a Raspberry Pi Zero, you'll need a converter for the Mini HDMI socket and for the Micro USB socket (see Figure 1-4). You can get these cables from an electrical components retailer, and you may be able to buy them at the same time as you buy your Raspberry Pi. Any other cables you need (for example, to connect to PC speakers or a USB hub) should come with those devices.

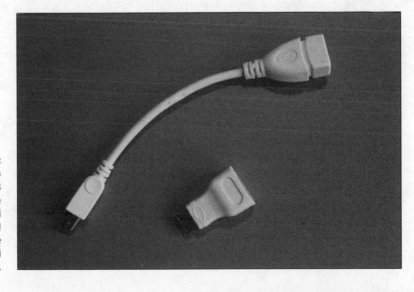

FIGURE 1-4: The Micro USB–to–USB converter cable and the Mini HDMI–to–HDMI converter for the Raspberry Pi Zero.

Optional extras

There are a few additional items you may want to get for your Raspberry Pi. They can make your Raspberry Pi easier to use and enable new applications.

» **USB hub:** The Raspberry Pi has one, two, or four USB sockets (depending on the model you get). Consider using a powered USB hub, for two reasons. Firstly (and especially if you have a Model A, A+, B, or Zero), you're going to want to connect other devices to your Pi at the same time as your keyboard and mouse, which need two sockets. And secondly, a USB hub provides external power to your devices and minimizes the likelihood of experiencing problems using your Raspberry Pi, especially if connecting relatively power-intensive devices such as hard drives. Make sure your USB hub has its own power source, independent of the Raspberry Pi.

» **External hard drive:** If you want lots of storage, perhaps so that you can use your music or video collection with the Raspberry Pi, you can connect an external hard drive to it over USB. You'll need to connect your hard drive through a powered USB hub, or use a hard drive that has its own external power source.

» **Raspberry Pi Camera:** The Raspberry Pi has stimulated entrepreneurs to create all kinds of add-ons for it, but the Camera Module is a product that originated at the Raspberry Pi Foundation. This fixed-focus camera can be used to shoot HD video and take still photos. The standard camera has 8-megapixel resolution, and the Raspberry Pi High Quality Camera offers 12-megapixel resolution. There is also a version of the standard camera without an infrared filter (the PiNoIR Camera), which can be used for wildlife photography at night or weird special effects by day.

» **Speakers:** Raspberry Pis (excluding the Pi 400) have a standard audio out socket, compatible with headphones and PC speakers that use a 3.5mm audio jack. You can plug headphones directly into it, or use the audio jack to connect to speakers, a stereo, or a TV. If you're using a TV or stereo for sound, you can get a cable that connects the 3.5mm audio jack and the audio input(s) on your television or stereo. You won't always need speakers: If you're using an HDMI connection, the audio is sent to the screen with the video signal, so you won't need separate speakers. If you're using a DVI monitor, you can get an HDMI-to-DVI adapter that includes audio extraction, so you can connect the audio separately. Some adapters can also convert from HDMI to VGA, with sound extracted separately.

» **Case:** It's safe to operate your Raspberry Pi as is, but many people prefer to protect it from spills and precariously stacked desk clutter by getting a case for it. The Pibow Coupe (`https://shop.pimoroni.com/collections/pibow`) is one of the most attractively designed cases, assembled from layers of colored plastic (see Figure 1-5). It's designed by Paul Beech, who designed the Raspberry Pi logo. There are also official red-and-white cases for current Raspberry Pi models. The case for the Pi Zero includes three different tops, so you can either seal it, leave a camera hole, or have access to the GPIO pins. You don't have to buy a case, though. You can go without or make your own using cardboard or Lego bricks. Whatever case you go with, make sure you can still access the GPIO pins so that you can experiment with connecting your Pi to electronic circuits and try the projects in Part 5 of this book.

FIGURE 1-5:
The Pibow Coupe case on the Raspberry Pi 4.

>> **Raspberry Pi 4 Case Fan:** If you're really pushing the performance of your Raspberry Pi 4, you might find it gets a bit hot. The Raspberry Pi 4 Case Fan (see Figure 1-6) is an official accessory that fits inside the official Raspberry Pi case. It connects to your GPIO pins, and the fan spins to keep air flowing through the case. It's useful for power users, but most people won't need one.

FIGURE 1-6:
The Raspberry Pi
4 Case Fan.

IN THIS CHAPTER

» Introducing Linux

» Using Raspberry Pi Imager to set up
your microSD card

» Choosing an operating system for
your Raspberry Pi

Chapter **2**

Downloading the Operating System

B efore you can do anything with your Raspberry Pi, you need to provide it
with an operating system (OS). The operating system software enables you
to use the computer's basic functions and looks after activities such as
managing files and running applications, like word processors or web browsers.
Those applications use the operating system as an intermediary to talk to the
hardware, and they won't work without it. This concept isn't unique to the
Raspberry Pi. On your laptop, the operating system might be Microsoft Windows
or macOS. On iPads it's iPadOS, on iPhones it's iOS, and on other devices it might
be Android.

In this chapter, we introduce you to Linux, the operating system most frequently
used on the Raspberry Pi, and we show you how to create a microSD card with an
operating system on it. You'll need to use another computer to set up the microSD
card. It doesn't matter whether you use a Windows, macOS, or Linux machine, but
the computer needs to be able to write to microSD cards, and must have a connection to the Internet.

Introducing Linux

The operating system used on the Raspberry Pi is GNU/Linux, or often just Linux. The Raspberry Pi might be the first Linux computer you've used, but the operating system has a long and honorable history.

Richard Stallman created the GNU Project in 1984 with the goal of building an operating system that users were free to copy, study, and modify. Such software is known as *free software,* and although this software is often given away, the ideology is about free as in "free speech" rather than free as in "free beer." Thousands of people have joined the GNU Project, creating software packages that include tools, applications, and even games. Stallman aimed to make his operating system compatible with Unix, an operating system that was created by AT&T's Bell Labs and that started to gain popularity in the 1970s. That would make it easy for existing Unix users to switch to using the GNU Project.

In 1991, Linus Torvalds released the central component of Linux, the *kernel,* which acts as a conduit between the applications software and the hardware resources, including the memory and processor. He still works on the Linux kernel, sponsored by the Linux Foundation, which is the nonprofit consortium that promotes Linux and supports its development. The Linux Foundation reports that 1,730 different organizations contributed to the kernel between 2007 and 2019.

GNU/Linux brings together the Linux kernel with the GNU components it needs to be a complete operating system, reflecting the work of thousands of people on both the GNU and Linux projects. That so many people could cooperate to build something as complex as an operating system, and then give it away for anyone to use, is a modern miracle.

Because GNU/Linux can be modified and distributed by anyone, lots of different versions of it exist. They're called *distributions,* or *distros,* but not all of them are suitable for the Raspberry Pi. The recommended distribution of Linux for the Raspberry Pi is Raspberry Pi OS. Software created for one version of Linux usually works on another version, but Linux isn't designed to run Windows or macOS software.

Strictly speaking, Linux is just the kernel in the operating system, but as is commonly done, we refer to GNU/Linux as *Linux* in the rest of this book.

REMEMBER

Older Raspberry Pi models use SD cards instead of microSD cards. When we say "microSD card" in this book, the same applies to an SD card.

Imaging a microSD Card for Your Raspberry Pi

It's possible to buy a microSD card with Raspberry Pi OS already installed. If you've already got a microSD card, you can skip to Chapter 3 now.

If you want to use a different operating system on your microSD card, or want to reuse an old microSD card that has no software on it yet, you'll need to set it up first.

To set up a microSD card for your Raspberry Pi, there are two steps. You carry these out on another computer, not your Raspberry Pi.

>> Download the image file of the operating system you want to use. The image file is a special format that describes all the different files that need to be created on the microSD card.

>> To convert the image file into a microSD card that will work on the Raspberry Pi, you need to *flash* the card. You can't just copy the file across. (Flashing is a process for copying an operating system onto the microSD card. During the process, the many files required by the operating system are extracted from the single image file you download.)

Raspberry Pi Imager is simple software that downloads the operating system and flashes it to the microSD card for you. It's available for the Windows, macOS, and Ubuntu operating systems. You can download it from www.raspberrypi.org/software and install it in the same way as any other software for your computer.

When you run Raspberry Pi Imager, you see a simple user interface, as shown in Figure 2-1. When it runs, you need to give the software permission to make changes on your computer, even though it will only be changing your microSD card.

TIP

If you press Ctrl+Shift+X on Windows, you can open the advanced settings. Here, you can set up the Wi-Fi for your new SD card, enable SSH for remote access to the Raspberry Pi, set the hostname for the Pi on your local network, and change the locale settings (time zone and keyboard layout). On macOS, press Cmd+Shift+X to access options to disable overscan, set the hostname, and enable SSH.

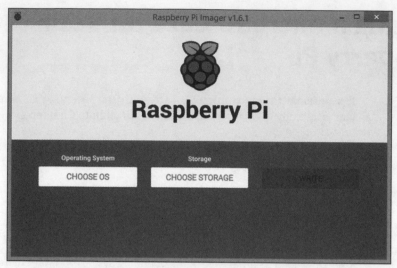

FIGURE 2-1:
The Raspberry Pi
Imager software.

To use the software, follow these steps:

1. **Click Choose OS.**

   Here you choose which operating system you'd like to use. We offer advice on your choice in the next section. If you're eager to get started straight away, click Raspberry Pi OS (other), and choose Raspberry Pi OS Full, including recommended applications. This will give you the software you need for the rest of this book. Your operating system is stored (or cached) on your computer, so it can be flashed to another card later without needing to download it again.

2. **Click Choose Storage.**

   You need to tell your computer where your SD card is. Take care here: The selected drive will be wiped, and the software can show options that include USB drives that are plugged in or even your watch when it's plugged in to charge. To be safe, you could disconnect other drives you're not using. Select your microSD card. Remember to eject or unmount the drives before disconnecting them to avoid losing data.

3. **Click Write.**

   The operating system is downloaded (if necessary) and written to your microSD card.

When it finishes, you're ready to insert your microSD card into your Raspberry Pi, and connect it up as described in Chapter 3.

Choosing the Right Operating System for Your Raspberry Pi

A number of operating systems are supported by the Raspberry Pi Imager software (see Figure 2-2). They're all available for free forever, except for TLXOS, which offers only a free trial. They're grouped into categories, such as general-purpose operating systems, media players, and gaming operating systems.

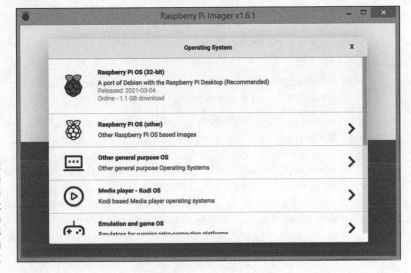

FIGURE 2-2:
Some of the operating system options that appear when you click Choose OS in Raspberry Pi Imager.

Here's an overview of the options:

>> **Raspberry Pi OS:** The distribution that the Raspberry Pi Foundation recommends is called Raspberry Pi OS. It's a version of a Linux distribution called Debian that has been optimized for the Raspberry Pi. It includes graphical desktop software (see Chapter 4), a web browser (see Chapter 4), and various development and programming tools. Raspberry Pi OS is the quickest way to get up and running with your Raspberry Pi, and for most users, it's the one you'll want to use. There are three versions available: one with the desktop and recommended applications, one with the desktop but no recommended applications, and one that does not have the desktop environment or recommended applications. We recommend that you choose the one with the desktop and recommended applications (in this book, we assume that's what you're using). Click Raspberry Pi OS (Other) and select the operating system with the recommended applications.

>> **Ubuntu:** Ubuntu is a popular Linux distribution, and it's available here in three versions. The desktop version gives you a familiar desktop environment, the server edition enables you to use your Raspberry Pi as a server, and the core version is for if you're using your Raspberry Pi as part of an Internet of Things setup. If you don't know how to use the server and core versions, the desktop is the one you need!

>> **Manjaro ARM Linux:** This is another Linux distribution for the desktop, with built-in features to make it easy to customize.

>> **RISC OS Pi:** Most people run Linux on the Raspberry Pi, but you can also use an alternative operating system called RISC OS, which has a graphical user interface (GUI). RISC OS dates back to 1987, when Acorn Computers created it for use with the upmarket Archimedes home computer. You can find documentation at www.riscosopen.org.

>> **LibreELEC:** This is a version of the Kodi media center for playing music and video (see Chapter 8 for a guide to using LibreELEC).

>> **RetroPie:** This retrogaming system includes emulators for a range of vintage home computers (including the Commodore 64, Amiga, Amstrad CPC, various Atari machines, and the ZX Spectrum), as well as game consoles (including a number of Nintendo machines and the Sony PlayStation). You can use the Multi Arcade Machine Emulator (MAME) option to play games from classic coin-operated arcade machines. You'll need to find the game files separately. A number of games have been released by their creators for free distribution online (including games for MAME at http://mamedev.org/roms, and Sean's Amstrad games at www.sean.co.uk/books/amstrad/index.shtm). You can transfer games to RetroPie using a USB key, over your home network, or over the Internet. When you first start RetroPie, you can configure your keys, including a USB game controller if you have one. Use the key you set as A (East) to confirm. Press F4 to exit RetroPie, and then you can type in sudo raspi-config to get into the Raspberry Pi settings and set up your Wi-Fi. See Chapter 21 for information on an arcade cabinet you can build to work with RetroPie. The documentation for RetroPie is at https://retropie.org.uk/docs.

>> **Recalbox:** This is another games system, with emulators for a huge range of classic home computers and video game systems. All the emulators include demonstration games, so an afternoon's arcade action is built in. In the menu system, use the cursor keys to move through the options; press A to confirm and S to go back. If you don't have a USB games controller, the Enter key on your keyboard replaces the Start button, and the Spacebar is the Select button. Tap the Enter key (if your keyboard has one) to enter the settings. In the games, press Esc to quit.

- >> **OctoPi:** The OctoPi operating system includes OctoPrint, a web interface for using 3D printers.

- >> **TLXOS:** This is a trial version of ThinLinX's thin client software, which enables a Raspberry Pi to work as a virtual desktop, interacting with software that is running on a different computer. The ThinLinX Management Software also enables one or more Raspberry Pis to be centrally managed. If you're using lots of Raspberry Pis for a project such as digital signage or to implement a number of virtual desktops, this could help to streamline the process of managing them. When the trial expires, you'll need to buy a license for the software, currently priced at $10.

- >> **Misc utility images:** Here you can find settings to change the boot priority, so you can set your Raspberry Pi to boot from USB, for example.

- >> **Erase:** This option will format your SD card, deleting all the data on it.

- >> **Use Custom:** Other operating systems may be distributed online, without testing by the Raspberry Pi Foundation. Using this option, you can flash any other image files you've downloaded.

TIP

If you fall in love with Raspberry Pi OS, it's also available for Windows and Mac computers. You can download it at www.raspberrypi.org/software/operating-systems.

You can also download the operating system images at that link, to image using any microSD card imaging software.

IN THIS CHAPTER

» **Inserting the SD or microSD card**

» **Connecting a monitor or TV, keyboard, and mouse**

» **Connecting to your router or Wi-Fi**

» **Connecting and testing the Raspberry Pi Camera Module**

» **Using the desktop and raspi-config to change the settings on your Raspberry Pi**

Chapter **3**

Connecting Your Raspberry Pi

N ow you've got a Raspberry Pi, a small pile of cables, and various accessories. In this chapter, we show you how to connect the cables and accessories to your Raspberry Pi and how to change its settings. We also show you how to connect to it remotely.

Chapter 1 lists everything you might need in order to use your Raspberry Pi, including the various cables.

TIP

Connecting Your Raspberry Pi

Here's a guide to setting up your Raspberry Pi. The most important thing to note here is that the power is connected last.

1. **Insert your SD or microSD card.**

   Your SD or microSD card contains the software and data for your Raspberry Pi (see Chapter 2). On the Raspberry Pi 400, insert the microSD card into the slot on the back, with the label facing up. The slot is between the general-purpose input/output (GPIO) pins and the HDMI sockets. On the Raspberry Pi Zero, your microSD card slides into the slot on the top of the board, with the card label facing up. On other models, your SD or microSD card goes into the slot on the bottom of the board. When you put your Raspberry Pi back on the desk, the card label should be facing the desk.

Gently press the card home to make sure it's well connected. The card will stick out from the side of the board. On some models (including the Raspberry Pi 400), you can press the card to pop it out again. On others, you can remove the card by just pulling it.

To avoid data loss, you should only insert and remove cards with the power switched off, and you should shut down your Raspberry Pi properly when you've finished using it (see Chapters 4 and 5).

2. **Connect your keyboard and mouse.**

Your keyboard and mouse can be connected directly to the USB sockets on your Raspberry Pi, and they should work fine on current models. If your Raspberry Pi has blue USB ports, they're the USB 3.0 ones, so save them for other devices that need the extra speed. Connect your mouse and keyboard to the other ports. For earlier models and those with a single USB socket, we recommend connecting the keyboard and mouse to an externally powered USB hub that is connected to the Pi. It reduces the risk of problems caused by the devices drawing too much power from the Pi, and gives you more sockets to play with. If you're using an official Raspberry Pi keyboard or another keyboard with a built-in hub, you can plug the mouse into the keyboard.

The Pi Zero models use Micro USB sockets. Before you can connect your USB hub, you need to plug in a converter that will enable you to connect to standard USB devices. The USB converter goes into the Micro USB socket, labeled as *USB* on the board. Take care with this one because the socket is the same shape and size as the the power socket.

You can set up Bluetooth devices after you've entered the desktop environment. See the "Configuring Bluetooth devices" section, later in this chapter, for more info.

3. **Connect an HDMI or DVI monitor.**

On the Raspberry Pi Zero models, first plug the HDMI converter into the Mini HDMI socket. On other models, you can connect your HDMI or micro HDMI cable directly to your Raspberry Pi. The Raspberry Pi 4 and Raspberry Pi 400 have two micro HDMI ports, so you can use two screens.

If you have a DVI display rather than an HDMI display, you need to use an adapter on the screen end of the cable. The adapter itself is a simple plug, so you just plug the HDMI cable into the adapter and then plug the adapter into your monitor and turn the screws on the adapter to hold the cable in place.

4. Connect a composite video screen.

If your TV has an HDMI socket, use that socket for optimal results. Alternatively, you can use the composite video socket if your Raspberry Pi has a socket. On the original Model A and B, it's a round, yellow-and-silver socket. On later (and current) models, it's the same socket as the audio output. You'll need to use a special RCA cable for this socket — you can't just connect an audio cable. Connect one end of your RCA cable to the socket and the other end to the Video In socket on your TV, which is likely to be silver and yellow.

You may need to use your TV's remote control to switch your TV over to view the external signal coming from the Raspberry Pi. If you're using HDMI on your TV, you may need to turn on the TV so that the Raspberry Pi can detect it when you switch it on.

Note that the Pi Zero models do not have a composite video socket, but they do have composite video output. You can solder your own connector to the board where it's labelled *TV*. For instructions, see `https://magpi.raspberrypi.org/articles/rca-pi-zero`.

There is no composite video support on the Raspberry Pi 400.

On a Raspberry Pi 4, you need to enable composite output. Enable `enable_tvout=1` in `config.txt`. See the appendix of this book for guidance on editing `config.txt`. The easiest solution is to connect another screen temporarily to make this change.

5. Connect to the network.

The Raspberry Pi Model A, A+, and Zero have no wired network connection on the board. The other Raspberry Pi models have an Ethernet socket on the right edge of the board. Use this socket to connect your Raspberry Pi to your Internet router with a standard Ethernet cable. The Raspberry Pi automatically connects to the Internet when used with a router that supports the Dynamic Host Configuration Protocol (DHCP), which means it works with most domestic routers. For advice on troubleshooting your Internet connection, see the appendix.

If you're using a Wi-Fi adapter, you can plug it into a USB socket so that it's ready for when you switch on your Raspberry Pi. Many models, including the Raspberry Pi 4, Raspberry Pi 400, and Raspberry Pi Zero W and WH, have built-in Wi-Fi.

6. Connect the audio.

If you're using an HDMI TV, the sound is routed through the HDMI cable to the screen, so you don't need to connect a separate audio cable. Otherwise, the audio socket of your Raspberry Pi is a small black or blue box stuck along the top edge of the board on the Model A and B, and on the bottom edge of the board on later and current models. If you have earphones or headphones from a portable music player, you can plug them directly into this socket. Alternatively, you can plug a suitable cable into this socket to feed the audio into a TV, stereo, or PC speakers for a more impressive sound. If you're using PC speakers, note that they need to have their own power supply. There is no audio socket on the Raspberry Pi 400 and Pi Zero models.

7. Connect the power.

The last thing you should do is connect the power. Take particular care with the Raspberry Pi Zero, because the power socket looks the same as the USB socket to its left.

WARNING

The Raspberry Pi Foundation warns against using battery power unless you know what you're doing, because it's easy to damage your Pi unless you provide a steady 5 volts (5V) of power. Some cellphone emergency battery chargers can be used to provide that steady power, but proceed with caution.

The Raspberry Pi has no on/off switch, so when you connect the power, it starts working. To turn it off again, you disconnect the power. To avoid losing data, you should shut down first (see Chapters 4 and 5) and wait for that process to finish.

Setting Up Your Raspberry Pi

When you switch on your Raspberry Pi for the first time using Raspberry Pi OS, you're guided through the basic settings.

First you set your country, language, and time zone. Then, you're prompted to change the password. The default Raspberry Pi OS username is pi and the password is raspberry. Both of these are case-sensitive, so you can't use PI instead, for example. It's a good idea to change the password, but you can click Next to skip any step in the setup process. If your screen display has a black border around it, the Set Up Screen option helps fix this.

Next, you can set up your wireless network. Start by choosing your network from those that the Raspberry Pi has detected. Click Next, and you're prompted for the password. When you click Next again, your Pi connects to the network. There's an option to update the software, which checks whether any of the software on your card needs updating. Even if you've just created your microSD card, there may be updates ready to install. The microSD card images are updated less frequently than the software updates available.

When the setup is complete, you have the option to restart now, so any changes you've made to the settings take effect, or to restart later. We recommend you restart now.

Configuring Your Raspberry Pi in Raspberry Pi OS

For most of the rest of this book, we assume that you're using Raspberry Pi OS with the desktop. It's the most user-friendly option and the best way to get started with the Pi.

When your Pi has finished booting, you should be in the desktop environment. You'll learn more about this topic in Chapter 4, but for now, let's take a look at how you use it to finish setting up your Pi or adjust its settings in the future.

Click the button in the top left, with the Raspberry Pi logo on it, to open the menu. Move down to Preferences and choose Raspberry Pi Configuration. The tool that opens is shown in Figure 3-1.

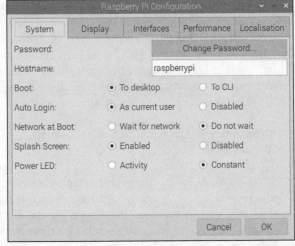

FIGURE 3-1:
The Raspberry Pi Configuration tool in the desktop.

Sean McManus

By default, the tool opens to its System tab. The options here include

>> Change the password. The default password for the username pi is raspberry.

>> Change the hostname (which is the name used for this Raspberry Pi on the network).

>> Control whether it boots into the desktop or the command line interface (CLI), which is explained in Chapter 5.

>> Set whether the pi user is automatically logged in.

- » Set the Pi to wait for the network at the start.

- » Display the graphical splash screen that shows when the Pi is booting.

- » Change whether the power LED on the board is constantly lit, or only when there is activity on the microSD card.

The tool's Display tab allows you to enable pixel doubling, so you can better see the screen output on very high-resolution displays and to enable or disable the screen blanking screensaver. You can also enable or disable underscan. Underscan and overscan change the size of the displayed screen image to optimally fill the screen. Underscan should be disabled to fill the screen if you see a black border around your screen and enabled if the desktop doesn't all fit on the screen.

The tool's Interfaces tab enables you to enable or disable various connection options on your Raspberry Pi, including enabling the Raspberry Pi camera. Other options here include SSH (short for Secure Shell), which is a way of setting up a secure connection between computers, usually so that you can control one computer from another computer. The VNC software enables remote access to your Raspberry Pi with a graphical interface and is also enabled here. (For more on SSH and VNC, see "Connecting Using SSH" and "Connecting Using VNC," later in this chapter.) The other interfaces are SPI, I2C, Serial Port, Serial Console, 1-Wire, and Remote GPIO (which enables another machine on the network to access the Pi's GPIO pins). In most cases, you only need to change these settings if you're using a particular add-on or working on a project that requires them.

The tool's Performance tab gives you access to options for overclocking and changing the GPU memory.

So, what is *overclocking*, anyway? It's when you make a computer work faster than the manufacturer recommends, by changing some of its settings. That said, the options offered to you within this tool have been chosen by the Raspberry Pi Foundation, and they have previously said they don't expect overclocking to cause any measurable reduction in your Pi's lifetime. The speed of the CPU is measured in MHz, and the highest overclocking setting increases the speed to 1000 MHz. You won't necessarily be able to use the top setting: It depends on your Pi and your power supply. Overclocking is not currently supported on the Raspberry Pi 3, 4, or 400.

As for changing the GPU memory, here's the lowdown on that particular option: Your Raspberry Pi's memory is shared between the central processing unit (CPU) and the graphics processing unit (GPU). These processors work together to run the programs on your Raspberry Pi, but some programs are more demanding of the CPU, and others rely more heavily on the GPU. If you plan to do lots of graphics-intensive work, including playing videos and 3D games, you can improve your

Raspberry Pi's performance by giving more of the memory to the GPU. Otherwise, you may be able to improve performance by stealing some memory from the GPU and handing it over to the CPU. Raspberry Pi OS allocates 76MB to the graphics processor and gives the rest to the CPU. In most cases, this setting will work fine, but if you experience problems, you can change how the memory is shared between the two processors. The configuration menu asks how much memory you want to give to the GPU and fills the entry box with the current value as a guide. The rest of the memory is allocated to the CPU. You can safely experiment with the memory split to see which works best for the kind of applications you like to use.

If you're using the Raspberry Pi Case Fan (see Chapter 1), you need to enable it in the Performance tab.

The options on the tool's Localisation tab enable you to set the character set used in your language (the locale), your time zone, the keyboard setup you want to use, and your Wi-Fi country. If you're using the Raspberry Pi outside its home country of the U.K., you may find you need to adjust settings here, especially if you see unexpected results when using the keyboard.

TIP

You can adjust the mouse and keyboard sensitivity separately by going through the main menu to the mouse and keyboard settings, also in the `Preferences` folder.

TIP

If you're using Raspberry Pi OS without the desktop, you can find an alternative tool for configuration options by typing **sudo raspi-config** on the command line. Note that you can't use the mouse to move through its menus. You use up- and down-arrow keys to select different options on the screen, and left- and right-arrow keys (or Tab, which is usually above the Caps Lock key) to select actions such as OK, Cancel, Select, and Finish. Press Enter to confirm a choice.

There's a screen configuration tool for adjusting the screen resolution and orientation. Open the menu, go to the Preferences category, and choose Screen Configuration. In the tool, click Configure on the menu, select Screens, and choose your screen. Alternatively, you can right-click the large screen name (for example, HDMI-1). You can then adjust the resolution as shown in Figure 3-2. After choosing your settings, click the green tick to confirm and then click OK if it looks fine. If the changes stop your screen from working properly, just wait and the previous settings will revert after 10 seconds.

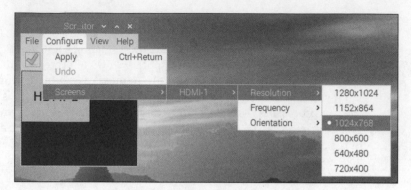

FIGURE 3-2:
Adjusting the screen resolution.

Changing Your Wi-Fi Settings

To change your Wi-Fi settings, click the fan-shaped icon in the top right, shown in Figure 3-3. It opens a menu that shows you the available networks, together with an option to turn off Wi-Fi.

FIGURE 3-3:
The Bluetooth and Wi-Fi buttons, beside the volume control and clock.

Click a network to select it and you're prompted to enter the Pre-shared Key, which is the Wi-Fi password. If the connection fails or drops, the Wi-Fi icon changes to an icon with two red crosses on it.

You can test whether your connection is working by opening the web browser and visiting a web page with it (see Chapter 4).

After you've set up a Wi-Fi network using the tool, your Raspberry Pi remembers it, so it automatically reconnects whenever you restart your Raspberry Pi or reconnect after previously turning off Wi-Fi. You can also use the network connection you have set up from the command line.

REMEMBER

When you connect to a new network, the Pi doesn't remember the previous network's password, so you need to reenter it if you want to reconnect to it later.

Configuring Bluetooth Devices

The Raspberry Pi 4, 400, 3, and Pi Zero W are Bluetooth-enabled, so you can use a wireless Bluetooth keyboard and/or mouse with your Pi.

WARNING

Not all wireless devices are Bluetooth-enabled: Keyboards and mice that come with their own USB dongles typically don't use Bluetooth.

The process of getting two Bluetooth devices to work together is called *pairing* them. Check the instructions for your device to see how you make it discoverable so that your Raspberry Pi can pair with it. This isn't always obvious: On Sean's mouse, the process involved pushing and holding a button and then pushing the two mouse buttons together and holding them until the mouse started flashing.

Once you have made your device discoverable, click the Bluetooth menu at the top of the screen (refer to Figure 3-3) and choose Add Device. Your Raspberry Pi will search for devices. When it finds your device, click it and then click the Pair button. When setting up a keyboard, we had to enter a code shown on the screen on the new keyboard. With the mouse, we were asked to confirm that a code was showing on the mouse (which it couldn't be, because the mouse has no display), but we confirmed that it was, to complete the setup.

Connecting the Raspberry Pi Camera Module

Lots of accessories and add-ons are available for the Raspberry Pi, but the Raspberry Pi Camera Module devices are official products from the Raspberry Pi Foundation. The camera module is a small circuit board with a strip of ribbon cable that plugs directly into the Raspberry Pi board. The Raspberry Pi High Quality Camera is a heavier and larger board, with a built-in screw mount for a tripod. You need to buy a lens separately and fix it to the camera. You can see both in Figure 3-4.

TIP

It's easiest to connect your camera before you plug your Raspberry Pi into any cables.

The Raspberry Pi 400 does not have a camera connector.

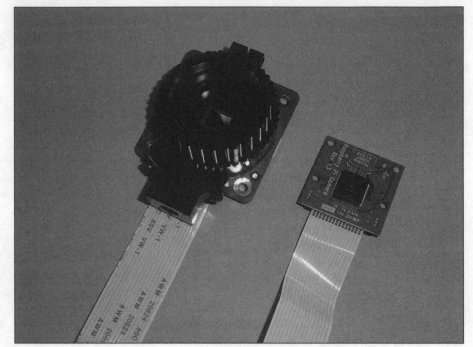

Connecting the camera on a Pi Zero

The Pi Zero camera socket uses a different width of cable to the main Raspberry Pi boards. You can buy the cable separately or get it with the official Pi Zero case.

The camera and the Pi board have similar sockets for the cable. You press the connector between your thumb and finger and gently pull to open the connector. The parts don't separate, but there's enough of a gap to remove and insert the cable. On the Raspberry Pi, the camera connector is on the right of the board.

Figure 3-5 shows the camera connected to a Raspberry Pi Zero. On the camera, insert the cable with the shiny contacts facing the camera front, and then press the socket closed again. On the Raspberry Pi Zero, insert the cable with the shiny contacts facing the bottom of the board (the flat side). When the cable is flat, the camera will be facing down, but you can bend the cable so that the camera sits on top of the board and faces up. One of the covers for the official Pi Zero case has a hole in it for the camera lens.

FIGURE 3-5:
Connecting the
Raspberry Pi
Camera Module
to a Raspberry Pi
Zero.

Connecting the camera on other Raspberry Pi models

The camera connector socket on the other Raspberry Pi models is about 1 inch long, and it touches the bottom edge of the board.

To open the camera connector on your Raspberry Pi board, hold the ends between your finger and thumb and gently lift. The plastic parts don't separate, but they move apart to make a gap. This is where you insert the camera's cable.

At the end of the camera's cable are silver connectors on one side. Hold the cable so that this side faces to the left, away from the side with the USB socket(s). Insert the cable into the connector on the board and press it gently home, and then press the socket back together again.

If the cable needs to be connected at the camera end, the shiny side of the cable should point towards the camera front. Figure 3-6 shows a camera connected to a Raspberry Pi 4.

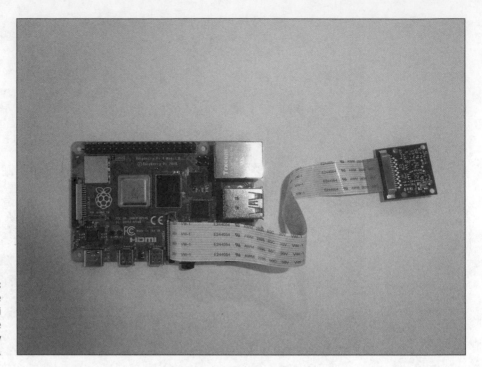

FIGURE 3-6:
Connecting the
Raspberry Pi
Camera Module
to a Raspberry
Pi 4.

Testing the Camera Module

Let's test whether your camera is working correctly.

WARNING

Make sure the camera is enabled: Go into the Raspberry Pi Configuration tool, click Interfaces, and select Enabled beside the Camera option. For more information, see the "Configuring Your Raspberry Pi in Raspberry Pi OS" section, earlier in this chapter. You'll need to reboot after enabling the camera.

We'll test the camera from the command line interface, which is covered in more depth in Chapter 5. Click the Terminal icon at the top of the screen to start. It has a >_ symbol on it. To take a still photo, type in this command:

```
raspistill -o testshot.jpg
```

You should see what the camera sees onscreen for a moment before it takes the photo. The picture is saved with the filename testshot.jpg. You can verify that the image was created by looking at the files in your directory with this command:

```
ls
```

You can use lots of different options to take still photos, too. This example takes a shot with the pastel filter and flips the picture horizontally (-hf) and vertically (-vf):

```
raspistill -ifx pastel -hf -vf -o testshot2.jpg
```

All those hyphens and letter combinations might seem a bit random to you now, but after you read Chapter 5, they should make more sense. To see the documentation for raspistill, type

```
raspistill | less
```

Use the down-arrow key to move through the information, and press Q to finish.

Your photos are stored in your pi directory. See Chapter 4 for instructions on how to use File Manager to find your files and Image Viewer to see them.

To shoot video, you use raspivid. Enter this command to shoot a 5-second film:

```
raspivid -o testvideo.h264 -t 5000
```

The video is saved with the filename testvideo.h264 and is 5000 milliseconds (5 seconds) long. You can view the video you made using

```
omxplayer testvideo.h264
```

The video footage is captured as a raw H264 video stream. For greater compatibility with media players, it's a good idea to convert it to an MP4 file. Start by installing MP4Box using this command:

```
sudo apt install gpac
```

Then you can convert your video file (called testvideo.h264) into an MP4 file (called testvideo.mp4) like this:

```
MP4Box -add testvideo.h264 testvideo.mp4
```

You can get help on using raspivid with

```
raspivid | less
```

There is also a library called picamera that you can use in Python to access the camera from your own Python programs. By adding the Video Sensing extension in Scratch, you can use the camera to make onscreen characters react to movement in the video.

For more information on using the Raspberry Pi Camera Module, see the documentation at www.raspberrypi.org/documentation/usage/camera.

Connecting Using SSH

If your Raspberry Pi has a network connection, you should be able to access it with another computer on the same network using Secure Shell (also known as SSH), which is a way to make a secure connection between computers. This can be helpful if you have set up your Pi to use it *headerless* (without a screen). Here's how to set it up:

1. Change your password.

To prevent unauthorized access to your Raspberry Pi after a remote connection is enabled, start by changing your password. You can use the Raspberry Pi Configuration tool (see "Configuring Your Raspberry Pi in Raspberry Pi OS," earlier in this chapter), or see Chapter 5 for instructions on changing the password in the command line.

2. Enable SSH on your Pi.

Use the Raspberry Pi Configuration tool to enable SSH on the Interfaces tab. Alternatively, use this instruction to create an empty file in the boot directory called ssh and then reboot your Pi:

sudo touch /boot/ssh

3. Get your Pi's IP address.

Use ifconfig in the command line to get your Raspberry Pi's IP address (shown as inet on wlan0 for a WiFi connection or on eth0 for an Ethernet connection).

4. Connect from your other device.

In Windows 10, macOS, and Linux, the SSH software is already installed. You can use it from the command line by typing in **ssh pi@198.51.100.0**. You replace the numbers with your Raspberry Pi's IP address. If you changed the username on your Raspberry Pi, change **pi** too. If you're using a computer that doesn't have SSH software, or you prefer to use a graphical interface, you can download an SSH client such as Putty. In Putty, you input the IP address of your

Raspberry Pi and click the Open button. After you enter your Pi's password, you can use the command line on the Raspberry Pi to manage and fix files, viewing it through your PC screen and using your PC keyboard. SSH apps are also available for the iPhone/iPad and Android. You can find fuller instructions for using SSH on your other machine at www.raspberrypi.org/documentation/remote-access/ssh.

REMEMBER

You can only use SSH to access the Raspberry Pi command line, not the desktop.

Connecting Using VNC

Using VNC, you can control your Raspberry Pi's desktop environment from another computer. If you register for a RealVNC account, you can access your Raspberry Pi over the public Internet. Just follow these steps:

1. Change your password.

To prevent unauthorized access to your Raspberry Pi after a remote connection is enabled, start by changing your password. You can use the Raspberry Pi Configuration tool (see "Configuring Your Raspberry Pi in Raspberry Pi OS," earlier in this chapter), or see Chapter 5 for instructions on changing the password in the command line.

2. Enable VNC on your Pi.

Use the Raspberry Pi Configuration tool to enable VNC on the Interfaces tab.

3. Get your Pi's IP address.

Click the VNC icon, which appears in the upper right of your desktop. The window that opens shows you the IP address you need to connect to your Pi.

4. Install VNC Viewer.

You'll need to install VNC Viewer on the device you want to connect from. You can download it at www.realvnc.com/en/connect/download/viewer. Apps are available for Windows, macOS, and Linux. If you want to manage one Raspberry Pi from another, there's a version of VNC Viewer for the Raspberry Pi, too. Apps for Android and iOS are available, but they're quite hard to use without a physical mouse and keyboard.

5. **Connect from your other device.**

Run VNC Viewer, and you'll be prompted to enter a VNC server address. This is the IP address displayed by your Raspberry Pi. The first time you connect, VNC Viewer warns you that it hasn't connected to this device before. Click Continue. When you enter your username and password, a window opens that shows your Raspberry Pi screen. You can now control your Raspberry Pi remotely.

TIP

Using VNC, you can share a keyboard and mouse between your PC and your Raspberry Pi. Connect them to your PC; then use VNC to control your Raspberry Pi from your PC.

2
Getting Started with Linux

Use the Raspberry Pi desktop to manage the files and start the programs on your Raspberry Pi.

Discover some of the games and applications provided in Raspberry Pi OS.

Surf the web and manage bookmarks for your favorite sites.

Watch slide shows with the Image Viewer and use it to rotate your photos.

Explore your Linux system and get to know the directory tree and file structure.

Back up your Raspberry Pi's microSD card.

Use the Linux shell to organize, copy, and delete files on your microSD card, and to manage user accounts.

Use the desktop or the shell to discover, download, and install new software.

IN THIS CHAPTER

» Using the Raspberry Pi Desktop to manage your Raspberry Pi

» Using external storage devices in the desktop environment

» Copying, moving, and managing files and their permissions

» Viewing web pages on the Raspberry Pi

» Using some of the built-in applications and games on your Raspberry Pi

» Customizing the desktop with your preferred settings

Chapter **4**

Using the Desktop Environment

The quickest way to start playing with your Raspberry Pi is to use the Raspberry Pi Desktop. It works in a similar way to the Windows and macOS operating systems, which let you use icons and the mouse to find and manage files and operate applications. That makes it relatively intuitive to navigate, and it means you can easily find and try out some of the software that comes with your Linux distribution.

In this chapter, we talk you through using the desktop and introduce you to some of its programs.

Navigating the Raspberry Pi Desktop

Figure 4-1 shows the Raspberry Pi Desktop. The photo in the middle of the screen is just a *wallpaper* (a decorative background image on the screen), so don't worry if you see a different image there.

REMEMBER

The strip along the top of the screen is called the *taskbar*, and this is usually visible in whatever application you're using.

FIGURE 4-1:
The Raspberry Pi
Desktop.

LXDE Foundation e.V. / Raspberry Pi Foundation; wallpaper photo by Greg Annandale of the Raspberry Pi Foundation

Using the Applications menu

For most of the applications you might want to run, you use the Applications menu. At the top left of the screen is the Raspberry Pi icon. Click it and you'll see the menu appear, similar to the one shown in Figure 4-2.

As you move the mouse cursor over the categories of applications, a submenu appears on the right, showing you the applications in that category. Click one of these once to start it.

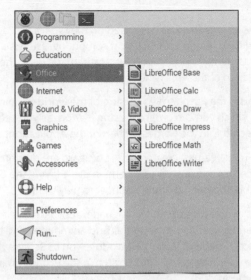

FIGURE 4-2:
The Applications
menu.

LXDE Foundation e.V. / Raspberry Pi Foundation

TIP

If you right-click an application on the menu, you can add its icon to the desktop so that you can start it more quickly in the future.

Buried among the submenus in the Applications menu, you'll find a wealth of applications including the following:

>> **Bookshelf:** This app gives you access to free magazines and books from the Raspberry Pi Foundation that you can download. You'll find it in the Help section of the Applications menu.

>> **Claws Mail:** You can use this email package for sending and receiving messages on your Raspberry Pi. We tell you more about it later in this chapter. It's in the Internet section of the Applications menu.

>> **Debian Reference:** The Raspberry Pi OS version of Linux is a Pi-specific version of the Debian distribution, so this icon gives you a guide to using Linux on your Pi. The documentation is stored on your SD card, but appears in a web browser, like a website. To get started, click the icon and then click the HTML (Multi-Files) link at the top of the screen. You probably won't need to use this resource often, but it's good to know it's there if you get stuck. To find this reference guide, go through the Help section of the Applications menu.

>> **LibreOffice:** This popular suite of productivity applications includes word processing, spreadsheets, and presentations. See Chapter 6 for a guide to getting started with them, and you can find them in the Office section of the Applications menu.

» **Mathematica:** Mathematica, which is based on the Wolfram programming language, is used for scientific and technical computing. There's a short introduction to Mathematica in Chapter 19. Mathematica is in the Programming section of the Applications menu.

» **Minecraft Pi:** This is the Raspberry Pi version of the world-building game Minecraft, which you can program using Python, as you see in Chapter 13. Minecraft is in the Games section of the Applications menu.

» **Python games:** These games, created by Al Sweigart, are demonstrations of Python, but they also provide entertainment. In the "Playing the Games" section, later in this chapter, we show you how to play them.

» **Scratch:** This is a simple programming language, approachable for people of all ages, which can be used to create games and animations and to manage electronics projects. Chapters 9 and 10 introduce you to Scratch and show you how to make your own game. Find Scratch in the Programming section of the Applications menu.

» **Sense HAT emulator:** The Sense HAT is a hardware add-on for the Raspberry Pi that you can use for creating experiments and other projects based on its built-in sensors. See Chapter 15 for more information on it. This emulator is also in the Programming section of the Applications menu.

» **Shutdown:** When you've finished using the Raspberry Pi, use this icon to switch it off before you remove the power. There are also options here to log out and restart (reboot) your Pi. This is a top-level option in the Applications menu.

» **Sonic Pi:** This is a programming language for creating music. See Chapter 14 for a guide to making your own tunes with it. You can find Sonic Pi in the Programming section of the Applications menu.

» **Terminal:** Terminal opens a window you can use to issue instructions from a command line (see Chapter 5). You can find the terminal in the Accessories section of the Applications menu, and there is also a button on the taskbar to go straight to the terminal.

» **Thonny Python IDE:** Thonny enables you to create and run programs in the Python programming language. See Chapters 11 and 12 for advice on getting started with Python. Find Thonny under Programming in the Applications menu.

» **Wolfram:** This is a programming language that aims to incorporate knowledge into it so that programmers can get results more quickly. You can find out more about it at www.wolfram.com/language. Wolfram is filed under Programming in the Applications menu.

The top-left corner of the screen also includes some buttons (refer to Figure 4-2) that you can use to gain quick access to (from left to right) the web browser, File Manager, and Terminal.

Running applications that are not on the menu

Some applications will install but won't appear on the Applications menu. In that case, you can run them using the Run option on the menu. Here's how:

1. Click the icon in the top left of the desktop to open the Applications menu.

2. Select the menu's Run option.

3. In the Run dialog box that appears, type the name of the application and press Enter.

You might prefer to try running the application using the command line interface (the shell), so you can see any error messages. See Chapter 5 for advice on using the shell. To run an application from the shell, type in its name.

Resizing and closing application windows

You'll probably want to use more than one application in a desktop session, so you need to know how to close applications when you've finished with them and how to rearrange windows on the screen.

The application windows have controls similar to the ones in Microsoft Windows that enable you to resize and close them. Figure 4-3 shows the Task Manager application, with these controls in the top right:

>> **X button:** Closes the window.

>> **Maximize button:** Enlarges the application window so that it fills the screen. After you click this button, you can click the new button that appears in its place to return the window to its original size (just like in Windows).

>> **Minimize button:** Hides the application from view but doesn't stop it from running. You can return to the application by clicking its name on the taskbar at the top of the screen.

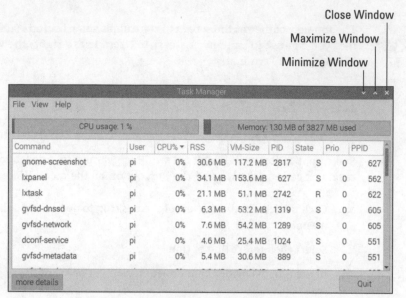

Close Window

Maximize Window

Minimize Window

FIGURE 4-3:
The Task
Manager.

LXTask, written by Hong Jen Yee, Jan Dlabal; derived from Xfce4 Task Manager, by Johannes Zellner

It's easy to change the size of windows — so that you can see more than one at a time, for example. Move the mouse cursor to one of the edges until the Mouse icon changes, and you can click and drag it inward or outward to reshape the window. You can also click and drag a corner to change the window's height and width at the same time. To reposition windows on the screen, click and drag the title bars at the top of them to move them. For example, you can arrange two windows side by side.

Using the Task Manager

You can see which applications are running on your Raspberry Pi by running the Task Manager (refer to Figure 4-3). You can find it on the Applications menu in the Accessories folder, but you can also go straight to it by holding down the Shift and Ctrl keys and pressing Esc.

If you have an application that is not responding, you can stop it by using the Task Manager. To terminate the application, right-click it in the task list and choose Term from the menu that appears. This sends a request to the application and gives it a chance to shut down safely, closing any files or other applications it uses. Alternatively, you can choose Kill. It terminates the application immediately, with the possible loss of data. We recommend you try using Term first, and then try Kill if Term doesn't work.

WARNING

You should use the Task Manager to close applications only as a last resort. Most of the tasks you see in the Task Manager are system tasks, which need to be running for the desktop to work properly. Avoid closing applications you don't recognize — that might crash the desktop and result in losing data in any open applications.

Using File Manager

You can manage your files using the command line (see Chapter 5), but it's often easier to do it in the desktop. File Manager (see Figure 4-4) is used to browse, copy, delete, rename, and otherwise manage the files on your Raspberry Pi or connected storage devices.

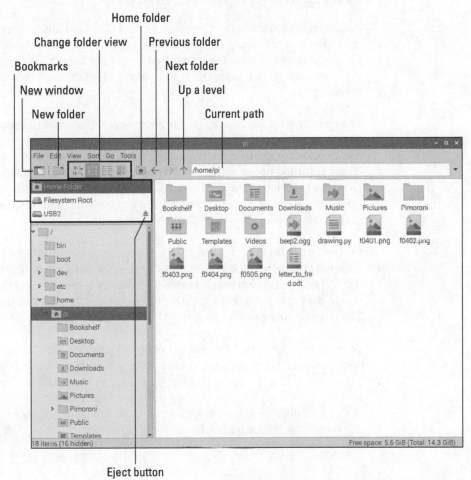

FIGURE 4-4:
The File Manager on the Raspberry Pi Desktop.

LXDE File Manager, written by Hong Jen Yee / Raspberry Pi Foundation

You start File Manager by either clicking its button at the top left of the desktop or using the Applications menu, where it is among the Accessories.

In Linux, people usually talk of storing files in directories, but the Raspberry Pi Desktop uses the term *folders* instead, which is probably familiar to you from other computers you've used. A folder is just a way of grouping a collection of files or applications and giving that collection a name. You can put folders inside other folders too.

TIP

Sometimes you might need to refresh the view of File Manager to reflect your latest changes. To do that, press the F5 key on the keyboard or choose View⇨ Reload Folder.

Navigating File Manager

On the right of File Manager, you can see the files (and any folders) that are inside the folder you're looking at. Each file has an icon indicating the type of file it is. In Figure 4-4, you can see the icons used for sound files (beep2.ogg), Python programs (drawing.py), Libre Office documents (letter_to_fred.odt), and images (f0401.png).

You can double-click a folder in this area to open it, and you can double-click a file to open it with the default application for that file type, if one is set up. An image file opens using the Image Viewer, for example, and a Scratch file opens in Scratch. If you want to choose which application to open a file in instead, you can right-click the file's icon to open a menu with an option called Open With. Select it to bring up a menu of all of the applications available on your Raspberry Pi, and then make your choice.

On the left is the directory tree, which you can use to navigate to any folder on your Raspberry Pi. Click a folder here to view its contents on the right. If a folder has a right-pointing triangle beside it, it means there is at least one other folder inside that folder. Click the triangle to see these subfolders in the directory tree. The triangle rotates to point down. You can click it again to hide the subfolder(s).

Your home folder is where you are expected to store most of your files, such as documents and photos. It is the only place you have permission to write and edit files as an ordinary user. In Chapter 5, we look at Linux and its directory structure in more detail, but for now the key thing is to store your files and folders only in the home folder, or in any folder inside it. You can go to your home folder by clicking the Home Folder button in the bookmarks box in the top left, or clicking the icon on the navbar.

Rather confusingly, your home folder is actually a directory called pi. There is a directory called home, but that's used for storing the home folders for all of the computer's users. Your home folder has the name pi because the default username is pi.

The Desktop folder, inside your home folder, shows you the applications and files that are on the desktop. If you repeatedly edit a document and you want it to be on the desktop for easy access, simply move it into the Desktop folder. The Downloads folder is where you'll find files you download with the browser. The Bookshelf folder is where the Bookshelf app puts the books it downloads. There are empty folders called Documents, Music, Pictures, Public, Templates, and Videos, too. In Figure 4-4, you can also see a folder called Pimoroni that was created when Sean installed some software from the company of that name for one of its add-ons (see Chapter 21).

When you're using the desktop, you can plug in external USB storage devices, such as external hard drives or USB keys (also known as flash drives), and the Raspberry Pi automatically recognizes them. Figure 4-5 shows you the window that appears when you connect a device. You can then view the device in File Manager to access its files. In Figure 4-4, Sean's USB key is shown as the folder USB2. Before removing an external storage device, you should use the Eject button. There's one beside the device in the File Manager (see Figure 4-4), and another on the right of the taskbar. As you may know from other computers you've used, this curiously named process doesn't propel your drive across the room: it makes it safe to remove without data loss.

Chapter 5 tells you more about the different folders on your Raspberry Pi.

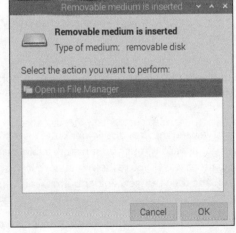

FIGURE 4-5:
Removable storage attached to your Raspberry Pi is automatically detected.

LXDE Foundation e.V. / Raspberry Pi Foundation

Across the top of File Manager is a menu bar, including File, Edit, View, Sort, Go, and Tools menus. Many of the activities in these menus can be carried out in other ways with File Manager, as we show you, but if you get stuck, this menu is a good way to quickly get back on track.

TIP

If there are folders you use particularly often, you can bookmark them — an idea borrowed from web browsers, and from (in the dim, distant past) print books before that. A bookmark makes it easy for you to go straight back to where you were. To add a bookmark to the folder you're viewing, right-click it, and then select Add to Bookmarks. Your bookmarks are shown above the directory tree. Click one of these bookmarks to go straight to its folder.

Underneath File Manager's menu bar is an icon bar that includes a number of useful shortcuts (refer to Figure 4-4):

>> **New Window:** You can have several File Manager windows open at the same time, showing different folders. If you arrange the windows side by side on your screen, you can easily move and copy files between different folders.

>> **New Folder:** Click this button to quickly create a new folder inside the folder you're viewing.

>> **View as Thumbnails, View as Icons, View as Small Icons, and View as Detailed List:** Click these four buttons to change how the current folder is displayed. We particularly like the View as Thumbnails option, where the icons for some image files are replaced with the pictures themselves. It makes it easier to manage folders with lots of images in them. It works for GIF and PNG files, which are often used for artwork like logos and screenshots. Photos in JPG format still appear as icons. The detailed list view shows you the file sizes and when the files were last modified.

>> **Home:** This button takes you back to your home folder so that you have quick access to your work.

>> **Previous Folder:** File Manager keeps a history of the folders you view, and the Previous Folder button works a bit like a web browser's Back button. It takes you back to the last folder you accessed. You can click it repeatedly to keep going back.

>> **Next Folder:** After you've used the Previous Folder button, you can use the Next Folder button to go forward through your history again, taking you back to a folder you visited after the one you're looking at now. If you click the Previous Folder button and then the Next Folder button, you'll end up where you started.

>> **Up a Level:** A folder might be inside another folder, known as a *parent folder*. The Desktop folder is inside your pi folder, for example, so pi is the parent

folder for Desktop. Click the Up a Level button to go to the parent folder. Pressing the Backspace key (usually used when typing to delete a single character to the left of the cursor) has the same effect as clicking this button.

>> **Path:** The path is a text description of the location of the folder you're looking at, including a list of the folders above it. Chapter 5 covers paths in depth, but if you know a path, you can type it and then press the Enter key to go straight to it in File Manager.

TIP

The Tools menu includes an option to Open Current Folder in Terminal, so you can use the shell to manage the files there if it's quicker. The keyboard shortcut is F4.

Copying and moving files and folders

File Manager makes it easy to copy and move your files and folders, without the need for any text commands.

When you right-click a file or folder in File Manager, a menu opens that enables you to rename the file, move it to the trash, or cut or copy it. (If you're using British English, you'll see Move to Wastebasket instead of Move to Trash.)

If you cut a file, it is *moved* to wherever you choose to paste it. If you copy the file, a *duplicate* copy of it is placed where you paste it. You paste by going to the folder where you want the file to be stored and then right-clicking an empty space inside a folder and choosing Paste from the menu that appears. (If you copy or cut a file without pasting it, nothing happens to it.)

TIP

You can also drag files onto a folder's icon to move them into it.

Selecting multiple files and folders

There are several ways to select more than one file at a time so that you can delete, copy, or move them all at the same time:

>> Hold down the Ctrl key and click each of the files in turn to add them to your selected files.

>> To select a group of consecutive icons (read from left to right, top to bottom), click the first icon, hold down the Shift key, and then click the last icon.

>> Click the mouse on the background of File Manager and hold the button down while you lasso the files you want to select.

After you've selected a group of files, you can drag them all into a different folder by clicking one of the selected files and dragging it into the folder. You can also right-click one of your selected files and choose to cut or copy the whole group, as shown in Figure 4-6.

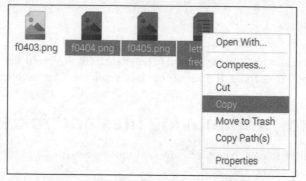

LXDE File Manager, written by Hong Jen Yee / Raspberry Pi Foundation

FIGURE 4-6:
Right-clicking a
file in File
Manager brings
up a menu of
options.

The Raspberry Pi Desktop supports some keyboard shortcuts that might be familiar to you from Microsoft Windows. You can use Ctrl+A to select all files and folders, Ctrl+C to copy, Ctrl+V to paste, and Ctrl+X to cut selected files and folders. It's worth remembering, however, that Ctrl+C is used to cancel an operation on the Linux command line (see Chapter 5), so the Copy shortcut isn't universal on your Raspberry Pi the way it is in Windows.

TIP

If you're selecting almost all the files, it's probably easiest to use Ctrl+A to select all and then hold down the Ctrl key and click to deselect the files you don't want. There's also an option on the Edit menu to invert your selection (also available with Ctrl+I), so you can select the files you don't want and then use this option to flip your choice so that everything else is selected instead.

Creating new folders and blank files

Organizing your files in folders makes it easier to manage them. You can more easily see what files you have where, go straight to a file when you need it, and back up a group of files by copying the folder to an external storage device.

It's easy to make a new folder. First go to the location where you want the new folder to be stored. Typically, it's in your home folder or one of its subfolders, such as your Desktop. Right-click a blank space in the right pane of File Manager and choose New Folder from the menu that opens. You'll be prompted to enter a name and click OK or press Enter to confirm. If you change your mind, click Cancel instead.

You can also use the button underneath File Manager's menu bar, as shown in Figure 4-4.

TIP

The right-click menu also has an option to create an empty file. If you want to practice creating folders and moving files around, you can create a few blank files so that you can do this safely, without worrying about moving anything you didn't intend to.

Deleting files and folders

To delete a file or a folder, right-click it in File Manager and choose Move to Wastebasket from the menu that appears. As you see when copying files, you can hold down the Ctrl key when you click to select several files or folders at the same time, and you can click the background of the File Manager window and drag the mouse to lasso files you want to delete, too. You can also send files to the wastebasket by selecting them and then pressing the Delete key on the keyboard (usually marked Del or Delete, and not to be confused with the Backspace key).

The *wastebasket* is used as a temporary place to put any folders or files you plan to remove. You can find it on the desktop (refer to Figure 4-1), and double-click its icon to see what's inside. You can empty the wastebasket, and delete any files or folders in it, by right-clicking it and choosing Empty Wastebasket.

If you put something in the wastebasket that you change your mind about, right-click its icon in the wastebasket and choose to restore it to where it was before. (This is especially useful if you've forgotten where it used to be!) You can also cut or copy it so that you can paste it wherever you want.

Sorting files

When you right-click an empty space in the right pane in File Manager, a menu opens with an option to change how the files there are sorted. You can sort files by name, modification time, size, or file type, in either ascending or descending order.

The detailed list view reveals more information about each file, showing a short description, its size, and when it was last modified. You can click the column headings to sort the view by the filename, description (which groups similar files), size, or modification date. If you click the column heading again, the sort order is reversed.

Exploring your Raspberry Pi

Linux has a rigorous permissions structure that governs who can access all its files and whether they have permission to modify them or run them. It's a good thing, because it means it's relatively difficult for you to do any real harm to your Raspberry Pi's operating system accidentally. You're free to use File Manager to explore all the files your operating system uses, but if you try to delete an essential file, you'll be told you don't have permission.

If you want to explore your system, go to your home folder, click the Up a Level button twice (refer to Figure 4-4), and then take a look in the folders there. Chapter 5 covers some of these folders in more depth.

Go to the usr folder and the share subfolder to find a lot of the installed software. The code-the-classics and python_games folders contain the games that come with Raspberry Pi OS. After you've learned some Python, you could come back here to tinker with the code.

Browsing the Web with Chromium

Raspberry Pi OS includes the Chromium web browser, which has been optimized for the Raspberry Pi. You access it from the Globe icon in the top left of the screen (see Figure 4-2), or through the Internet part of the Applications menu.

Figure 4-7 shows the browser in use. Its layout is similar to other browsers you might have used in the past, with a thin toolbar at the top and most of the screen given over to the web page you're viewing.

If you know the address of the website you want to visit, you can type it into the Address bar, as shown in Figure 4-7. When you start to type an address, a menu under the Address bar suggests pages you've previously visited that might match what you want. Click one of these to go straight to it, or carry on typing. When you've finished typing the address, press the Enter key.

You can scroll the page using the scroll bar on the right side of the browser or the scroll wheel on your mouse.

When the mouse pointer is over a link, the pointer changes to a small hand. You can then click the left mouse button to follow that link to another web page. The browser keeps a list of the web pages you visit (called your *history*), so you can click the Back button (refer to Figure 4-7) to retrace your steps and revisit the pages you browsed before the current one. The Forward button beside it takes you forward through your history again.

Forward   Menu

Back   Reload   Address bar   New tab   Ad blocker settings

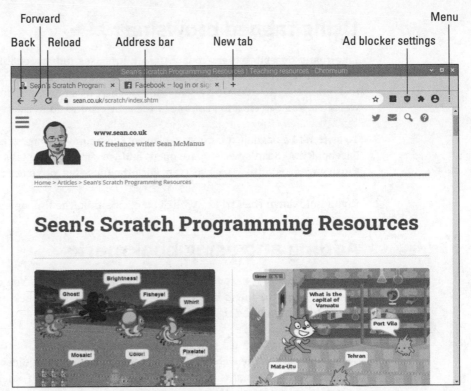

FIGURE 4-7:
The Chromium
browser.

©2017 The Chromium Authors

Some web pages update frequently with new information, so you can click the Reload button to download the current page again and see any updates since you first opened it. While a page is downloading, this button becomes a Stop button. Click it to halt the download.

REMEMBER

Chromium includes an ad-blocker to strip advertising from the pages you visit. You can change the settings using the button shown in Figure 4-7.

Searching within web pages

To find a word or phrase within a web page, press Ctrl+F after the page has loaded. The Find bar opens at the top of the screen, with a box for you to type into. The first occurrence of the text you're looking for is highlighted on the page in orange, and you can press the Enter key or click the Down button in the Find bar to move to the next one. You can close the Find bar again by clicking the Close button (an X) on the far right end of the Find bar or pressing the ESC key.

Using tabbed browsing

Like many other browsers today, Chromium uses tabs to enable you to switch between several websites you have open at the same time. Click the + button (refer to Figure 4-7) to add a new tab, which opens to show your most often visited websites. You can click to visit one of these or type an address on the Address bar.

To switch to a page, just click its tab above the main web page area. In Figure 4-7, Facebook and Sean's website are open, and we can click the tabs to flick between those pages instantly. To close a tab, click the Close button to the right of its name.

TIP

If you hold down the Ctrl key while you click a link, the link opens in a new tab.

Adding and using bookmarks

Bookmarks make it easy to revisit your favorite web pages. You can add a bookmark by using the menu in the top right, using Ctrl+D, or clicking the star inside the Address bar on the right.

The window for adding a new bookmark looks like Figure 4-8. The default name for a bookmark is the web page's title, but you can edit it. Folders can be used to organize your bookmarks so that they're easier to use. The Folder menu includes an option to choose another folder. It enables you to create a new folder. You can also choose to store a bookmark in the Bookmarks Bar folder. You can show the *Bookmarks bar* underneath the Address bar, giving you one-click access to your favorite websites at all times. To display the Bookmarks bar, click the button in the top right to open the Chromium menu, hover over Bookmarks, and then select Show Bookmarks Bar from the options that appear. There's a keyboard shortcut too: Ctrl+Shift+B.

To add the bookmark, click the Done button (refer to Figure 4-8).

To access your bookmarks while you're browsing, click the Menu button in the top right (refer to Figure 4-7) and choose Bookmarks. The bookmarks on the Bookmarks bar are shown on the menu that opens, and others can be found through the Bookmark manager on this menu. You can visit a website on the Bookmarks bar by displaying the bar (Ctrl+Shift+B) and clicking its entry on the bar, and there's easy access to your other bookmarks on the right of this bar too, in the Other Bookmarks folder.

To manage your bookmarks, go to the Bookmarks manager with Ctrl+Shift+O. Hover over a bookmark there, and click the menu button on the right of it to see options to edit or delete the bookmark.

FIGURE 4-8:
Adding a bookmark in Chromium.

If you sign in to your Google account while using Chromium on your Raspberry Pi, it can synchronize your bookmarks across your different devices.

TIP

One of the best Chromium features is the ability to create bookmarks for all open web pages, in their own folder. It's handy if you're doing some research in different tabs to be able to store all the pages you're looking at in one place. To do this, open the Chromium menu, hover over *Bookmarks,* and then select Bookmark All Tabs from the menu that appears. You can enter a new name for the folder, and then click Save.

Protecting your privacy

As you know, your browser stores the history of web pages you visit. If you want to make a visit to a website without any traces being left in the browser — perhaps to plan your Christmas shopping without the risk of other family members coming across the websites you've visited — open a new, incognito window first. You do this from the menu in the top right. When you close the private browsing window, your secret session stops.

When information has already been stored in the browser, you can delete it by opening the Chromium menu in the top right and clicking Settings. Click Privacy and Security on the left (if shown) or scroll down to Privacy and Security, and then click Clear Browsing Data. From the menu in the top right, you can also visit your browser history and delete any entries.

Sending and Receiving Email with Claws Mail

Claws Mail is an open source email application that is preinstalled on your Raspberry Pi. Find it in the Internet category of the Applications menu.

If you want to use email on your Raspberry Pi, you need to know the details of the server for sending and receiving your email. Your email provider most likely publishes this information on its website. You also need to know your user ID and password, which are likely to be the same as you use when logging on with webmail.

When you start Claws Mail for the first time, it walks you through a configuration wizard to add an account. If you experience any difficulties, you can edit your email accounts (including adding or removing them) by using the Configuration menu. There is an Auto-configure option, but this didn't work for our account when we tried it, so be prepared to do the extra work of putting all the information in the boxes manually if it doesn't work for you, either.

When you're set up, click the Get Mail button in the top left to download your email. Claws Mail is similar to many other email clients, including Thunderbird and Outlook. Your mail folders are shown on the left, and your messages are listed on the right, at the top. You can use the message preview pane at the bottom right to read messages, or you can double-click a message to open it in its own window.

Across the top is a menu bar with options for composing a new message, replying to a message, replying to all people copied on that message, and forwarding the message. There's also a Trash or Wastebin button you can use to delete a message.

Using the Image Viewer

It's easy to look at your digital photos and other images. Among the accessories on the Applications menu is the Image Viewer. You can start it from the menu (in the Graphics folder) or by double-clicking or right-clicking an image file.

The Image Viewer displays the picture, with a toolbar underneath it, as you can see in Figure 4-9. From left to right, this is what the buttons do:

IMG_3584.JPG (4032x3024) 22%

FIGURE 4-9:
Gnomes on
phones, as seen
through the
Image Viewer.

GpicView, written by Hong Jen Yee / Raspberry Pi Foundation

» **Previous:** Goes to the previous photo in the folder. Note that any unsaved changes (such as rotation) are lost. You can also use the left-arrow key on the keyboard.

» **Next:** Goes to the next photo in the folder. As with the Previous button, clicking this discards any unsaved changes you've made to the current photo. You can also use the right-arrow key on the keyboard.

» **Start Slide show:** Begins a slide show of all photos in the folder. The interval between photos is set at 5 seconds, but you can change it in the preferences. You can also press the W key to start a slide show. There might be a short delay before the slide show begins.

» **Zoom Out:** Reduces the magnification of the image. The keyboard shortcut is the Minus (–) key.

» **Zoom In:** Increases the magnification of the image. Scroll bars appear if the image becomes too big to fit in the Image Viewer, and you can use these to see different parts of the picture. The keyboard shortcut is the plus sign (+) key, with no need to use Shift.

» **Fit Image to Window:** Shrinks a large image to make it fit the Image Viewer snugly. If an image is smaller than the Image Viewer window, it won't be blown up to fill it, though. This button (or its keyboard shortcut, F) is a good way to recover if you get lost zooming in or out.

» **Go to Original Size:** Resets any zooming by showing the image at its full original size. This might be bigger than the Image Viewer window, in which case scroll bars appear, to enable you to move around the image. The keyboard shortcut is G.

» **Full Screen:** Expands the image to fill the monitor, so you lose the Image Viewer controls. Right-click the image to open a menu with all the same options. To revert to using the Image Viewer in a window, choose Full Screen from the menu or press ESC. You can also use the F11 key to switch the full screen view on and off. On the Raspberry Pi keyboard, use Fn+F1 in place of F11.

» **Rotate Left:** Rotates the image 90 degrees counterclockwise. The keyboard shortcut is L.

» **Rotate Right:** Rotates the image 90 degrees clockwise. The keyboard shortcut is R.

» **Flip Horizontally:** Mirrors the image horizontally and can also be done with the H key.

» **Flip Vertically:** Turns the image upside down. The V key does the same.

» **Open File:** Opens a new image file. You can also drag and drop an image on the Image Viewer from a folder in File Manager. This doesn't move the file — it just opens it.

» **Save File:** Saves the image (including any rotations or mirroring you have done) and replaces the original image. You get a warning before it happens. Keyboard shortcut: S.

» **Save File As:** Saves the image with a new filename so that it doesn't overwrite the original image. (You can also press the A key to do this.) Use the menu at the bottom of the Save File As window to choose the image format.

» **Delete:** Deletes an image from your storage device. If you delete an image, it's not sent to the wastebasket: It's deleted and cannot be recovered. You get one warning, but then it's toast! You can also use the Delete key.

» **Preferences:** Holds the settings you can change for Image Viewer so that you can customize it for your needs. You can turn off the warnings you get before overwriting or deleting an image, set Image Viewer to automatically save rotated images, change the background colors of Image Viewer, and change the slide show interval. There's also an option to rotate images by changing their orientation value in the EXIF tag, which changes some of the information stored with the image to say which way up the camera was, instead of actually rotating the image content itself. It's okay to keep this selected, but this is where you disable it, if you prefer.

» **Exit Image Viewer:** Closes the Image Viewer application. You can also close the window by clicking the Close button in the top right as you would with any other window.

Using the Text Editor

Among the accessories on the Applications menu is Mousepad, which is a simple text editor. To find it, click Text Editor in the Accessories part of the Applications menu. You can use Mousepad for writing and word processing, but it's not ideal for creating print-ready documents. It's most useful for editing documents intended to be read by computers, such as web pages and configuration files.

The menus are logically organized, and if you've ever used a text editor on another computer, you'll find your way around in Mousepad easily.

The File menu is used to start new documents and open, save, and print files. There's also an option to close the window here, although you can just use the Close button in the upper right of the window to close it.

The Edit menu gives you tools for undoing and redoing your work and for cutting, copying, pasting, deleting, and selecting all your text. Mousepad uses Windows shortcuts too, so you can use Ctrl+C to copy, Ctrl+V to paste, Ctrl+X to cut, and Ctrl+A to select all text.

The Search menu has options to find a particular word or phrase, go to a particular line in the document, or replace a chosen word or phrase with an alternative. When using Find and Replace, you can check the box beside Replace All to change all occurrences of your search term in one go, or you can step through them individually using the Replace button.

The Document menu has an option to switch on *word wrap* (which means text starts a new line when it reaches the edge of the window, instead of a horizontal scroll bar appearing). The auto-indent feature means that any indentation used on one line is automatically applied to the next line when you press Enter.

Under the View menu, you can switch on line numbers and change the color scheme.

Configuring Printers

In the Preferences part of the Applications menu, you can find the Print Settings. Here you can see the print queue and add and remove printers. On Sean's Raspberry Pi, Print Settings automatically detected and set up his network-connected printer.

The underlying Common Unix Printing System software is open source and depends on the contributions of volunteers, so not every printer is supported. If you have a reasonably old network-connected printer, it should work fine. Newer models and USB-connected models are less likely to be well supported.

Customizing the Desktop

You can do quite a few things to stamp your identity on the desktop and make it easier to use. As with other desktop computers you might have used, you can change the look and feel of it. To find the options for this, click Appearance Settings in the Preferences section of the Applications menu.

In the desktop options, you can change the picture used as a backdrop (the *wallpaper*), change the desktop color if you're not using wallpaper, and change the color of icon descriptions (the text color). You can tick a box to display your Documents folder and mounted disks on the desktop, which makes it easier to find your files. The Taskbar tab gives you options for changing the size, position, and color of the menu bar that is usually at the top of the screen. The System tab enables you to change the default font used throughout the desktop environment and the colors used in the title bars of windows.

TIP

You can also get to the appearance settings by right-clicking the desktop and choosing Desktop Preferences from the menu that appears.

To adjust the sensitivity of the keyboard and mouse, use the mouse and keyboard settings in the Preferences section of the Applications menu. For left-handers, you can swap the left and right mouse buttons, too.

Playing the Games

A number of games are included in Raspberry Pi OS. You'll find them all in the Games section of the Applications menu. Here's a quick roundup:

>> **Boing!:** This is a version of Pong, a bit like digital tennis. Two players bounce a ball, left and right, between them. If one player misses, the other player scores a point. If you play alone, your bat is on the left. You can choose one or two players on the start screen using the arrow keys. Press the Spacebar to start the game or get back to the menu when it ends. Player 1 (left) uses the A and Z or Up and Down keys to move. Player 2 (right) uses the K and M keys. The computer can be your Player 2. The first player to score 10 points wins.

>> **Bunner:** This a clone of Frogger, where your challenge is to move safely up the screen, dodging traffic and crossing rivers by leaping onto logs. Press the Spacebar to start the game and get back to the title screen at the end, and press the arrow keys to move.

>> **Cavern (see Figure 4-10):** This game is inspired by Bubble Bobble, another classic from the '80s. Move your monkey character with the arrow keys, avoiding the robots. Use the Spacebar to fire a bubble to entrap the robots, as you zip around collecting fruit. When you fall off the bottom of the screen, you come back at the top.

>> **Minecraft Pi:** See Chapter 13 for information on playing and programming this game.

>> **Myriapod:** Use the arrow keys to move and the Spacebar to fire in this Centipede clone. Your gun is at the bottom of the screen. Your aim is to shoot and destroy the centipede that crawls down the screen and to avoid getting hit by the fly, which you can also zap. Destroy rocks to make it easier to shoot the centipede. We use a screen resolution of 1,024 x 768 usually, but we had to change it to 1,152 x 864 to see the whole game arena.

>> **Substitute Soccer:** This football game has three difficulty levels. Move using the arrow keys and Spacebar to pass or shoot. If you have a friend over, they can play using the W, A, S, and D keys to move and the left Shift key to pass/shoot. A triangle above a player shows which one is active.

These games were coded by Raspberry Pi cofounder Eben Upton, and come from the book *Code the Classics, Volume 1.* Find it in the Bookshelf app.

There is also an entry called Python Games, which gives you access to 12 further games, written by Al Sweigart. Click a game to select it. The games are as follows:

>> **Flippy:** A version of Reversi. Click to place your counter. Any counters in a line between your new counter and any of your other counters will switch to your color. The winner is the one with the most counters in their color.

>> **Four in a Row:** Click and drag the pieces to drop them into the frame. The winner is the first to get four pieces in a row, in any direction. The computer AI is pretty smart!

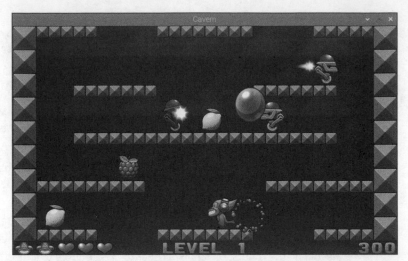

FIGURE 4-10:
Cavern, one of the games in Raspberry Pi OS.

Raspberry Pi Foundation

>> **Gemgem:** In this game, there's a grid of gems. You can swap any two neighboring gems to make a row of three, which then disappear. Use the mouse to click the two gems you want to swap. It's against the clock. Achieving a row of three extends your time.

>> **Inkspill:** This is a favorite of Sean's. You have a grid of colored squares. Starting at the top left, you change the color of the ink. Any squares of the same color that are touching the square join your ink spill, making it bigger. The idea is to keep changing colors to try to fill the screen with a single color.

>> **Memory Puzzle:** This is a version of Pairs or Concentration, a classic card game. You click a card to turn it over, and then click another card to see if they match. Your aim is to pair up all the cards.

>> **Pentomino:** This version of Tetris uses the arrow keys to move the blocks and the Q key to rotate them. The aim is to stop the blocks from piling up to the top of the screen. When you complete a row, it disappears. This game has some different tile designs from the original game.

>> **Simulate:** This game is a pattern repeating one. Click the panels to repeat the sequence that's played to you.

>> **Slide Puzzle:** This sliding puzzle game uses the arrow keys or mouse to move tiles. Can you unscramble the puzzle?

>> **Squirrel:** Your aim is to eat the smaller squirrels, while avoiding the bigger ones. Each time you eat, you grow. Move using the arrow keys.

>> **Starpusher:** This puzzle game starts easy but gets harder. The game challenges you to move stars onto the target spaces on the floor. You'll need to think clearly to avoid getting any stuck in corners. Arrow keys move your

character, and you'll probably need to use the Backspace key to restart the level sometimes.

>> **Tetromino:** This game is like Tetris with the original tile designs. Arrow keys move the pieces and the Q key rotates them.

>> **Wormy:** This version of the classic mobile phone game Snake uses the arrow keys to move the snake. The aim is to eat the red blocks, without bumping into your own tail or the edge of the screen. Each time you eat, you grow, so the game gets harder and harder.

Finding and Installing New Applications

You can discover new software to install using the command line (see Chapter 5), but there's also a friendly tool you can use in the desktop environment. On the Applications menu, hover over the Preferences option and click Add/Remove Software to get started. You need to have an active Internet connection.

Figure 4-11 shows you the tool. In the top left is a search box, where you can enter the name of an application you're looking for, or a phrase such as *puzzle games* to explore what's available. On the left are categories you can click to see your options.

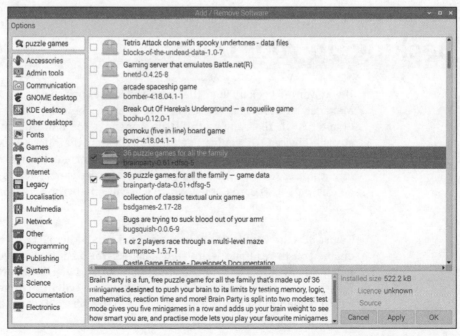

FIGURE 4-11:
The Add/Remove
Software menu.

Raspberry Pi Foundation

The main pane shows you the packages, with a scroll bar on the right that you can use to see the full list. Those that are already checked (or ticked) and shown in bold are already installed on your Raspberry Pi. You can click a package to see its description below. To select a package for installation, tick the box beside it. To remove it, untick it.

When you've finished choosing your software, click the OK button to install and remove the applications. You will be prompted to enter your password (which is *raspberry* if you haven't changed it). It can take some time to download and install the software, so it's a good idea to choose a few applications and leave them to install in one batch while you do something else.

The menu ensures that any applications that your chosen application requires also get installed. When I installed Brain Party (see Chapter 19), for example, the menu automatically installed its separate data package for me.

REMEMBER

The menu makes it easy to install software, but you might find that not all the software works well on the Raspberry Pi. It's easy enough to try something, though, and remove it if it doesn't do what you need. It's all free.

TIP

The Recommended Software app in the Preferences part of the Applications menu is used to install or remove recommended software. If you chose to install a version of Raspberry Pi OS without these recommended applications, you can install them from here.

Backing Up Your Data

If you want to back up your files, you can easily copy them to a USB key using File Manager as described earlier in this chapter, or using shell commands. (See Chapter 5 on file copying, and the Appendix for more on mounting external storage devices.) If you've got a lot of files on the MicroSD card, though, and you've spent time customizing it with your preferred settings and software, you might prefer to make a backup copy of the entire card. There's an application to do this, called SD Card Copier, which you can find in the Accessories section of the Applications menu.

To use SD Card Copier effectively, you need a USB MicroSD card reader, which will enable you to read and write additional MicroSD cards from your Raspberry Pi, using a device plugged into one of the USB ports. If you don't have a MicroSD card reader, you can use the application to back up to a USB flash drive; you would need a card reader, though, to restore the backup to a MicroSD card so that you can use it in your Raspberry Pi.

The MicroSD card or the USB key that you back up to will be totally erased. Ensure that there's nothing on it you need before you begin.

The application is shown in Figure 4-12. It has two menus, where you choose which device to copy from and which device to copy to. If you're not sure which MicroSD card contains your operating system (the one the Pi is currently using), check the Copy To Device menu: It won't be listed there, because you can't use this application to write to the card the Raspberry Pi is using for its operating system.

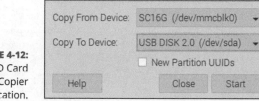

FIGURE 4-12:
The SD Card
Copier
application.

Raspberry Pi Foundation

The backup might take some time, during which it can look like nothing's happening, so be patient.

Logging Out and Shutting Down

When you've finished using your Raspberry Pi, shut it down before removing the power supply. The options to shut down or restart (reboot) your Pi are on the Applications menu, under Shutdown, but you can also use Ctrl+Alt+Delete.

Instead of shutting down, you can log out, which will prompt you to log in again. The default username and password are *pi* and *raspberry*. Chapter 5 shows you how to add additional users with their own logins and home folders.

After your Pi has shut down, you can disconnect the power. When you reconnect it, your Pi will start up again.

If you have a Raspberry Pi 400, you can use the Fn+F10 combination to power down or switch on again.

IN THIS CHAPTER

» **Exploring the Linux file system**

» **Creating, removing, and browsing files and directories**

» **Discovering and installing great free software**

» **Managing user accounts on your Raspberry Pi**

» **Customizing the shell with your own commands**

Chapter **5**

Using the Linux Shell

Hollywood loves a screenful of text. If directors want to show how computer-savvy someone is, or how they've managed to break into a computer system, they flood the screen with writing. It's funny to see the actors staring agog at the screen when often all they've done is list a bunch of files.

This movie cliché contributes to a mystique around the command line interface, though, that can make it seem harder than it is. We both worked with computers before graphical interfaces were the norm, so we're here to reassure you: It's going to be okay. You'll find the text-based interface gives you fast and effective control over your Raspberry Pi, and it's often quicker than using the graphical desktop. It also gives you some understanding of what's going on behind the scenes on your Raspberry Pi.

The Linux shell is the text-based way of issuing instructions to your Raspberry Pi. The shell on the Raspberry Pi is called Bash, which is used in most other Linux distributions too. Its name is short for *Bourne Again Shell*, a pun because it was created to replace the Bourne shell. The Bourne shell gets its name from its creator, Stephen Bourne. In this chapter, you learn how to use the shell to manage your Raspberry Pi.

To open a shell window, click the Terminal icon at the top of the screen, which has a >_ prompt on it. Alternatively, you can find the Terminal in the Accessories section of the Applications menu. Either approach opens a window on the desktop that you can use to access the shell.

TIP

If the screen goes blank while you're using the shell, don't worry: You can get it back again by pressing any key on the keyboard.

Understanding the Prompt

When you log in to your Raspberry Pi, you see a prompt that looks like this, with a cursor beside it that's ready for you to enter a command:

```
pi@raspberrypi:~ $
```

At first glance, that prompt can look quite foreign and unnecessarily complicated (why doesn't it just say *OK* or *Ready?*), but it actually contains a lot of information. This is what the different bits mean:

» `pi`: This is the name of the user who is logged in. Later in this chapter, we show you how to add different users to your Raspberry Pi, and if you log in as a different user, you see that user's name here instead.

» `raspberrypi`: This is the hostname of the machine, which is the name other computers might use to identify the machine when connecting to it.

» `~`: In Linux, people talk about organizing files in *directories* rather than *folders*, but it means the same thing. This part of the prompt tells you which directory you're looking at (the current working directory). The tilde symbol (a horizontal wiggly line) is shorthand for what is known as your *home directory*, and its presence in the prompt here shows that you're currently working in that directory. As we explain in Chapter 4, this is where you should store your work and other files. An ordinary user doesn't have permission to put files anywhere except for their home directory or any directories inside that home directory.

» `$`: The dollar sign means that you're a humble, ordinary user and not an all-powerful superuser. If you were a superuser, you would see a # symbol instead.

Exploring Your Linux System

It's perfectly safe to take a look at any of the files and directories on your Raspberry Pi. As an ordinary user, you're blocked from deleting or damaging any essential files in any case, so you can explore without fear of deleting anything important.

Listing files and directories

The command for listing files and directories is ls. Because you start in your home directory, if you enter it now, you see the folders and files (if any) in your home directory. Here's what the output looks like on Sean's Raspberry Pi — in this chapter, we use bold text for the bits you type and use normal text for the computer's output:

```
pi@raspberrypi:~ $ ls
Bookshelf  Desktop  Documents  Downloads  Music  Pictures  Public  Templates
    Videos
```

WARNING

Linux is case-sensitive, which means LS, ls, Ls, and lS are completely different instructions. Linux doesn't see that uppercase and lowercase letters are related to each other, so an S and an s look like completely different symbols to the computer, in the same way that an A and a Z look different to humans. If you get the capitalization wrong in your command, it won't work, and that applies to everything in the shell. If you misplace a capital letter in a filename, Linux thinks the file you want doesn't exist. When you come to use more-advanced command options later, you'll find that some commands use uppercase and lowercase options to mean different things.

Changing directories

The output is all blue, which means these are all directories, so you can go into them to take a look at the files they have inside. The command to change a directory is cd, and you use it together with the name of the directory you would like to go into, like this:

```
pi@raspberrypi:~ $ cd Bookshelf
```

The prompt changes to show the directory you have changed to after the tilde character, and you can double-check that the current directory has changed by using ls to view the files there, if any. The Bookshelf directory contains a copy of the manual, but the other folders are empty.

Changing to the parent directory

So far, we've used cd to change into a directory that's inside the current working directory. However, you will often want to change into the directory *above* the current working directory, which is known as its *parent* directory. The Bookshelf directory is inside the pi directory, for example, so the pi directory is the parent directory for it.

To change to the parent directory, you use cd with two dots. You can use that command while in Bookshelf to change your home directory (indicated by a ~ symbol in the command prompt):

```
pi@raspberrypi:~/Bookshelf $ cd ..
pi@raspberrypi:~ $
```

The ~ symbol is really just shorthand for your home directory. The directory's real name is the same as your username, which means it is usually *pi*, the default user-name. The parent directory of your home directory is, rather confusingly, called *home*, and it's used to store the home directories of all users of the computer.

When you're in your home directory, try using cd .. to go into the directory called *home*. If you use it again, you will find yourself at the highest directory of your operating system, known as the *root* and indicated with a / in the command prompt. Try navigating through the parent directories to get to the root and then listing what's there, like this:

```
pi@raspberrypi:~ $ cd ..
pi@raspberrypi:/home $ cd ..
pi@raspberrypi:/ $ ls
bin  boot  dev  etc  home  lib  lost+found  media  mnt  opt  proc  root  run
   sbin  srv  sys  tmp  usr  var
```

Feel free to use the cd command to nose around these directories. You can use ls to see what's in the directory, and cd to change into any directory you come across. A lot of directories are empty, but in some you'll find files. When you've finished in a directory, use cd .. to go back to its parent.

Understanding the directory tree

When people think about how the directories are organized on a computer, they often use the metaphor of a tree. A tree has a single trunk with many branches that come off it, secondary branches that sprout from those branches, and so on until you get down to twigs.

Your Raspberry Pi has a single root at the top, with directories that come off it, and subdirectories inside those, and maybe subdirectories inside those too.

Figure 5-1 shows a partial picture of the directory tree on your Raspberry Pi. It doesn't show all the subdirectories in the root, and it doesn't show all their sub-directories either, but it does show you where your home directory is, relative to other directories and the root. You can think of it as a map. If you're at the root and you want to get to the Bookshelf directory, the tree shows you need to go through the home and pi directories to get there.

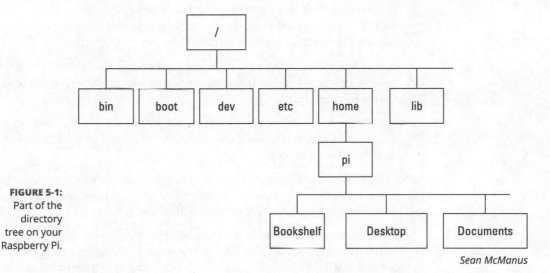

FIGURE 5-1:
Part of the directory tree on your Raspberry Pi.

When you get to the root, you see approximately 20 directories there. All the programs, files, and operating system data on your Raspberry Pi are stored in these directories, or in their subdirectories. It's safe to go into the various directories and have a look around, and to use file to investigate any files you find.

TECHNICAL STUFF

You will rarely need to use any of these directories, but in case you're curious, here's what some of them are used for:

>> bin: This is short for *binaries*, and it contains small programs that behave like commands in the shell, including ls and mkdir, which you will use to make directories later.

>> boot: This contains the Linux kernel, the heart of the operating system, and also contains configuration files that store various technical settings for the Raspberry Pi. The appendix shows you how you can edit the config.txt file here to change some of your computer's settings.

» `dev`: This stores a list of devices (such as disks and network connections) that the operating system understands.

» `etc`: This is used for various configuration files that apply to all users on the computer.

» `home`: As already discussed, this directory contains a directory for each user, and that is the only place a user is allowed to store or write files by default.

» `lib`: This directory contains *libraries* (shared programs) that are used by different operating system programs.

» `lost+found`: The name looks intriguing, but hopefully you'll never have to deal with this directory. It's used if the file system gets corrupted and recovers partially. You don't usually have permission to enter this directory.

» `media`: When you connect a removable storage device like a USB key and it is automatically recognized in the desktop environment, its details are stored in the media directory.

» `mnt`: This directory is used to store the details of removable storage devices that you mount yourself. (See the section about mounting external storage devices in the appendix.)

» `opt`: This directory is used for optional software on your Raspberry Pi. Usually in Linux, this directory is used for software you install yourself, but on the Raspberry Pi, many programs install into `/usr/share` instead.

» `proc`: This directory is used by the Linux kernel to give you information about its view of the system. Most of this information requires expertise to interpret, but it's fun to take a peek anyway. Try entering **less /proc/cpuinfo** to see how the kernel views the Raspberry Pi's processors, or **less /proc/meminfo** to see how much memory your Raspberry Pi has and how it's being used. (You'll learn how to use `less` fully later, but for now, you just need to know that you press Q to quit.) Shortly, you'll see how to use the `file` command to check the type of a file. If you use the `file` command on these files, they appear to be empty, which is a peculiarity that arises because they're being constantly updated.

» `root`: You don't have permission to change into this directory as an ordinary user. It's reserved for the use of the root user, which in Linux is the all-powerful user account that can do anything on the computer. The Raspberry Pi discourages the use of the root account and instead encourages you to use `sudo` to issue specific commands with the authority of the root user (sometimes called the *superuser*). The command is short for "superuser do." Later in this chapter, we show you how `sudo` is used to install software (see "Installing software").

- **run**: This directory provides a place where programs can store data they need and have confidence it will be available when the operating system starts up. Data in the `tmp` folder is vulnerable to being removed by disk cleanup programs, and the `usr` directory might not always be available at start-up on all Linux systems. (It can be on a different file system.)

- **sbin**: This directory contains software that is typically reserved for the use of the root user.

- **srv**: This is empty by default, and is sometimes used in Linux for storing data directories for services such as FTP, which is used to copy files over the Internet.

- **sys**: This directory is used for Linux operating system files.

- **tmp**: This directory is used for temporary files.

- **usr**: This directory is used for the programs and files that ordinary users can access and run.

- **var**: This directory stores files that fluctuate in size (or are variable), such as databases and log files. You can see the system message log with the command `less /var/log/messages`. (Use the arrow key to move down, and press Q to quit.)

Using relative and absolute paths

We've been discussing how to move between directories that are immediately above or below each other on the directory tree, a bit like the way you might work in a desktop environment. You click to open one folder, click to open the folder inside it, and click to open the folder inside that. It's easy (which is why it's popular), but if you've got a complex directory structure, it soon gets tedious.

If you know where you're going, the shell enables you to go straight there by specifying a *path*, which is a description of a file's location. There are two types of paths: relative and absolute. A *relative* path is a bit like giving directions to the directory from where you are now (go up a directory, down through the `Desktop` directory, and there it is!). An *absolute* path is more like a street address: It's exactly the same wherever you are.

The root is represented by a / symbol. Absolute paths are usually measured from the root, so they start with a / and then they list the directories you go through to find the one you want. For example, the absolute path to the `pi` directory is `/home/pi`. Whichever directory you're in, you can go straight to the `pi` directory using

```
cd /home/pi
```

If you wanted to go straight to the Desktop directory, you would use

```
cd /home/pi/Desktop
```

To go straight to the root, just use a slash by itself, like this:

```
cd /
```

Besides using the root as a reference point for an absolute path, you can also use your home directory, which you represent with a tilde (~). You can use it by itself to jump back to your home directory:

```
cd ~
```

Alternatively, you can use it as the start of an absolute path to another directory that's inside your home directory, like this:

```
cd ~/Desktop
```

Relative paths use your current working directory as the starting point. It's shown in the command prompt, but you can also check it by entering the command

```
pwd
```

The pwd command is short for print working directory. As in Python (see Chapter 11), *print* in this case means "display on the screen," not "write onto paper."

Whereas the command prompt uses the tilde (~) character if you're in your home directory, pwd tells you where that actually is on the directory tree and reports it as /home/pi.

A relative path that refers to a subdirectory below the current one just lists the path through the subdirectories in order, separating them with a slash. For example, in Figure 5-1, you can see a directory called home, with a directory called pi inside it, and a directory called Desktop inside that. When you're in the directory with the name home, you can change into the Desktop directory by specifying a path of pi/Desktop, like this:

```
pi@raspberrypi:/home $ cd pi/Desktop
pi@raspberrypi:~/Desktop $
```

You can change into any directory below the current one in this way. You can also have a relative path that goes up the directory tree by using .. to represent the parent directory. Referring to Figure 5-1 again, imagine that you want to go from the Desktop directory into the Bookshelf directory. You can do that by going through the pi directory using this command:

```
pi@raspberrypi:~/Desktop $ cd ../Bookshelf
pi@raspberrypi:~/Bookshelf $
```

As the prompt shows, you've moved from the Desktop directory into the Book-shelf directory. You started in Desktop, went into its parent directory (pi), and then changed into the Bookshelf directory there. You can go through multiple parent directories to navigate the tree. If you wanted to go from the pi directory to the boot directory, you could use

```
pi@raspberrypi:~ $ cd ../../boot
pi@raspberrypi:/boot $
```

That takes you into the parent directory of pi (the directory called home), takes you up one more level to the root, and then changes into the boot directory.

You can choose to use an absolute or relative path, depending on which is most convenient. If the file or directory you're referring to is relatively close to your current directory, it might be simplest to use a relative path. Otherwise, it might be less confusing to use an absolute path. It's up to you. Paths like this aren't used only for changing directories. You can also use them with other commands and to refer to a specific file by adding the filename at the end of the path. For example, you can use the less command like this:

```
less /boot/config.txt
```

It shows you the contents of the config.txt file, no matter which directory you're in. Press Q to finish.

As you discover more commands in this chapter that work with files, you'll be able to use your knowledge of paths to refer to files that aren't in the same directory as your current working directory.

Be careful not to confuse absolute and relative paths. In particular, pay attention to where you use a slash. You should only use a / at the start of the path if you intend to use an absolute path starting at the root.

WARNING

If you want to change into a directory for a quick look around and then go back again, you can use a shortcut to change back to the previous directory:

TIP

```
cd -
```

If you enter this, the shell shows you the previous directory you were in and then changes your current working directory to that.

You can also change to your home directory quickly by using the cd command alone, like this:

```
pi@raspberrypi:/boot $ cd
pi@raspberrypi:~ $
```

Checking file types

If you want to find out more about a particular file, you can use the file command. A good place to experiment with this is the python_games folder.

You can find it like this:

```
pi@raspberrypi:~ $ cd /usr/share/python_games
pi@raspberrypi:/usr/share/python_games $
```

In this directory, there are lots of images, sounds, and Python files that make up some of the games that are preinstalled on the system. You can use the file command to find out more about those files. After the command name, put the name of the file you'd like more information on.

TIP

You can list several files in one command by separating them with spaces, like this:

```
pi@raspberrypi:/usr/share/python_games $ file boy.png match0.wav wormy.py
boy.png:    PNG image data, 50 x 85, 8-bit/color RGBA, non-interlaced
match0.wav: RIFF (little-endian) data, WAVE audio, Microsoft PCM, 16 bit, mono
    44100 Hz
wormy.py:   Python script, ASCII text executable
```

As you can see, the file command can tell you quite a lot about a file. You not only learn what kind of data is in the first two files (an image and an audio recording), but also how big the image is (50 x 85 pixels) and that the audio is mono.

TECHNICAL
STUFF

If you're an experienced computer user, you may have been able to guess what kind of files those were from the file extensions (the .png, .wav, and .py on the ends of the filenames). Linux doesn't require files to have extensions like that, however, so the file command can sometimes be a huge help. (In practice, many applications choose to use file extensions, and users often prefer to do so because it's more user-friendly than having filenames without any context for the file type.)

You can also use the `file` command on a directory. For example, when you're in your `pi` directory, you can find out about `Desktop` and `Bookshelf` like this:

```
pi@raspberrypi:~ $ file Desktop Bookshelf
Desktop:      directory
Bookshelf:    directory
```

That confirms that both of these are directories. Using a command called `file` to find out about a directory may seem counterintuitive, but it illustrates an important feature of Linux: Linux considers everything to be a file, including hard drives and network connections. It's all just a bunch of files, according to Linux.

Investigating more advanced listing options

You can use `ls` to look inside any directory outside the current working directory by specifying its path, like this:

```
pi@raspberrypi:~ $ ls /boot
```

Although you're in your home directory, that command gives you a listing from the `/boot` directory.

When we provide information for a command to process like this, such as a filename or a path, it's called an *argument*. Many Linux commands can accept arguments in this way (including the `cd` and `file` commands).

Some commands can also accept options. Options tell the command how to do its work, and they have the format of a hyphen followed by a code that tells the command which option(s) to use.

There are several options you can use with `ls` to change its results, shown in Table 5-1. For example, use

```
pi@raspberrypi:~ $ cd /boot
pi@raspberrypi:/boot $ ls -R
```

This lists all the contents in the `boot` directory, and then all the contents in the `overlays` directory that is inside the `boot` directory. Use the scroll bar on the right side of the shell window if you can't see them all.

When you're using options and arguments together, the options come before the arguments, so the format of the typical Linux command is

```
command -options arguments
```

TABLE 5-1   Options for the ls Command

Option	Description
–1	Outputs the results in a single column instead of a row. Note that this option is a number 1 and not a letter l.
–a	Displays all files, including hidden files. The names of hidden files start with a single period (full stop). Hidden files are usually put there (and required) by the operating system, so they're best left alone. You can create your own hidden files by using filenames that start with a period.
–F	Adds a symbol beside a filename to indicate its type. When you use this option, directories have a / after their names, and executables have a ∗ after their names.
–h	In the long format, expresses file sizes using kilobytes, megabytes, and gigabytes to save you the mental arithmetic of working them out. It's short for *human*-readable.
–l	Displays results in the long format, which shows information about the permissions of files, when they were last modified, and their size. Note that this option uses a letter *l*, short for *long*.
–m	Lists the results as a list separated by commas.
–R	The recursive option; as well as listing files and directories in the current working directory, opens any subdirectories and lists their results too, and keeps opening subdirectories and listing their results, working its way down the directory tree. You can look at all the files on your Raspberry Pi using ls –R from the root. Be warned: It takes a while. To cancel when you get bored, use Ctrl+C.
–r	The reverse option; displays results in reverse order. By default, results are in alphabetical order, so this shows them in reverse alphabetical order. If your directory listing has multiple columns, the files are sorted by column, not row. Read down the first column, then the second column, and so on. Note that –r and –R are completely different options.
–s	Shows the file size.
–S	Sorts the results by their size.
–t	Sorts the results according to the date and time they were last modified.
–X	Sorts the results according to the file extension.

For example, try using the –X option to list the contents of the python_games folder. All the .png, .py, and .wav files will be grouped together, so it's easier to see what's there. The command to use is

```
pi@raspberrypi:~ $ ls -X /usr/share/python_games
```

You can use several options together by adding all the option codes after a single hyphen. For example, if you want to look in all your directories under your current directory (option R) and you want to group the results by file type (option X) and use symbols to indicate directories and executables beside their filenames (option F), you would use

```
pi@raspberrypi:~ $ ls -RXF
```

This is what it looks like, if you run the command from the /opt/minecraft-pi directory:

```
pi@raspberrypi:/opt/minecraft-pi $ ls -RXF
.:
api/  data/  lib/  minecraft-pi*  CONTROLS.txt  HOW_TO_RUN.txt  LICENSE.txt
    VERSION.txt

./api:
java/  python/  spec/

./api/java:
doc/  lib/  src-api/  src-demos/  McPiDemos.jar  McPi.jar  HOW_TO_RUN_DEMOS.txt
```

The output continues down the screen. This only shows the start of it. One thing you might notice is that a single period (*full stop*) is used to refer to the current directory in the pathnames, so the path for the first set of results is simply a period. This short code for the current directory is similar to the two periods used to refer to the parent directory. This directory shows how results are grouped. We see the directories first, then the executable file minecraft-pi, then the text files. You can also see how symbols are used to show what's a directory (/) and an executable (*).

TIP

When you're experimenting with ls (or at any other time, come to that), use the command clear to empty the screen if it gets messy and hard to follow.

Understanding the Long Listing Format and Permissions

One of the most useful ls options is long format, which provides additional information on a file, compared to a standard listing. You trigger it using the option -l (the letter *l*) after the ls command, like this:

```
pi@raspberrypi:~ $ ls -l
total 152
-rw-r--r-- 1 pi pi 256 Nov 18 13:53 booknotes.txt
drwxr-xr-x 2 pi pi 4096 Oct 28 22:54 Bookshelf
drwxr-xr-x 2 pi pi 4096 Oct 28 22:54 Desktop
drwxr-xr-x 5 pi pi 4096 Oct 28 22:35 Documents
drwxr-xr-x 2 pi pi 4096 Oct 28 22:54 Downloads
drwxr-xr-x 2 pi pi 4096 Oct 28 22:54 Music
```

```
drwxr-xr-x 2 pi pi 4096 Oct 28 22:54 Pictures
drwxr-xr-x 2 pi pi 4096 Oct 28 22:54 Public
drwxr-xr-x 2 pi pi 4096 Nov 3 17:43 seanwork
-rw-r--r-- 1 pi pi 20855 Nov 12 2020 spacegame.sb3
drwxr-xr-x 2 pi pi 4096 Oct 28 22:54 Templates
drwxr-xr-x 2 pi pi 4096 Oct 28 22:54 Videos
```

This listing includes some of Sean's work files on the Raspberry Pi (`booknotes. txt`, `spacegame.sb3`, and the directory `seanwork`), which we can use to show you how different files are described.

This layout might look a bit eccentric, but it's easier to follow if you read it from right to left. Each line relates to one file or directory, with its name on the right and the time and date it was last modified before the name. For older files, the date's year appears in place of the modification time, as you can see for the file `spacegame.sb3` in the preceding list.

The number in the middle of the line is the size of the file. Most of the entries (including `Bookshelf`, `Desktop`, and `seanwork`) are directories that have the same file size (4096 bytes), although they have vastly different contents. That's because directories are files too, and the number here is telling you how big the file is that describes the directory, and not how big the directory's contents are. The file size is measured in bytes, but you can add the `-h` option to give you more meaningful numbers — translating 4096 bytes into 4K, for example.

The rest of the information concerns *permissions*, which refer to who is allowed to use the file and what they are allowed to do with it. Linux was designed from the start to offer a secure way for multiple users to share the same system, and so permissions are an essential part of how Linux works.

TECHNICAL
STUFF

Many people find they can use their Raspberry Pi without needing to know too much about permissions, but permissions tell you a lot about how Linux works, and you might find the knowledge useful if you want to be a bit more adventurous.

The permissions on a file are divided into three categories: things the file's *owner* can do (who is usually the person who created the file), things that *group owners* can do (people who belong to a group that has permission to use the file), and things that *everyone* can do (known as the *world* permissions).

In the long listing, you can see that the word `pi` is shown twice for each file. These two columns represent the owner of the file or directory (the leftmost of the two columns) and the group that owns the file. These both have the same name here because Linux creates a group for each user with just that user in it, and with the same name as the user. In theory, the group could be called something like *students* and include all students who have usernames for the computer.

The leftmost column contains a code that explains what type of file each file is, and what the permissions are on that file. To make sense of the code, you need to break it down into four chunks, as in Table 5-2, which represents the code shared by booknotes.txt and spacegame.sb3 in our long listing.

TABLE 5-2

Understanding Permissions

File type	Owner	Group	World
–	rw–	r––	r––

The two main file types you're likely to come across are regular files and directories. Regular files have a hyphen (–) for their file type at the start of their code, and directories have a d. You can see both of these symbols used in our long directory listing.

Next come the permissions for the owner, group, and world. These are the three different types of permission someone can have:

>> **Read permission:** The ability to open and look at the contents of a file, or to list a directory

>> **Write permission:** The ability to change a file's contents, or to create or delete files in a directory

>> **Execute permission:** The ability to treat a file as a program and run it, or to enter a directory using the cd command

WARNING

That probably seems logical and intuitive, but there are two potential catches: First, you can only read or write in a directory if you also have execute permission for that directory; and, second, you can rename or delete a file only if the permissions of its *directory* allow you to do so, even if you have write permission for the file.

The permissions are expressed using the letters r (for read), w (for write), and x (for execute), and these make up a 3-letter code in that order. Where permission has not been granted, the letter is replaced with a hyphen. So in Table 5-2, you can see that the owner can read and write the file, but the group owner and world (everyone else) can only read it.

The code for the Desktop folder in our long listing is drwxr-xr-x. The first letter tells you it's a directory. The next three letters (rwx) tell you that the owner can read it, write to it, and execute it, which means they have freedom to list its contents (read), add or delete files (write), and enter the directory in the first place to carry out those actions (execute). The next three characters (r–x) tell you that group owners may enter the directory (execute) and list its contents (read), but may not create or delete files. The final three characters (r–x) tell you that everyone else (the world) has been granted those same read-only permissions.

Several commands are used to change the permissions of a file (including chmod to change the permissions (or *mode*), chown to change a file's owner, and chgrp to change the file's group owner). We don't have space to go into detail here, but see "Learning More about Linux Commands," later in this chapter, for guidance on how to get help with them. The easiest way to change permissions, in any case, is through the desktop environment. Right-click a file in File Manager (see Chapter 4) and choose Properties from the menu that appears. You can then use the Permissions tab in the File Properties window that appears (see Figure 5-2) to change the permissions associated with a file.

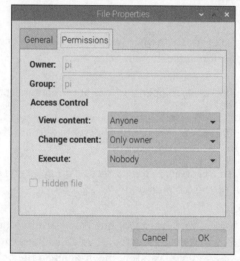

FIGURE 5-2:
Changing file permissions using File Manager.

Raspberry Pi Foundation

Slowing Down the Listing and Reading Files with the Less Command

The problem with ls is that it can deluge you with information that flies past your eyes faster than you can see it. If you open a shell window from the desktop environment, you can use a scroll bar to review information that has scrolled off the screen.

The more usual solution, however, is to use a command called less, which takes your listing and enables you to page through it, one screen at a time. To send the listing to the less command, you use a pipe character (|) after your listing command, like this:

```
ls -RXF | less
```

When you're using less, you can move through the listing one line at a time using the up- and down-arrow keys, or one page at a time using the Page Up (or b) and Page Down (or space) keys. To page up or down on a Raspberry Pi keyboard, hold down the Fn key and press the Up or Down arrow key.

You can search by pressing / and then typing what you'd like to search for and pressing Enter. When you've finished, press the Q key (upper- or lowercase) to quit.

TIP

You can cancel a Linux command, including an overwhelming listing, by pressing Ctrl+C.

TIP

You can also use less to view the contents of a text file by giving it the filename as an argument, like this:

```
less /boot/config.txt
```

This is a great way to read files you find as you explore Linux. The less command warns you if the file you want to read might be a binary file, which means it's computer code and likely to be unintelligible, so you can try using the less command on anything and bow out gracefully if you get the warning. Displaying binary code onscreen can result in some strange results, including distorting the character set in the shell.

TIP

If you want to see the first ten lines of a file, perhaps just to check which version it is, you can use the command head followed by the filename.

Now you have all the tools you need to explore your Linux operating system!

Speeding Up Entering Commands

Now that you've learned a few basic commands, we can teach you a few tricks to speed up your use of the shell.

First of all, the shell keeps a record of the commands you enter, called your *history*, so you can save retyping if you want to reuse a command. If you want to reuse the last command, just type in two exclamation marks (!!) and press Enter. If you want to use an earlier command, tapping the up arrow brings back your previous commands in order (most recent first) and puts them after your prompt. The down arrow moves through your history in the other direction if you overshoot the command you want. You can edit the command before pressing Enter to issue it.

The shell also tries to guess what you want to type and automatically completes it for you if you tap the Tab key. You can use it for commands and files. For example, type

```
cd /bo
```

and then press the Tab key, and the path is completed as /boot/.

This technique is particularly helpful if you're dealing with long and complicated filenames. If it doesn't work, you haven't given the shell enough of a hint, so you need to give it more letters to be sure what you mean.

Using Redirection to Create Files

Before you look at how you delete files and copy them, you should prepare some files to play with.

It's possible to send the results from a command to a file instead of to the screen; in other words, to *redirect* them. You could keep some listing results in a file, for example, so you have a permanent record of them or so you can analyze them using a text editor. You turn screen output into a file by using a greater-than sign and the filename you'd like to send the output to, like this:

```
ls > ~/listing.txt
```

REMEMBER

You don't need to have the file extension of .txt for it to work in Linux, but it's a useful reminder for yourself, and it helps if you ever copy the file back to a Windows machine.

WARNING

Try using this command twice from two different directories and then looking at the contents of `listing.txt` with the `less` command. You'll see just how unforgiving Linux is. The first time you run the command, the file `listing.txt` is created. The second time you do it, it's replaced without warning. Linux trusts you to know what you're doing, so you need to be careful not to overwrite files.

If you want a bit of variety, you can use other commands to display content onscreen:

>> **echo:** This displays whatever you write after it onscreen. You can use it to solve mathematics problems if you put them between two pairs of brackets (parentheses) and put a dollar sign in front, for example:

```
echo $((5*5))
```

>> **date:** This shows the current time and date.

>> **cal:** This shows the current month's calendar, with today highlighted. You can see the whole year's calendar using the option -y.

If you want to add something to the end of an existing file, you use two greater-than signs, as you can see in this example:

```
pi@raspberrypi:~ $ echo I made this file on > testfile.txt
pi@raspberrypi:~ $ date >> testfile.txt
pi@raspberrypi:~ $ cal >> testfile.txt
pi@raspberrypi:~ $ echo $((5+31+5)) Days until my birthday! >> testfile.txt
pi@raspberrypi:~ $ less testfile.txt
I made this file on

I made this file on
Wed 25 Nov 10:07:40 GMT 2020
    November 2020
Su Mo Tu We Th Fr Sa
 1  2  3  4  5  6  7
 8  9 10 11 12 13 14
15 16 17 18 19 20 21
22 23 24 25 26 27 28
29 30

41 Days until my birthday
```

You can use redirection like this to create some files you can use to practice copying and deleting. If you don't want to spend time creating the file contents, you can make some empty files by redirecting nothing, like this:

```
> testfile1.txt
```

Creating Directories

As you may know from other computers you've used, it's a lot easier to manage the files on your computer if they're organized into directories (or folders). You can create a directory using the command `mkdir`:

```
mkdir work
```

To save time, use one command to create several directories, like this:

```
pi@raspberrypi:~ $ mkdir work college games
pi@raspberrypi:~ $ ls
Bookshelf  college  Desktop  Documents  games  Music  Pictures  Public
    Templates  Videos  work
```

You might see additional files here, especially if you followed the earlier examples to make some text files, but the important thing is that one command made three new directories for you.

TIP

The mkdir command's ability to make several directories at the same time isn't unusual: Many other commands can also take several arguments and process them in order. You can see the listing of two different directories like this, for example:

```
ls ~ /boot
```

TIP

The mkdir command doesn't give you a lot of insight into what it's doing by default, but you can add the –v option (short for verbose), which gives you a running commentary as each directory is created. You can see what that looks like in the next example.

If you want to make some directories with subdirectories inside them, it would be a nuisance to have to create a directory, go inside it, create another directory, go inside that, and so on. Instead, use the –p option, like this:

```
pi@raspberrypi:~ $ mkdir –vp work/writing/books
mkdir : created directory 'work'
mkdir : created directory 'work/writing'
mkdir : created directory 'work/writing/books'
```

The command keeps you informed of any changes it makes, but if the work directory already exists, you won't see the first line of output shown here.

Deleting Files

After experimenting with creating files and directories, you probably have odd bits of file and meaningless directories all over the place, so it's time to tidy up.

To delete files in Linux, you use the rm command, short for *remove*. Use it *very* carefully. There's no trash can or recycle bin to recover your file from again, so when it's gone, it's gone. Actually, expert Linux users *might* be able to get it back using specialized software and huge amounts of time and patience, so it's not a secure deletion. But for an average user without access to such software and expertise, when you tell Linux to remove a file, it acts fast and decisively.

The rm command has this format:

```
rm options filename
```

As with mkdir, the command doesn't tell you what it's doing unless you use the verbose option (–v). As an example, you could remove a file called letter.txt using

```
pi@raspberrypi:~ $ rm –v letter.txt
removed 'letter.txt'
```

Like mkdir, running the rm command can take several arguments, which means it can remove several files at once if you list all their names, for example:

```
pi@raspberrypi:~ $ rm –v letter.txt letter2.txt
removed 'letter.txt'
removed 'letter2.txt'
```

This is where you need to be extremely careful. Imagine that you have two files called *old index.html* and *index.html*. The latter is your new website home page, which you've toiled over all weekend. (You can see where this is going, can't you?) You want to clear out the old development file, so you issue this command:

```
pi@raspberrypi:~ $ rm –v old index.html
rm : cannot remove 'old': No such file or directory
removed 'index.html'
```

Arrrggh! Because of that space in the *old index.html* filename, the rm command thinks that you wanted to remove two files — one called *old* and the other called *index.html*. It tells you it can't find the file called *old*, but goes right ahead and wipes out *index.html*. Nasty!

To pin up a safety net, use the –i option (for interactive), which tells you which file(s) will be deleted and prompts you to confirm each deletion. Using that would have avoided this mistake, as shown here:

```
pi@raspberrypi:~ $ rm -vi old index.html
rm : cannot remove 'old': No such file or directory
rm : remove regular file 'index.html'?
```

No, no, no! When prompted, you enter **Y** to confirm the deletion or **N** to keep the file and move on to the next one (if any).

TIP

The risk of deleting the wrong file is one reason you should avoid files with spaces in their names. For the record, the correct way to remove a file whose name contains a space would be to enclose it in quotes:

```
pi@raspberrypi:~ $ rm -vi 'old index.html'
```

Using Wildcards to Select Multiple Files

Often, a directory contains lots of files that have similar filenames. Sean's digital camera, for example, creates files with names like these:

```
img_8474.jpg
img_8475.jpg
img_8476.jpg
img_8477.jpg
img_8478.jpg
```

If you want to delete a group of them, or to copy them or do anything else with them, you don't want to repeat the command by typing out each filename in turn. Computers are good at repetition, so it's better to leave the donkey work to the shell.

Wildcards enable you to do that. Instead of giving a specific filename to a command, you can give it a pattern to match, such as all files that begin with img or all the files that have the extension .jpg.

The asterisk wildcard replaces any number of any character, so *.jpg returns any filenames that end with .jpg, no matter how long they are and no matter how many of them there are. The question mark asterisk replaces just one character, so img?.jpg would select img1.jpg, img2.jpg, and imgb.jpg but ignore img11.jpg or any other longer filenames.

If you want to choose files that begin with a particular letter, you can use the square brackets wildcard. To choose any files beginning with the letters *a*, *b*, or *c*, you would use [abc]*. To narrow that down to just those that end with .jpg, you would use [abc]*.jpg.

Table 5-3 provides a quick reference to the wildcards you can use, with examples.

TABLE 5-3   ## Raspberry Pi Wildcards

Wildcard	What It Means	Usage Example	What Is Selected in the Example
?	Any single character	photo?.jpg	Any files that start with *photo* and have exactly one character after it before the .jpg extension. For example, photo1.jpg or photox.jpg but not photo10.jpg.
*	Any number of characters (including no characters)	*photo*	Any files that have the word *photo* in their filenames.
[...]	Matches any one of the characters in brackets	[abc]*	All files that start with the letter *a, b,* or *c*.
[^...]	Matches any single character that isn't between the brackets	[^abc]*	Any files that do not start with the letter *a, b,* or *c*.
[a–z]	Matches any single character in the range specified	[a-c]*.jpg	Any files that start with a letter *a, b,* or *c* and end with the .jpg extension.
[0–9]	Matches any single character in the range specified	photo[2–5].jpg	Matches photo2.jpg, photo3.jpg, photo4.jpg, and photo5.jpg.

You can use wildcards anywhere you would usually use a filename. For example, you can delete all your files starting with the letters img, like this:

```
rm -vi img*
```

To delete all files ending with the extension .txt, use

```
rm -vi *.txt
```

WARNING

Be especially careful about where you insert spaces when you're using wildcards. Imagine that you add a sneaky space in the previous example, like this:

```
rm -vi * .txt
```

Doh! The shell thinks you want it to delete *, which is a wildcard for every file, and then to delete a file called .txt. Luckily, you've used the -i option, so you'll be prompted before deleting each file — though people often omit that when they're deleting a lot of files, because otherwise they spend a long time confirming each deletion, which is almost as tedious as not using wildcards in the first place.

One way you can test which files match a wildcard is to use the `file` command with it before you delete using it. For example:

```
file *.txt | less
```

Take care that you don't introduce any spaces between testing with `file` and removing with `rm`! You can tap the Up arrow key to get the last command back, and then edit it carefully to change the command, leaving the wildcard alone.

Another thing to be careful about is using wildcards with hidden files. Hidden files begin with a period (full stop), so you might think that .* would match all the hidden files. It does, but it also matches the current directory (.) and its parent directory (..), so .* matches everything in the current directory and the directory above it.

Removing Directories

You can use two commands for removing directories. The first one, `rmdir`, is the safer of the two because it refuses to remove directories that still have files or directories inside them. Use it with the name of the directory you want to remove — for example, `books`, — like this:

```
rmdir books
```

If you want to prune a whole branch of the directory tree, you can use the `rm` command to remove a directory and delete anything inside it and its subdirectories. Used with the recursive option (-R), it works its way down the directory tree, and with the force option (-f), it deletes any files in its way. It's a rampaging beast of a command. Here's an example:

```
rm -Rf books
```

It acts silently and swiftly, deleting the `books` directory and anything in it.

You can add the interactive option to cut the risk, which prompts you for confirmation of each deletion, as you can see in this example where we've left a file in the folder work/writing/books:

```
pi@raspberrypi:~ $ rm -Rfi work
rm: descend into directory 'work'? Y
rm: descend into directory 'work/writing'? Y
rm: descend into directory 'work/writing/books'? Y
rm: remove regular file 'work/writing/books/rapidplan.txt'? Y
rm: remove directory 'work/writing/books'? Y
rm: remove directory 'work/writing'? Y
rm: remove directory 'work'? Y
```

WARNING

You can use wildcards when removing directories, but take special care with them and make sure you don't introduce any unwanted spaces that result in your removing * (everything). If you use rm -Rf .* to try to remove hidden directories, you also match the current directory (.) and the parent directory (..). That means it deletes every file in the current directory (hidden or not), all its subdirectories and their contents (hidden or not), and everything in the parent directory, including its subdirectories (again, whether or not they are hidden).

WARNING

Our own experience of the Linux community has been that it's friendly and supportive, and people welcome newcomers who want to join. But occasionally, you might come across some joker online advising inexperienced users that the solution to their problems is to issue the command rm -Rf /* as a user with root permissions. This attempts to delete everything, starting at the root.

Copying and Renaming Files

One of the fundamental things you'll want to do with your files is copy them, so let's take a look at how to do that. The command you need to use is cp, and it takes this form:

```
cp -options copy_from copy_to
```

Replace copy_from with the file you want to copy, and copy_to for where you want to copy it to.

For example, if you wanted to copy the file config.txt from the /boot directory to your home directory (~) so that you can safely play with it, you would use

```
cp /boot/config.txt ~
```

If you wanted to copy the file into your current working directory, wherever that is, you could use

```
cp /boot/config.txt .
```

You can also specify a path to an existing folder to send the file to by using

```
cp /boot/config.txt ~/files/
```

Your original file and the copy don't have to have the same name. If you specify a different filename, the copy takes that name. For example:

```
cp /boot/config.txt ~/oldconfig.txt
```

That copies config.txt from the /boot directory to your home directory and renames it as oldconfig.txt. This same technique enables you to keep a safe copy of a file you're working on, in case you want to revert to an old version later. The paths are optional, so you could create a backup copy of the file timeplan.txt in the same directory by going into the directory that stores timeplan.txt and using

```
cp timeplan.txt timeplan.bak
```

You can use several options with cp, some of them familiar from the rm command. The cp command overwrites any files in its way without asking you, so use the -i (interactive) option to force it to ask you before it overwrites any existing files with the new copies. The -v (verbose) option gives you an insight into what the command has done, as it does with rm.

You can use wildcards, so you can quickly copy all your files, or all your files that match a particular pattern. If you want it to copy subdirectories too, however, you need to use the recursive option, like this:

```
cp -R ~/Documents/* ~/homebak
```

That command copies everything in your Documents directory (including any sub-directories) into a folder called homebak in your home directory. For advice on using the shell to copy to external storage devices, see the appendix.

If you don't want to make a copy of a file, but instead want to move it from one place to another, use the mv command. For example, if you misfiled one of your files and you want to move it from the australia directory to the japan directory (both in your home directory), you would use

```
mv ~/australia/itinerary.txt ~/japan
```

That works as long as the destination directory exists. If it doesn't, the command assumes that you want the file to have the new filename of japan, and so the file stops being itinerary.txt in the *australia* directory, and becomes a file called *japan* in the home directory. It's confusing if you do it by mistake, but this quirk is how you rename files in Linux. You move them from being the old name into being the new name, usually in the same folder, like this:

```
mv oldname newname
```

TIP

There's no recursive option with the mv command because it moves directories as easily as it moves files by default.

Finding Files on Your Raspberry Pi

Handily enough, there's a command called find that you can use to track down files on your Raspberry Pi. You can use it with wildcards or a filename. To avoid filling the screen with warnings about folders you can't access, it's best used with sudo if you're searching from the root. You can use it like this:

```
sudo find / -name filename
```

For example, to find where listing.txt is stored, use:

```
sudo find / -name listing.txt
```

Installing and Managing Software on Your Raspberry Pi

Linux distributions come with thousands of packages, which are software programs that are ready to download from the Internet and install on your computer. In this chapter, we show you how to use the command line to install software. There is also a simple tool for installing software on the Raspberry Pi within the desktop environment, as described in Chapter 4. You might find it useful to know what's going on in the background, though, and the command line gives you more of a sense of that.

Some packages require other packages to work successfully (known as *dependencies*), but luckily a program called a *package manager* untangles all these dependencies and takes responsibility for downloading and installing the software you want, together with any other software it needs to work correctly. On the Raspberry Pi, the package manager is called apt. It's short for "Advanced Package Tool."

Installing software requires the authority of the root user or superuser of the computer. The Raspberry Pi doesn't come with a root account enabled, in common with some other Linux distributions. One school of thought says that a root account is a security threat because people are inclined to use it all the time rather than log in and out of it when they need it. That leaves the whole system and its files vulnerable, including to any malicious software that might get in. Instead of using a root account, you use the word sudo before a command on the Raspberry Pi to indicate that you want to carry it out with the authority of the root user. You can't use it before all commands, but it's essential for installing software.

TIP

If you ever get an error message that tells you something can be done only with the authority of the root, try repeating the command but putting sudo in front of it.

Updating the cache

The first step in installing software is to update the cache, which is the list of packages the package manager knows about. You do that by entering the following command:

```
sudo apt update
```

This command needs the permission of the root user, so we put sudo in front of it.

WARNING

You need to have a working Internet connection for this to work, and it may take some time. Consider leaving the Raspberry Pi to get on with it while you have a cup of tea — or a slice of raspberry pie, perhaps.

Finding the package name

The package manager cache (the apt cache, in Linux terminology) contains an index of all the software packages available. You can search it to find the software you want using a tool called apt-cache. For example, you can find all the games by using

```
apt-cache search game
```

You don't need to use sudo this time, but it still works if you do.

The list is huge, so you might want to use `less` to browse it, like this:

```
apt-cache search game | less
```

The screen output looks like this (I left out a couple of additional packages near the top of this list relating to 0ad):

```
pi@raspberrypi:~ $ apt-cache search game
0ad - Real-time strategy game of ancient warfare
2048-qt - mathematics based puzzle game
3dchess - Play chess across 3 boards!
4digits - guess the-number game, aka Bulls and Cows
7kaa - Seven Kingdoms Ancient Adversaries: real-time strategy game
7kaa-data - Seven Kingdoms Ancient Adversaries - game data
a7xpg - chase action game
a7xpg-data - chase action game - game data
abe - Side-scrolling game named "Abe's Amazing Adventure"
abe-data - Side-scrolling game named "Abe's Amazing Adventure"
[list continues...]
```

The bit before the hyphen tells you the name of the package, which is what you need to be able to install it. That might not be the same as the game's title or its popular name. For example, there are lots of solitaire card games you can install, but none of them has the package name `solitaire`. To find the package name for a solitaire game, you would use

```
apt-cache search solitaire
```

This search returns 22 results, and the first one is

```
ace-of-penguins - penguin-themed solitaire games
```

You can read a longer description of a package and see its download size using `apt show`. For example, to find out more about `ace-of-penguins`, use:

```
apt show ace-of-penguins
```

Installing software

If you know the name of the package you would like to install, the following command downloads it from the Internet and installs it, together with any other packages it needs to work correctly:

```
sudo apt install ace-of-penguins
```

The last bit is the name of the package we found by searching the cache.

WARNING

When you're searching the cache, you use `apt-cache` in the command, and when you're installing software, you use `apt`. It's easy to get these two commands mixed up, so if your instruction doesn't work, double-check that you're using the right one. Installing software and updating the cache requires `sudo`, but searching the cache doesn't.

Note that not all of the software available in the packages works well on the Raspberry Pi. It's easy enough to try, though, and remove it again if it doesn't work for you.

Running software

Some programs can be run directly from the command line by just typing their names, such as

```
htop
```

which enables you to see all the processes running on your Raspberry Pi. It's already installed for you. Press Q to quit when you've seen enough.

Most end-user applications require you to be in the desktop environment to run them. After installing them, you can find them on the Applications menu. The Ace of Penguins games, for example, install into the Games folder in your Applications menu.

Whether a program should be run from the command line or in the desktop environment depends on the program, so consult its instructions if you can't work out how to start it.

Upgrading the software

The package manager's responsibility doesn't end once it has installed software. It can also be used to keep that software up to date, installing the latest enhancements and security improvements. You can issue a single command to update all the software on your Raspberry Pi:

```
sudo apt upgrade
```

Update the cache first to make sure apt installs the latest updates to your installed packages. You can combine both commands into a single line, like this:

```
sudo apt update && sudo apt upgrade
```

The && means that the second command should be carried out only if the first one succeeds. If the update to the cache doesn't work, it won't attempt to upgrade all the software.

The upgrading process often ties up your Raspberry Pi for some time.

If you want to update just one application, you do that by issuing its install command again. Imagine that you've already installed Ace of Penguins and you enter

```
sudo apt install ace-of-penguins
```

That prompts apt to check for any updates to that package and install them. If there are none, it tells you that you're already running the newest version.

Removing software and freeing up space

The package manager can also be used to remove software from your Raspberry Pi. For example:

```
sudo apt remove ace-of-penguins
```

This particular command leaves traces of the applications, which might include user files and any files containing settings. If you're sure you won't need any of this information, you can completely remove and clean up after an application using

```
sudo apt purge ace-of-penguins
```

You can do two other things to free up some precious space on your SD or microSD card and clean up your system. First, you can automatically remove packages that are no longer required. When a package is installed, other packages it requires are usually installed alongside it. These packages can remain after the original program has been removed, so there's a command to automatically remove packages that are no longer required. It is

```
sudo apt autoremove
```

It lists the packages that will be removed and tells you how much space it will free up before prompting you to enter a Y to confirm that you want to continue.

When you install a package, the first step is to download its installation file to your Raspberry Pi. After the package has been installed, its installation file remains in the directory /var/cache/apt/archives. Over time, as you try out more and more packages, this can amount to quite a lot of space on your SD or MicroSD card. Take a look in that directory to see what's built up there. These files aren't doing much. If you reinstall a program, you can always download the installation file again.

The second thing you can do to clean up your SD card is remove these files using

```
sudo apt clean
```

Finding out what's installed

To find out what software is installed on your Raspberry Pi, you can use

```
dpkg --list
```

The dpkg command is short for "Debian Package." Debian is the Linux distribution that Raspberry Pi OS is based on.

This command doesn't need root authority to run, so it doesn't require you to put sudo at the start.

If you want to find out whether a specific package is installed, use

```
dpkg --status packagename
```

For applications that are installed, this also provides a longer description than the short apt-cache description, which might include a web link for further documentation.

WARNING

The Raspberry Pi includes many packages that come with the Linux operating system and are required for its operation. If you didn't deliberately install a package, exercise caution before removing it.

Managing User Accounts on Your Raspberry Pi

If you want to share the Raspberry Pi with family members, you could create a user account for each one so that they all have their own home directory. The robust permissions in Linux help to ensure that people can't accidentally delete each other's files, too.

When we looked at the long listing format earlier in this chapter, we discussed permissions. You might remember that users can be members of groups. On the Raspberry Pi, groups control access to resources like the audio and video hardware, so before you can create a new user account, you need to understand which groups that user should belong to. To find out, use the groups command to see which groups the default pi user is a member of:

```
pi@raspberrypi:~ $ groups pi
pi adm dialout cdrom sudo audio video plugdev games users input netdev spi i2c
    gpio lpadmin
```

When you create a new user, you want to make him a member of most of these groups, except for the group pi (which is the group for the user pi).

WARNING

If you give users membership of the sudo group, they will be able to install software, change passwords, and do pretty much anything on the machine (if they know how). In a home or family setting, that should be fine, however. The permissions system still protects users from accidentally deleting data they shouldn't, as long as they steer clear of the sudo command.

To add a user, you use the useradd command with the –m option to create a home directory for him and use the –G option to list the groups the user should be a member of, like this:

```
sudo useradd –m –G [list of groups] [username]
```

For example:

```
sudo useradd –m –G adm,dialout,cdrom,sudo,audio,video,plugdev,games,users,
    netdev,input,spi,gpio leo
```

WARNING

Make sure the list of groups is separated with a comma and there are no spaces in there.

You can do a quick check to confirm that a new home directory has been created with the user's name in `directory /home`, alongside the home directory for the `pi` user:

```
ls /home
```

You also need to set a password for the account, like this:

```
sudo passwd [username]
```

For example,

```
sudo passwd leo
```

Usernames are case sensitive, so if you use any capital letters, you must do so consistently. You're prompted to enter the password twice, to make sure you don't mistype it, and you can use this command to change the password for any user. There is no output on the screen as you type the password, which can be a bit off-putting, but keep typing and it should work fine.

You can test whether it's worked and log in as the new user without restarting your Pi by logging out from your current user account. Close the shell window and select Applications ⇨ Logout. Choose Logout from the options, and you'll be presented with the login screen, where you can test that the new username is working. When you're ready to return to the `pi` account, log out of the new account and log back in as `pi`. The default password for the pi account is *raspberry*.

WARNING

If you use the `passwd` command to set a password for the username root, you will be able to log on as the superuser, who has the power to do anything on the machine. As a last resort, this might enable you to get some types of software working, but we advise you against using it. It's safer to take on the mantle of the superuser only when you need it, by using `sudo`.

TIP

If you want to share the Raspberry Pi with different family members, you could just give each user their own microSD card to insert when they're using the machine, and let them log on with the `pi` username and password.

TIP

Raspberry Pi OS automatically logs in as the user `pi`, unless you disable that feature. You can turn it on or off using the Raspberry Pi Configuration tool (see Chapter 3).

Learning More about Linux Commands

Lots of information about Linux is available on the Internet, but plenty of documentation is also hidden inside the operating system itself. If you want to dig further into what Linux can do, this documentation can point you in the right direction, although some of it is phrased in quite a technical way.

Commands in Linux can take several different forms. They might be built into the shell itself, they might be separate programs in the /bin directory, or they could be aliases (which are explained in the next section). If you want to look up the documentation for a command, first find out what kind of command it is, using the type command, like this:

```
pi@raspberrypi:~ $ type cd
cd is a shell builtin
pi@raspberrypi:~ $ type mkdir
mkdir is /bin/mkdir
pi@raspberrypi:~ $ type ls
ls is aliased to 'ls --color=auto'
```

If you want to find out where a particular program is installed, use the which command together with the program name:

```
which mkdir
```

To get documentation for shell built-ins, you can use the shell's help facility. Just enter **help** followed by the filename you're looking for help with:

```
help cd
```

TIP

The help command's documentation uses square brackets for different options (which you may omit), and uses a pipe (|) character between items that are mutually exclusive, such as options that mean the opposite of each other.

For commands that are programs, such as mkdir, you can try using the command with --help after it. Many programs are designed to accept this and to display help information when it's used. A usage example is

```
mkdir --help
```

TIP

When we used this approach on apt, the help page told us that "APT has Super Cow Powers." Try apt moo to see what it means!

There is also a more comprehensive manual (or man page) for most programs, including program-based Linux commands and some applications such as Libre-Office (see Chapter 6). To view the manual for a program, use

```
man program_name
```

For example:

```
man ls
```

The manual is displayed using less, so you can use the controls you're familiar with to page through it. This documentation can have a technical bent, so it's not as approachable to beginners as the help pages.

If you don't know which command you need to use, you can search across all the manual pages using the apropos command, like this:

```
pi@raspberrypi:~ $ apropos delete
argz_delete (3)        - functions to handle an argz list
delete_module (2)      - unload a kernel module
git-branch (1)         - List, create, or delete branches
git-replace (1)        - Create, list, delete refs to replace objects
git-symbolic-ref (1)   - Read, modify and delete symbolic refs
git-tag (1)            - Create, list, delete or verify a tag object signed
                           with GPG
groupdel (8)           - delete a group
ntfsundelete (8)       - recover a deleted file from an NTSF volume.
rmdir (2)              - delete a directory
shred (1)              - overwrite a file to hide its contents, and optionally
                           delete it
[list continues ...]
```

You can then investigate any of these programs further by looking at their man pages or checking to see whether they can accept a --help request. The number in brackets (parentheses) tells you which section of the man page contains the word you searched for.

For a 1-line summary of a program, taken from its man page, use whatis:

```
pi@raspberrypi:~ $ whatis ls
ls (1) - list directory contents
```

If you're not yet drowning in documentation, there's an alternative to the man page, which is the info page. Info pages are structured a bit like a website, with a directory of all the pages at the top, and links between the various pages. You use `info` like this:

```
info ls
```

Move around using the cursor keys or Page Up/Down. When you go off the bottom of a page, you move to the top of the next page. You can go back up a page by moving your cursor off the top of the current page, too. When your cursor is on a link (underlined text), press Enter to follow the link. Press l (a lowercase letter L) to go back. Press Q to quit. There's a lot more you can do with `info`, so if you're curious take a look at the documentation at www.gnu.org/software/texinfo.

Customizing the Shell with Your Own Linux Commands

If you want to stamp your identity on your Raspberry Pi, you can make up your own Linux commands for it. You can have fun inventing a command that shows a special message if someone enters your name (use the `echo` command for this), but it's genuinely useful for making more memorable shortcuts so that you don't have to remember all the different options you might want to use. We show you how to make a command for deleting files that uses the recommended options to confirm each file that will be deleted, and to report on what's been removed. We call it `pidel`, a mashup of *Pi* and *delete*.

The first step is to test whether your preferred command name is already in use. If the `type` command tells you anything other than *not found*, you need to think up another command name, or risk upsetting an existing command. Here's our test:

```
pi@raspberrypi:~ $ type pidel
bash: type: pidel: not found
```

Now that you know that the command `pidel` is not yet taken, you can create your command. To do that, make an alias, like this:

```
alias pidel='rm -vi'
```

Between the quote marks, put the Linux command that you want to execute when you enter the `pidel` command. As you can see from this alias instruction, when you use `pidel`, it behaves like `rm -vi`, but you will no longer have to remember the letters for those options. For example:

```
pi@raspberrypi:~ $ pidel *.txt
rm: remove regular file 'fm.txt'? y
removed 'fm.txt'
rm: remove regular file 'toc.txt'? n
pi@raspberrypi:~ $
```

You can combine lists of commands in your alias definition by separating them with semicolons. For example:

```
alias pidel='clear;echo This command removes files with the interactive and
    verbose options on.;rm –vi'↵
```

Your alias only lasts until the computer is rebooted, but you can make it permanent by putting the alias instruction into the file .bashrc in your home directory. To edit that file, use

```
nano ~/.bashrc
```

Nano is a simple text editor that is covered in more detail in the appendix, but in brief, you can edit your file, use Ctrl+O to save, and Ctrl+X to exit.

Your alias can go anywhere in the .bashrc file. For convenience, and to avoid the risk of disturbing important information there, we suggest you add your aliases at the top. Each one should be on its own line.

Any commands added in .bashrc take effect when you next start up your Raspberry Pi. (See the next section, "Shutting Down and Rebooting Your Raspberry Pi.")

TIP

Sometimes you might want to replace an existing command with an alias so that your chosen options are enforced whenever you use it. If you look at the type for ls, for example, it's aliased so that it always uses the color option to classify files.

Shutting Down and Rebooting Your Raspberry Pi

Usually, you would shut down and reboot your Raspberry Pi using the Applications menu on the desktop, or the Fn+F10 combination on the Raspberry Pi 400. However, it's possible to set the Pi to boot straight into the command line (see

Chapter 3) so that you don't see the desktop. In that case, you can turn off your Raspberry Pi safely in this way:

```
sudo shutdown
```

You can set the shutdown to happen after a delay — for example, 3 minutes — like this:

```
sudo shutdown -h 3
```

You can cancel a scheduled shutdown using the –c option like this:

```
sudo shutdown -c
```

There is a shorter version you can use, which works fine on the Raspberry Pi, but may cause problems on other Linux systems you might use later:

```
sudo halt
```

To switch on your Raspberry Pi again, disconnect and reconnect the power, or use Fn + F10 on a Raspberry Pi 400.

You can reboot (or restart) your Raspberry Pi without disconnecting and reconnecting the power, like this:

```
sudo reboot
```

When you're in the Terminal window, you can close it by clicking the Close button in the top-right corner of the window, or by using this command:

```
exit
```

3

Using the Raspberry Pi for Both Work and Play

Chapter **6**

Being Productive with the Raspberry Pi

There comes a time in most people's lives when they have to get down to work, and when it's your turn to get down to work, the Raspberry Pi can help. Whether you need to do your homework or work from home, you can use LibreOffice, a fully featured office suite that's compatible with the Raspberry Pi.

If you haven't heard of LibreOffice, you might have heard of its ancestor, OpenOffice. A team of developers took OpenOffice as a starting point and developed LibreOffice based on it.

LibreOffice and Microsoft Office for Windows have lots of similarities between them, so LibreOffice will probably feel familiar to you. You can copy files between the two programs too, although you might lose some of the layout features when you do that.

In this chapter, we show you how to use four of the programs in LibreOffice for common household activities. You'll learn how to write a letter, use a spreadsheet to plan a budget, create a presentation, and design a simple party invitation.

LibreOffice is free to download and distribute. If you would like to support the project, you can donate to the charitable foundation that drives its development, The Document Foundation, or help to build or promote the software. Visit the website at www.libreoffice.org.

Installing LibreOffice on Your Raspberry Pi

If you're using the Raspberry Pi OS image that includes recommended applications, you should already have LibreOffice installed. If not, you can install it using the Recommended Software tool in the Preferences category of the Applications menu.

Alternatively, you can install or update the software using the following two commands in the Linux shell:

```
sudo apt update
sudo apt install libreoffice
```

For further guidance on installing software, and an explanation of how these commands work, see Chapter 5.

Working with LibreOffice on the Raspberry Pi

When you enter the desktop environment (see Chapter 4), you should find Libre-Office on the Applications menu, in the Office category, as shown in Figure 6-1. The menu has separate entries for LibreOffice Base (databases), LibreOffice Calc (spreadsheets), LibreOffice Draw (page layouts and drawings), LibreOffice Impress (presentations), LibreOffice Math (mathematical formulas), and LibreOffice Writer (word processing).

TIP

You can start a new LibreOffice file of any type from the File menu in one of the applications, irrespective of which application you're using. For example, you can create a new spreadsheet from the word processor's File menu. When you do this, the correct application opens (Calc, in this case) with a blank file ready for you to use.

In this chapter, we show you how to get started with Writer, Calc, Impress, and Draw.

TIP

You can also start LibreOffice and open a file in it by double-clicking a LibreOffice or Microsoft Office file in the desktop environment.

FIGURE 6-1:
The LibreOffice
entries, on the
Applications
menu.

If you're a student or an academic and you have to write scientific or mathematical formulae, the suite also includes LibreOffice Math, which is used to lay them out (but won't generate the answers for you, unfortunately). To use it, go to the File menu in LibreOffice and choose New ⇨ Formula or choose Applications ⇨ LibreOffice Math.

Saving your work

In all LibreOffice applications, you save your work from the File menu. You have a choice of formats. The OpenDocument Format (ODF) file types are the default, and can be read by other applications, including Microsoft Office. You can also save in the normal file formats of Microsoft Office.

REMEMBER

Save frequently. The applications save automatically from time to time and have some capabilities built in to recover files if the computer crashes, but it's better to catch the trapeze than to test the safety net.

Writing letters in LibreOffice Writer

LibreOffice Writer is a word processor, similar to Microsoft Word on Windows, which makes it the perfect application to use to write a letter.

It can open Microsoft Word files, and its default file format — the OpenDocument Text (ODT) file — can be opened and saved by Word, too. For anything but the most basic files, you're likely to experience some corruption of the document's appearance when you open a Word document in LibreOffice, however. You probably won't have the same fonts on your Raspberry Pi, for example, and more advanced layouts tend to get distorted.

Figure 6-2 shows LibreOffice Writer in action. If you've used other word processing packages, it won't take you long to find your feet here. The icons are similar to those used in Microsoft Office, and if you hover the mouse cursor over an icon, a tooltip appears and tells you what it does.

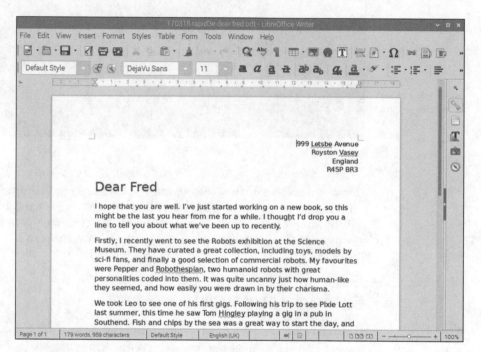

FIGURE 6-2:
Writing a letter with LibreOffice.

You can change the text appearance and the style using the icons and options on the toolbars above your document and then type on the page using your chosen formatting in the document. Alternatively, you can click and drag to highlight text in your document and then click the toolbar to apply different formatting to your selected text.

The pull-down menus at the top of the screen provide access to all of LibreOffice Writer's features. Browsing them is a good way to see what the application is capable of.

The Insert menu enables you to add special characters, manual breaks (including page breaks and line breaks), formatting marks (including nonbreaking hyphens), document headers and footers, footnotes, bookmarks (they don't appear onscreen, but can help you to navigate the document), comments (useful if you are collaborating on documents), frames (boxes for text that you can arrange where you want on the page), and tables.

The Format menu includes options for character formatting (which includes font and font effects, underlining, superscript, rotation, links, and background colors), paragraph formatting (which includes indents and spacing, alignment, text flow, and borders), bullets and numbering, page formatting (including paper size, background colors and images, headers, and footers), columns (for multicolumn layouts), and alignment.

Using those two menus, you can achieve most of what you need. The most common options are also replicated with icons on the toolbars at the top of the screen.

TIP

If you use styles to structure your document (using Heading 1 for the most important headings and Heading 2 for subheadings, for example), you can use the Navigator to jump to different parts of your document easily. Tap F5 to open it. The Navigator also enables you to jump to tables, links, and bookmarks.

TIP

Using the File menu, you can save your document as a PDF file (or *export* it). The great thing about this is that it preserves the formatting of the file, so you can share your document with people who might not have the same fonts or software as you, and guarantee they will see exactly what you see. Most people have software for reading PDF files. Microsoft Office can convert PDFs to an editable form, too, although it may not look the same as the PDF.

Managing your budget in LibreOffice Calc

LibreOffice Calc is a spreadsheet application, similar to Microsoft Excel. A good way to try it is to open one of your Excel spreadsheets using it. Your formulae should work fine and the cell formats should carry over. The interface is similar to LibreOffice Writer, with icons you can roll the mouse pointer over to find out what they do. Figure 6-3 shows Sean's holiday budget in LibreOffice Calc. We've used the slider at the bottom of the screen to magnify the content so that it's easy for you to read.

WARNING

More advanced Microsoft Excel spreadsheets that use macros might not be compatible with LibreOffice.

We don't have room to provide an in-depth guide to spreadsheets here, but we can show you how to work out a simple holiday budget.

Formula bar

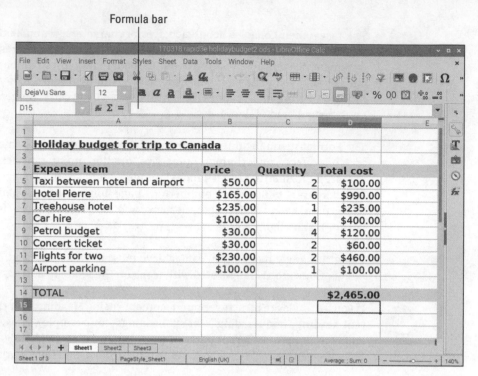

FIGURE 6-3:
How much?!
Planning a
holiday budget in
LibreOffice Calc.

A spreadsheet is basically a grid of information, but it's powerful because you can use it to perform calculations using that information. The boxes on the spreadsheet are called *cells*. To enter information into a cell, you just click it and then type what you want to enter. Alternatively, you can click a cell and then type into (or edit the contents of) the Formula bar at the top of the screen (refer to Figure 6-3).

Each cell has a grid reference, taken from the letter at the top of its column and the number on the left of its row. The top-left cell is A1, and the next cell to the right is B1, and the one below that is B2 (refer to Figure 6-3).

To start with, enter a list of the different expenses you'll incur, working your way down the screen in column A. Beside each item, in column B, enter how much it costs. In column C, enter how many of that item you will need. For example, one row of our spreadsheet shows the name of the hotel in column A, how much it costs per night (in column B on the same row), and then a 6 for the number of nights Sean will stay there in column C on that row. In Figure 6-3, you can see we've also written titles in the cells at the top of the columns of data so that we can easily see what is in each column.

You can make a column wider so that you can more easily enter the descriptions of your budget items. Click and drag the line between the letter at the top of the column and the letter at the top of the column to its right. If you double-click this line, the column will become the right width for its content.

To show a currency sign in a cell, click the Format menu, choose Cells, and then change the category to Currency and the format to the layout and currency symbol you would like to use. You can select a group of cells and format them at the same time by clicking and dragging the cells before you go into the Format menu.

You can enter formulae (or calculations) into cells, and the answers will appear in the spreadsheet. If you want to enter a formula into a cell, you type the equal sign (=), followed by the formula. You use an asterisk (*) for multiplication and a slash (/) for division. For example, click any empty cell and enter

```
=7*5
```

The result (35) appears in the cell on the spreadsheet where you entered the formula. You can view or edit the formula itself by clicking the cell and then clicking the formula bar above the spreadsheet, or by double-clicking the cell.

The magic happens when you start using the numbers in one cell to work out what should go in another one. You do that by using the grid reference of a cell in your formula. For the holiday budget, we want to multiply the cost of an item (such as a night in a hotel) by how many of them we buy (six nights' worth). The first of those values is stored in column B, and the second one is beside it in column C, both on the same row. After the titles and spacing at the top of the spreadsheet, the first expense is on row 5. In column D5, we enter

```
=B5*C5
```

This multiplies the values in cell B5 (the price) and cell C5 (the quantity) and puts the result (the total amount spent on that particular item) into cell D5. You can click cell D5 and then copy its contents and paste them into the cells below. There are options for copying and pasting on the Edit menu, but LibreOffice also supports Windows shortcuts, including Ctrl+C to copy and Ctrl+V to paste.

You might think the same number would go into those cells, but it actually copies the formula and not the result, and it updates it for the correct row number as it goes. If you copy the formula from cell D5 into cell D6 and then click D6 and look on the Formula bar, you'll see that it says

```
=B6*C6
```

After you've copied the formula down the column, you will have a column of results that shows the total cost of each expense item. The final step is to calculate the grand total, adding up the values in those cells. To do that, you use a special type of formula, called SUM, which adds up the values in a set of cells. To use that, follow these steps:

1. **Click a cell at the bottom of the cost column and type** =sum(. **Don't press Enter when you've finished.**

2. **Click the top cell in the column of expenses (D5), and hold down the mouse button.**

3. **Drag the mouse down the screen until the red box encloses all your cost entries.**

4. **Type a closing bracket — the right parenthesis — and then press Enter.**

The grand total appears in that cell, and your budget is complete. A spreadsheet is more than a glorified calculator because you can use it for planning and asking "What if?" For example, you can see what happens if you use a more expensive hotel. Just change the price of the hotel per night, and all the other cells that are calculated from that update automatically, including your total cost at the bottom. Similarly, you can double the length of your stay at the hotel by changing the number of nights in column B to see how that affects your budget total.

Creating presentations in LibreOffice Impress

If you're called upon to deliver a presentation, or if you want to force your holiday-photo slide show on your friends, you can use LibreOffice Impress to create your slides and play them back. You're probably realizing that most Libre-Office programs have a counterpart in the Microsoft Office suite, and Impress is a bit like Microsoft PowerPoint. You can open PowerPoint presentations using it, and although some of the nifty slide transitions are missing, we found that quite sophisticated layouts can be carried across without a problem.

Figure 6-4 shows Sean's holiday-photo slide show in Impress. To create a presentation, simply follow these steps:

1. **Start Impress, or choose to create a new presentation from the File menu in any of the LibreOffice applications.**

2. **Click the slide you want to edit in the Slides panel on the left. When you begin, this will be the single empty slide.**

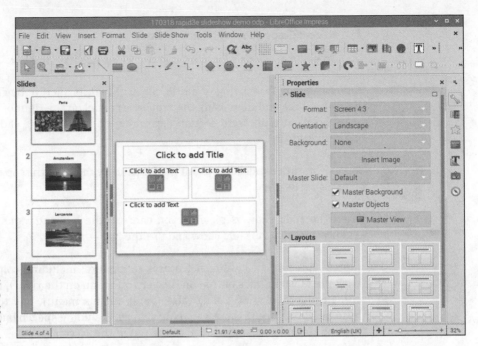

FIGURE 6-4:
Creating a photo slide show using LibreOffice Impress.

In the panel on the right, you can see 12 different slide layouts to choose from. You can also show the Layouts by clicking the Properties button from the vertical menu on the right.

3. **Click the slide layout you would like to use in the panel on the right.**

4. **In the panel in the middle, click the existing title text and replace it with the title you'd like to use for the slide.**

5. **Your slide has up to six boxes for content.**

Click one of these and start typing to add text in the box. Alternatively, in the center of the content box are often four buttons you can click to add different types of content, including a table, a chart, an image, or a video. If you want to add a picture, click the bottom-left button and then choose the picture you'd like to use. Note that if you click a different slide layout on the right, it is applied to the slide you're already working on. In the slide layouts, boxes with a blue line in them are for text only, and those without can be used for any content type.

6. **To add a new slide, right-click in an empty space in the Slides panel on the left, or use the Slide menu.**

You can also double-click an empty space in the Slides panel to add a new slide.

7. Repeat Steps 2 to 5 to fill in the slide.

8. To edit a previous slide, click it in the Slides panel on the left.

You change the formatting of a title, piece of text, or picture by clicking it in the main slide area and then using the Properties panel on the right of the screen. The panel opens at a different section depending on whether the slide content is an image or text.

TIP

You can start the slide show from the Slide Show menu at the top of the screen or by pressing F5.

When the slide show is playing, you can use the left and right cursor keys to advance through the slide show and use the Esc key to exit.

Impress has lots of additional features to explore, including colorful templates (click the Master Slides button on the vertical menu on the right), transitions that animate the display of a slide (also found on this menu), and tools (similar to those in LibreOffice Draw) for making shapes, including speech bubbles and stars. (See the second row of the menu at the top of the screen.)

If the panel on the right is taking up too much space, you can close it by clicking the close button (an X) at the upper-right corner of it. Reopen it again using one of the buttons in the vertical menu.

Creating a party invitation with LibreOffice Draw

LibreOffice Draw is used for designing simple page layouts and illustrations and can be used for making posters and invitations. Despite the application's arty name, the drawing tools are basic and are best suited to creating flowcharts and simple business graphics, although children might enjoy the ease with which they can add stars, smiley faces, and speech bubbles to their pictures.

We'll show you how to use Draw to make an invitation. Refer to Figure 6-5, which shows the LibreOffice Draw screen and our design, as you work through this quick guide. To make an invitation using LibreOffice Draw:

1. Start Draw or choose to create a new drawing from the File menu in any of the LibreOffice applications.

2. Use the toolbar on the left of the screen to select a drawing tool. As you move the cursor over the buttons, a short description appears.

For this example, click the Smiley Face drawing tool in the Symbol shapes on the toolbar.

3. **Move the mouse cursor to the page, and then click and hold the mouse button as you drag the mouse down and to the right.**

   As you move the mouse, you see the face fill the space you're making between where you clicked the button and where the cursor is. When you release the mouse button, the face is dropped in place. You might find it easier to simply place the face anywhere onscreen and then reposition and resize it afterward.

4. **After you have placed the face onscreen, you can reposition it by clicking and dragging it. To resize it, click it and then click and drag one of the blue boxes that appears on its edges.**

5. **Use a similar process to add a speech bubble from the group of items called Callouts on the toolbar. (Click the bubble on the toolbar to select it, and then click and drag the page to place it.)**

   When the speech bubble is on the page, you can resize and reposition it in the same way you arranged the face. To move the tail of the speech bubble, click and drag the yellow point at the end of it. Arrange it so that it points to the smiley face.

6. **Click the speech bubble and type some text.**

   The text spills out of the bubble if there is too much of it, so press the Enter key to start a new line when necessary, and resize the bubble to fit.

7. **Some of the buttons have menus you can open by clicking the small down arrow to the right of the icon. Click the menu beside the Star button to find the Vertical Scroll and position it on the page. Add text to it in the same way you added text to the speech bubble.**

8. **To change the background color of the scroll, face, or bubble, click it on the page and then change the color in the Area part of the Properties panel.**

   Click the color bar to choose a new color. Click the Fill menu that says Color in it and you can select a gradient, hatching pattern, or bitmap (colored pattern) instead of a solid color. Use the Properties button in the vertical menu on the right to show the Properties if you can't see them.

9. **To change the color of the text in the speech bubble or scroll, click it, press Ctrl+A to select it all, and then use the Character part of the Properties panel.**

   The Font Color option has an icon with red underlined text. You can also change the font and size of the text using the Character options.

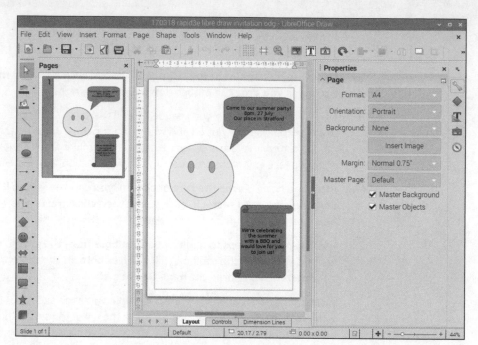

FIGURE 6-5:
Making a party
invitation using
LibreOffice Draw.

As you might expect, you can do lots more with Draw. The Curve option (on the left) enables you to draw freehand by clicking and dragging on the page, and it smoothes your lines for you. The Text option (the T icon on the toolbar at the top of the screen) enables you to place text boxes anywhere, so you can create poster-like layouts. The Fontwork gallery (also on the top toolbar) gives you a choice of different bulging, curved, bent, and circular text styles to choose from. After you've placed the Fontwork item, click its default Fontwork text and type your words. Click somewhere else on the page to have your words inserted in the eccentric style of your choice. If you want to use your own pictures or photos, the Insert Image button on the menu at the top enables you to choose an image file. When your image loads, you can resize and reposition it to fit your design.

IN THIS CHAPTER

» **Installing GIMP**

» **Understanding the GIMP screen layout**

» **Resizing, cropping, rotating, and flipping your photos**

» **Adjusting the colors and fixing imperfections**

» **Converting images between different formats**

Chapter **7**

Editing Photos on the Raspberry Pi with GIMP

We live in probably the best-documented era in our history. Not only do we write about our daily lives in blogs and on social media, but many people also carry cameras everywhere they go, built into their phones or other electronic devices. More serious photographers might have dedicated digital cameras. Whatever you use, and whatever you do with your day, photography is a great way to record your life and express yourself creatively.

The Raspberry Pi can play a part in this activity, enabling you to edit photos to improve their composition and quality. Although early Raspberry Pi models struggled with the large files generated by digital cameras, you can easily edit photos using current or recent models. We had no problems using a Raspberry Pi 2, 3, or 4.

In this chapter, we introduce you to GIMP, one of the most popular image-editing packages, and we give you some tips for editing your photos with it. You learn how to resize, crop, rotate, and flip your photos. We also tell you how to change colors and fix any imperfections, such as dust or unwanted details, in your shots.

Working with GIMP

The program we use is the GNU Image Manipulation Program, known as GIMP for short. It's a highly sophisticated tool, and it's available for free download not just on Linux but on Windows and Mac computers as well.

The easiest way to install GIMP on your Raspberry Pi is to use the Add/Remove Software tool. You'll find it in the Preferences category of the Applications menu. Search for "gimp" and select "GNU Image Manipulation Program" from the search results. Chapter 4 has more guidance on installing software using the Add/Remove Software tool.

Alternatively, you can enter the following at the shell:

```
sudo apt install gimp
```

If you experience any difficulties, consult Chapter 5 for advice on installing software using the shell.

After installation is complete, you can start GIMP from the Graphics category on the Applications menu in the desktop environment. (See Chapter 4 for more on the Applications menu.)

Understanding the GIMP screen layout

Figure 7-1 shows the screen layout of GIMP. GIMP can be used in such a way that each pane of tools or content is a separate window onscreen — though we find it easier to arrange everything in a single window, especially when we're using a smaller screen. If your layout looks different from the one shown in Figure 7-1, click to open the Windows menu at the top of the screen and select Single Window Mode.

When GIMP opens, the large area in the middle is empty, with a picture of Wilber, the GIMP mascot, in the background at the bottom. We've used the File menu in the top left to open a photo for editing, which you can see in the center pane.

Across the top of the screen is a bar with menus for File, Edit, Select, View, Image, Layer, Colors, Tools, Filters, Windows, and Help. You can browse these menus to get an idea of what the program can do, and to find options quickly if you don't know which icons they use on the toolbar.

New layer   Layers pane

FIGURE 7-1:
GIMP enables
you to edit
photos on your
Raspberry Pi.

On the left is a pane that contains icons for the tools at the top and the tool options at the bottom. When you roll the cursor over a tool's icon, a tooltip pops up to tell you what it does. When you click a tool to select it, the options at the bottom of the pane change, depending on the tool you're using. For example, if you're using the paintbrush, the options cover properties such as opacity and the brush type.

The pane on the right is also divided into halves. The bottom half has tabs for Layers, Channels, and Paths. Of these, the Layers tab (refer to Figure 7-1) is the most important because it enables you to edit your photos safely.

Layers are used for adding new elements to an image without disturbing whatever is underneath. For example, if you want to add text to an image, you do that in a new layer on top of the old one. If you change your mind, you can just remove the layer, and the picture underneath is unchanged. The Text tool (which has an *A* as its icon) automatically adds text in a new layer when you use it. If you intend to use the drawing tools, add a layer for each part of the drawing by clicking the New Layer button under this pane (refer to Figure 7-1). New layers appear on top of older layers, but you can change the order of layers by dragging them up or down in the pane on the right. Those near the top of this pane appear nearer the foreground in the image. To hide a layer temporarily, click the Eye icon next to it in the pane.

The top half of the right pane is for Brushes, Patterns, Fonts, and Document History. The brushes are used when you're drawing or painting on the image. The patterns and gradients are used for the Fill tool, which fills in a part of the image with a particular color or pattern. The Document History tab enables you to open files you've previously opened in GIMP.

You can change the width of the left and right panes to make it easier to see all the tabs. Put the mouse cursor at the edge of the pane adjoining the central image area. When the cursor turns into a two-headed arrow, click and drag left or right to resize the pane.

Resizing an image in GIMP

You can use GIMP to resize an image. All computer images are made up of *pixels*, which are tiny colored dots. Sean's camera produces images that are 4,272 pixels wide and 2,848 pixels high. High-quality images like these are great for printing photos, but they're probably too big if you want to use a photo in a game or a website you're developing.

Here's how you can resize an image using GIMP:

1. **Open the image by choosing File ⇨ Open.**

2. **Click to open the Image menu at the top of the screen, and click Scale Image.**

   A window like the one in Figure 7-2 opens.

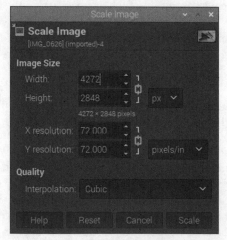

FIGURE 7-2:
The scale options in GIMP.

©1995–2012 Spencer Kimball, Peter Mattis, and the GIMP
Development Team

3. **In the Width box, enter in pixels the width you want the final image to be. Press Enter when you've finished entering the width.**

If you want to put a holiday snap on your website, you probably wouldn't want it to be more than 500 pixels wide. If you want to use a photo as a Scratch background (see Chapter 10), the ideal size is 480 by 360 pixels.

When you enter a new value for the width and press Enter or Tab, the height is updated automatically, so the image stays in proportion and doesn't become stretched. You can also enter a value for the height and have the width calculated automatically. If you want to be able to adjust the width and height independently, click the chain to the right of their boxes to break it.

4. **Alternatively, instead of using absolute values for the width and height, you can resize the image to a certain percentage. Click the Units drop-down list box (it says *px*) and choose Percent.**

The values in the Width and Height boxes will then be percentages. For example, you would enter 50% to shrink the image by half. The size of the image in pixels is shown under the Height box.

5. **When you've set the size, click the Scale button.**

TIP

At the bottom of the screen, underneath the Image pane, you can see some information about the file, including the current zoom level, which is how much the image has been magnified or reduced for you to view it. If you set this to 100%, you can get an idea of how much detail is in the image now, and it's easier to edit too.

WARNING

Resizing an image reduces its quality. This would be noticeable if you tried to create a high-quality print of it later. Don't overwrite an existing image with a resized version. Instead, save the resized image by choosing Save As from the File menu at the top of the screen and giving it a different filename.

Cropping your photo

If your photo has excessive space around an edge, or if you'd like to change the composition of the picture, you can cut off the sides, or *crop* it. To do that, follow these steps:

1. **Click the icon that looks like two overlapping right-angle tools, or press Shift+C to choose the Crop tool.**

2. **Click the image in the top left of the area you'd like to keep, hold down the mouse button, and drag the mouse down and to the right.**

When you release the mouse button, a box appears on the image, as you can see in Figure 7-3.

FIGURE 7-3:
Cropping a photo
in GIMP.

The inside of the box shows which bits of the image will be kept. Anything outside the box is cut off when you crop the image. You don't have to get the position or size of the box right first time, because it's easy to adjust.

3. **Click one of the corners and drag the mouse to change the size and shape of the box. You can also click and drag along an edge inside the box to adjust the width or height.**

4. **To reposition the box, click and drag in its center.**

5. **To crop the image, click inside the box or press the Enter key.**

TIP

If you make a mistake, you can use Ctrl+Z to undo.

Rotating and flipping your photo

If you rotate your camera sideways to take a picture, you might need to rotate the resulting image too. The easiest way to do this is to click to open the Image menu and then use the Rotation options on the Transform submenu there. You can rotate clockwise or anticlockwise (counterclockwise) by 90 degrees, rotate the image by 180 degrees, and flip it horizontally or vertically.

For a simple rotation like this, it's quicker to rotate a photo using the desktop's Image Viewer. (See Chapter 4.)

If you have a photo that's slightly wonky, you can manually adjust it in GIMP. Click the Rotate tool and then click the image (or press Shift+R) and you can enter an angle for rotation, or click and drag the image to rotate it. To change the pivot point about which the picture rotates, click the circle in the middle of the image and drag it.

This Rotate tool (unlike the options in the Transform menu) only works on the currently selected layer. (You can select a layer by clicking it in the bottom pane on the right.) If you plan to make other modifications to your photo, we recommend you correct its rotation before you add other layers.

Adjusting the colors

In common with other image editing programs, GIMP has options for adjusting the colors in a photo. You can find all these options on the Colors menu at the top of the screen. If your picture has a tint of color you don't want, or if you would like to add a tint, use the Color Balance settings to alter the amount of cyan, magenta, and yellow in the image. The Brightness and Contrast settings can help to bring out detail in shadows, or to give the image more impact.

There are also options on this menu (farther down, under Auto) to automatically adjust the colors using six different methods. These can give strange and undesirable results, but you can always undo them with Ctrl+Z if you don't like them. The Equalize option can be a quick fix for images that look wishy-washy, and the White Balance option can fix pictures that don't already have strong black and white areas.

The Color options only affect the currently selected layer. If you have added layers to your image, click the photo layer in the bottom pane on the right before adjusting the color.

Fixing imperfections

On Sean's holiday to Australia, he found a beautiful unspoiled beach in Darwin. He took a picture of it: a lone tree in the foreground, the shimmering sea, and wisps of cloud in a light blue sky. When he got home, he noticed that some idiot had left a crushed beer can in the foreground.

Thankfully, in GIMP, you can use a handy tool called the Clone tool to make little details like this vanish. It enables you to use part of the image as a pattern that you spray over another part of the image. In Sean's case, he can use a clean piece of beach as the pattern and spray it on top of the litter. Hey, presto! The rubbish vanishes.

Here's how you use the Clone tool:

1. **Zoom in to the image using the menu underneath it, and then use the scroll bars at the side of the Image pane to position the image so that you have a clear view of the imperfection.**

2. **Click the Clone tool, which looks like a rubber stamp, or press the C key.**

3. **Move the cursor to an unspoiled part of the image you would like to use as the pattern, or *clone source*.**

   This needs to be somewhere as plain as possible, more of a texture than a shape, with no obvious prominent details or lines. Sky, grass, or sand is perfect.

4. **Hold down the Ctrl key and click the mouse button.**

   A crosshair icon appears on your image at that spot.

5. **In the tool options, at the bottom of the left pane, you can see the brush that is being used. Click the shape (a circle, by default) if you want to change it.**

   For the best results, use a brush with a fading edge rather than a solid edge. You can change the size of the brush in this pane, too, by clicking the Size box and typing your preferred value. The bigger the brush you use, the bigger the pattern.

6. **Move the cursor to the imperfection in the image, and click the mouse button.**

   This copies an area the size of the brush from the clone source to the place where you clicked.

   If you've done it right, the imperfection should appear to vanish. If you see unwanted picture details included in the pattern, either reduce the size of the brush or move the clone source. Repeat this step until the imperfection is gone.

You can hold down the mouse button as you move the mouse to clear larger areas, but be aware that the clone source also moves as you move the mouse. If you replace a large area, you're likely to stray into a distinctive part of the image with the clone source, which will break the effect.

7. **Adjust the magnification at the bottom of the Image pane to view the image at 100%.**

   Check whether you can see any evidence of your handiwork. If so, you might need to try another clone source or brush size. Otherwise, you've succeeded!

Converting images between different formats

There are several different file formats that can be used for images, but not all programs can open all files. If you want to use a picture as a background in Scratch (see Chapter 10), for example, you need to use .jpg files, which usually deliver the best quality for photos, or .png or .gif files, which are optimal for illustrations.

The default format used by GIMP is .xcf, which stores additional information about your editing session along with the picture, but this format isn't widely used in other programs.

You can use GIMP to save the picture in a more widely used format, or to convert a picture between different file formats. First, open the image from the File menu, and then select that menu's Export As option. The Export As window looks a lot like the Save window, but here you can click Select File Type (By Extension) at the bottom and choose the file format you'd like to convert the image into.

WARNING

The conversion is quite memory-intensive, so on an older Raspberry Pi you might need to resize (shrink) a digital photo before you convert it.

Finding Out More about GIMP

There's much more you can do with GIMP, and you can find detailed documentation on its website. To access it, click Help at the top of the screen, and click User Manual to open a menu showing the different sections. When you select one, it opens in your browser. Alternatively, in any browser, go to https://docs.gimp.org/2.8/en.

Chapter **8**

Playing Audio and Video on the Raspberry Pi

n this chapter, we show you how you can turn your Raspberry Pi into a media center, capable of playing high-definition video files and music.

To do that, we use dedicated media player software based on the Kodi software that powers some set-top boxes and smart TVs. You can use it to play music and video you have on storage devices connected to your Raspberry Pi, or to play back media from other devices on your home network. To a more limited degree, you can also use it to play back streaming TV shows and radio stations from the Internet.

At the end of this chapter, we also show you how to play music on your Raspberry Pi in the desktop environment.

Setting Up Your Media Center

The Raspberry Pi can play back full HD 1080p video, which makes it ideal as the heart of a cheap and low-powered media center.

To start setting up your media center, create a microSD card with LibreELEC on it (see Chapter 2). LibreELEC is based on Kodi and is the recommended media player software for the Raspberry Pi. Different versions are available for the Raspberry Pi 4, Raspberry Pi 2/3, and Raspberry Pi Zero or original Raspberry Pi.

When you first boot up LibreELEC, you'll be guided through its main settings, including its name on the network, the network connection itself, and the remote access settings (so that you can access files on your device over the network).

Navigating the Media Center

The Kodi screen looks like Figure 8-1. LibreELEC uses Kodi's simple interface, which is designed to work with only a remote control. In this chapter, we assume you're using a mouse, but we give you some pointers on using remote controls in the section "Using a Remote Control," later in this chapter. If you're using a remote control, you should find the interface intuitive.

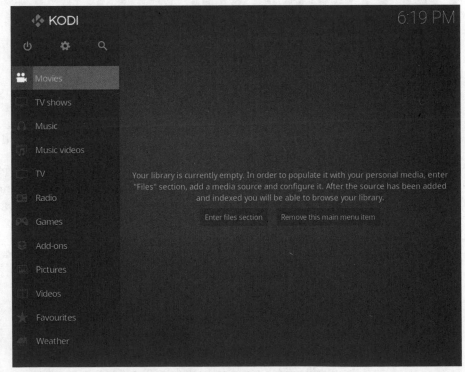

FIGURE 8-1:
The main menu on the left and the empty library area on the right. This shows how Kodi appears before media is added.

Sean McManus

The menu on the left side of the screen gives you access to the different content types, including movies, TV shows, music, and music videos. When you hover the mouse pointer over one of these options, the main screen area on the right shows you options and submenus for accessing your content. The Music section, for example, shows your top 100 songs and albums, recently played albums, recently added albums, and albums organized by genre, artist, and year. You can use the cursor keys to move around the options; or the mouse, with the scroll wheel moving between the options within a menu. To exit a menu, press the Escape key.

To select an item and start to play it, simply click it. You can pause playback by tapping the Spacebar. Tap Escape or left-click the mouse when the pointer is on the background to get back to the menu.

TECHNICAL STUFF

To use the TV and radio features, you'll need to have a separate tuner device and TV signal decoding software. For more information, see `https://kodi.wiki/ view/PVR`. The official Raspberry Pi TV HAT enables you to receive TV streams on your Raspberry Pi through a TV aerial. It costs about $28.

At the top of the menu on the left are buttons for switching off your Pi, accessing the settings menus (the Cogwheel icon), and searching for content (the magnifying glass).

Adding Media

Before you can play back content, you need to tell Kodi where it can find it. You have several options for providing it:

» **USB drive:** You can plug a USB drive that stores your movies or music directly into your Pi or its USB hub. A message appears in the top right, confirming that the USB device is being mounted, which means it's being prepared so that you can use it.

» **Networked media:** You can connect your Pi to your home network and then access other devices on the same network. Sean, for example, was able to connect his Pi to his Windows PC over the network and use Kodi on the Pi to play back the music and movies stored there. You might have a router with a built-in media server, so it can share any files on USB devices you connect to it. These networked devices most likely use the UPnP (Universal Plug and Play) standard.

Kodi can create a library of your media and index it to provide you with easy access to it. You'll need to add your content to the library first to benefit from this.

Adding music

To add music to your library, follow these steps:

1. **Click Music in the menu on the left.**

If you've already added music, you'll see options to browse it by artist, album, and year, among other categories.

2. **Choose Files from the menu.**

This will show you the folders you have added to your library already, and any connected storage devices.

3. **Choose Add Music.**

4. **Choose Browse in the Add Music Source dialog box that opens and use the options to find where your music is stored (see Figure 8-2).**

If you've connected a USB drive to your Pi, you can find it by clicking *Root filesystem*, *media*, and then the name of your USB device. To find media connected on your network, try the Windows Network (SMB) or Network File System options.

FIGURE 8-2: The browsing options for adding media to your Kodi installation.

Sean McManus

REMEMBER

You may need to enter your username and password, and you should enable the option to remember them for this path if you don't want to have to enter them whenever you access your media. Click folders to open them. The .. (two dots) option at the top takes you up a level in the folder structure.

5. **Click OK when you have navigated to the folder that contains all the music folders or files you want to add.**

6. **In the Add Music Source dialog box enter a name for the media source, if you want.**

   This will help you to identify the source on the menus.

7. **Click OK.**

8. **When prompted, confirm that you'd like to add the media to your library.**

TIP

Content that Kodi doesn't recognize, including home recordings, won't be added to the library, but you can still view them in the folder you added by using the browser. Start by choosing Music from the menu on the left, choose Files (as you did when adding media), and then select the folder that contains your added media. From there, you can open the folders and select the music files to play them.

Adding videos

To add movies to your library, follow these steps:

1. **Click Movies in the menu on the left.**

   If you can't see it, it's probably off the top of the screen. Hover the mouse cursor over the menu and use the scroll wheel to bring the Movies option back again.

   If you've already added some movies, you'll see a list of them with .. (two dots) at the top of the list. In that case, click .. twice to go up a level and then choose Files. This will show you the folders you have added to your library already.

2. **Choose Add Videos at the bottom of the folder list.**

3. **Choose Browse and use the options to find where your movies are stored.**

   If you've connected a USB drive to your Pi, you can find it by clicking *Root filesystem, media,* and then clicking the name of your USB device. To find media connected on your network, try the Windows Network (SMB) or Network File System options. Click folders to open them. The .. option at the top takes you up a level in the folder structure.

4. **Click OK when you have navigated to the folder that contains all the video folders or files you want to add.**

5. **In the Set Content options that open, click where it says *This directory contains* and select Movies.**

6. **(Optional) If your movies are stored in separate folders that match the movie title — a common way of organizing them — turn that option on.**

7. **Click OK.**

   If prompted, confirm that you'd like to refresh items in this path.

TIP

Content that Kodi doesn't recognize, such as your home movies, won't be added to the library, but you can still view them in the folder you added by using the browser. Start by clicking Movies on the left, clicking .. twice, choosing Files, and then selecting the folder that contains your added media. From there, you can open the folders and select the movie files to play them.

You can add TV shows or music videos using a similar process — just change the media type to TV Shows or Music Videos in Step 5.

Adding pictures

To add photos, you follow a similar process to adding music and movies. Start by clicking Pictures from the main menu on the left, and then choose Add Pictures from the menu that appears. You can then add a folder of photos.

To view the folder, hover over Pictures on the main menu on the left and then select the folder name. You can use the left- and right-arrow keys or the scroll wheel to navigate your photos and click Options in the bottom left to open a menu that enables you to display a slide show.

If you're in a folder and you want to view a different folder or add more pictures, click .. at the start of the photo folder.

Streaming media

Streaming media means that the content flows into your Raspberry Pi over the Internet as you watch it or listen to it. As a result, streaming services only work when you have a good Internet connection.

To enable streaming, you use *add-ons*, which are third-party applications that access sources of content online. Music add-ons, for example, enable you to listen to Internet radio stations and access some online music services. Video add-ons can give you access to online TV stations. The availability of add-ons varies over time as new services launch and older services disappear. The music and video add-ons can be particularly short-lived and unreliable because broadcasters are often keen to keep viewers using their own software and gateways.

To install an add-on, hover over Add-ons on the main menu on the left, and then choose Install from Repository from the menu at the top of the screen. From the new menu that appears, choose Kodi Add-On Repository to find add-ons from the official Kodi repository, which offers a range of add-ons that have undergone basic testing.

As well as music and video add-ons, there are options for a weather forecast provider, screensavers (in the Look and Feel category), and picture add-ons. Note that if you add a screensaver, you'll also need to enable it in the Interface settings. If you install a weather add-on, you'll need to select it in the Services settings.

Enjoying Your Media

After you've added your music, movies, and pictures to your Raspberry Pi, you can listen to, watch, and view them.

Playing music

To get started with playing music, hover over Music on the main menu to show your music collection in the main part of the screen. You can use the Categories menu at the top to browse by genre, artist, album, or song, among others. You can also choose Files from the Categories menu to go to the music folders you have added. In the main part of the screen are options for recently played or added albums, and random artists and albums too. Use the scroll wheel to go left and right through the media and options, and the up and down keys to move through the rows of options. Click an album to play it.

From the top of the main menu, click the icon showing four arrows, like a compass, to view the song full screen, where there are options to play, pause, stop, or skip the track. There are also options for repeating the track or setting a random song order. Note that there is a screensaver, so the screen display disappears after a moment or two. Press Escape to go back to the menu. You can tap the spacebar to pause or unpause the audio.

You can continue to browse your music while it is playing, and queue songs or albums to play next, by right-clicking them and choosing Queue Item from the menu that appears.

The Playlists option in the Categories section of the Music menu enables you to create lists of songs for playback that you can save and play whenever you want. Start by clicking New Playlist in the Playlists option to create a new playlist. To add a song from the Playlist browser, right-click its name and click Add from the menu that opens. When you've finished creating your playlist, choose Save from the menu on the left.

You can also make *smart* playlists, which are playlists that are generated from rules, such as songs that belong to particular years, genres, or artists. You need to add your rules before you can give your playlist a name. You use a simple menu system to set up your playlist rules.

REMEMBER

Smart playlists can only contain songs that have been added to the music library, but standard playlists can contain songs from any of your connected media devices.

Playing videos

Use the Movies, TV Shows, Music Videos, and Videos options to view your video content. What's available under each option will depend on the media you added in each section, and what type of media you told Kodi it was at the time (such as movie or TV show).

Note that content protected by digital rights management, including files bought from iTunes, can't be played in Kodi.

To play movies, hover over Movies on the main menu to see options for different categories, recently added movies, random movies, and unwatched movies. Click a movie to start watching it. You can also click Movies on the main menu to see a list of your movies, together with artwork and a synopsis downloaded from the Internet. Click the .. option at the top of this menu, and then click it again on the next menu to get to your files if you want to play media that isn't indexed in the library. As with audio, you can tap the spacebar to pause or unpause a movie.

Viewing photos

Click Pictures from the main menu to see a simple file browser that you can use to access your pictures, with your added folders listed. The media center software supports standard image formats, including JPEG, BMP, and TIFF, and generates thumbnails of your photos and folders. Navigate to a photo and click it. You use the left and right keys to move through your photos, and the Escape key (or right-click the mouse) to return to the menu.

Changing the Settings

The cogwheel at the top of the main menu gives you access to the settings for Kodi, divided into several sections. They include, among others:

>> **Player settings,** covering options such as whether the next song or video plays automatically, whether there is cross-fading between songs, and the use of subtitles.

>> **Media settings,** which enable you to manage your sources of media in the library and choose information providers for artist and album information.

>> **Interface settings,** which enable you to choose a skin (or design) for Kodi, your region, and a screensaver. If you install a screensaver add-on, you'll need to enable it here.

>> **System settings,** which can be used to choose the correct audio output device, configure notification sounds in Kodi, and configure power saving and the Internet connection.

>> **LibreELEC,** which includes options for managing the Internet connections available to the system and for enabling Secure Shell (SSH) for remote access to your Pi.

>> **File Manager,** which enables you to copy files between different folders and storage devices.

REMEMBER

By default, the sound is set to pass through your HDMI cable. If you want to use your Raspberry Pi's audio output (and speakers) but you're using HDMI for your screen, you'll need to change the audio setting from HDMI to Analog. You'll find this option on Kodi's System Settings menu. When you're viewing the settings menu, select Audio on the left and select Audio Output Device.

Using a Remote Control

Considering all the functionality we've covered in this chapter, you're not that far away from running a low-power home media center. To complete it, you can use a remote control.

There are many ways to remotely control your Raspberry Pi media center. You can use a USB device, a cheap infrared remote, a keyboard remote, or even your games console controller, if you have one. Or you can use the Kodi remote app (available for iOS and Android operating systems) to control Kodi on your Raspberry Pi.

You can find the remote-control settings on Kodi's System Settings menu, and then under Input. You might find that your remote control works without needing to change any settings, so try it first.

If you have a television that supports the HDMI CEC (Consumer Electronics Council) standard, a neat option is to enable your existing television remote to control your Pi. To do this, connect your networked Pi to your television's HDMI socket. Your Raspberry Pi appears as a new input. Use the TV's remote control to change to this input, and your Home screen then appears on the television.

Turning Off Your Media Center

To turn off your media center properly, click the Shutdown button in the top left of the Kodi main menu. You can just switch off the power supply, but you minimize the risk of corrupting your microSD card by shutting down properly.

Playing Music in the Desktop Environment

Looking for musical inspiration while you program? The good news is that you can also play music and video from the Raspberry Pi OS desktop environment (see Chapter 4). If you're still using LibreELEC, shut down your Raspberry Pi using the Shutdown icon in the top left and then reboot into Raspberry Pi OS. You will need to swap microSD cards.

VLC Media Player is a music and video player that works on the Raspberry Pi desktop. It's in the version of the desktop with recommended applications. If you don't have it, you can install VLC at the command line (see Chapter 5) using:

```
sudo apt install vlc
```

After VLC has been installed, you can find it in the Sound and Video category of the Applications menu and click its name there to start it up.

On the Media menu of VLC are options to open a file or directory. Usually, you'll want to open a directory so that you can play a whole album. By default, VLC shows you the album artwork (where available), but you can open the View menu and choose Playlist to see a list of songs so that you can pick another to play (as shown in Figure 8-3). In the box on the left, you can pick a device to play from, including several Internet services for streaming music. Choose Playlist from the View menu again to revert to the full-window artwork.

The playback controls to pause, play, skip, shuffle, and repeat songs are at the bottom left of the window. The volume control is at the bottom right.

FIGURE 8-3: Playing music in VLC Media Player.

Written by Hong Jen Yee and Juergen Hoetzel; icon by Arnaud Didry

4

Programming the Raspberry Pi

Get familiar with the Scratch interface and how you can use it to create your own simple animations or games programs.

Use Scratch to build an arcade game, which you can customize with your own artwork.

Learn how to use Python to create a times table calculator and Chatbot, a program that simulates basic artificial intelligence.

Use Pygame Zero with Python to create a simple arcade game that you can customize with your own sounds and artwork.

Discover how you can use Python to build worlds in Minecraft, including a maze you can explore.

Use Sonic Pi to compose your own computer music on your Raspberry Pi.

Chapter **9**

Introducing Programming with Scratch

The Raspberry Pi was created partly to inspire the next generation of programmers, and Scratch is the perfect place to start. With it, you can make your own cartoons and games and discover some of the concepts that professional programmers use every day.

Scratch is designed to be approachable for people of all ages. The visual interface makes it easy to see what you can do at any time, without having to remember any strange codes, and you can rapidly achieve great results. Scratch comes with a library of images and sounds, so it takes only a few minutes to write your first Scratch program.

In this chapter, we introduce you to Scratch so that you can start to experiment with it. In Chapter 10, we show you how to use Scratch to make a simple arcade game.

Understanding What Programming Is

Before we dip into Scratch, we should clear up some of the jargon surrounding it. A *program* is a repeatable set of instructions to make a computer do something, such as play a game. Those instructions can be extremely complicated because they have to describe what the computer should do in detail. Even a simple bouncing-ball game requires instructions for drawing the ball, moving it in different directions, detecting when it hits something, and then changing its direction to make it bounce.

Programming is the art and science of creating programs. You can create programs in lots of different ways, and Scratch is just one of them. In Chapter 11, you read about Python, another one.

Scratch and Python are both *programming languages*, different ways of writing instructions for the computer. Different programming languages are best suited for different tasks. Scratch is ideal for making games, for example, but it's not much use if you want to create a word processor. Using Python to create games takes longer, but it is more powerful than Scratch and gives you much more flexibility in the type of things you can get the computer to do.

Working with Scratch

There are two versions of Scratch in the recommended programs for Raspberry Pi OS:

>> **Scratch:** This is the original version of Scratch, widely known as Scratch 1.4. This version was optimized to perform well on the early Raspberry Pi models. If you're using an old Raspberry Pi model, you may prefer to use this version because it's faster, although it doesn't have all the features of the latest Scratch version. The screen layout is similar to Scratch 3, except that the Code Area is called the Scripts Area, and the tabs for Scripts, Costumes, and Sounds are above the Scripts Area. Additionally, the buttons to select different parts of the Blocks Palette are above it, instead of to its left.

>> **Scratch 3:** If you use the online version of Scratch (at https://scratch.mit.edu), Scratch 3 is the version you're familiar with. New features in this latest version of the language include Extensions. These enable you to add new capabilities to Scratch, including for text to speech, and for electronics on the Raspberry Pi. We recommend you use Scratch 3, so you can benefit from the latest improvements to Scratch.

In this chapter, we assume you're using Scratch 3.

To start Scratch, select your chosen version from the Applications menu in the upper left of the screen. You can find all three versions of Scratch in the Programming folder. If they aren't installed on your computer, you can add Scratch using the Recommended Software application in the Applications menu. It's in the Preferences folder.

Understanding the Scratch screen layout

Scratch divides the screen into four main areas, as you can see in Figure 9-1. The Stage is where you can see your game or animation take shape. There's a cat on it already, so you can get started straightaway by making it do things, as you see in a minute. The Stage is in the upper right.

Underneath the Stage is your Sprite List. You can think of sprites as the characters in your game. They're images that you can make do things, such as move around or change their appearance. For now, there's just the cat, which has the name Sprite1.

You create a Scratch program by snapping together *blocks*, which are short instructions. You find the blocks in the Blocks Palette, which is on the left. It displays the Motion blocks by default. They include instructions to move ten steps, rotate, go to a particular grid reference, and point in a particular direction.

The Code Area is where the magic happens! You assemble your program in this space by dragging blocks into it from the Blocks Palette. The Code Area is between the Blocks Palette and the Stage.

There are two buttons in the upper right of the screen (refer to Figure 9-1) to toggle the Stage between full size and small. When the Stage is small, the Code Area is bigger, so you may find that useful when you're writing scripts later in this chapter. You can also make the Stage fill the screen when you're running your program.

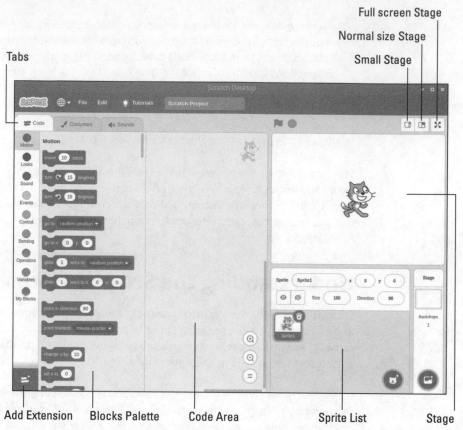

Full screen Stage

Normal size Stage

Small Stage

Tabs

FIGURE 9-1:
The screen layout in Scratch.

Add Extension   Blocks Palette   Code Area   Sprite List   Stage

Scratch is developed by the Lifelong Kindergarten Group at the MIT Media Lab. See https://scratch.mit.edu.

Making your sprite move

You can drag and drop your sprite (the cat) around the Stage to position it where you would like it to be at the start of your program.

Experimenting with Scratch is easy. To try out different blocks, just click them in the Blocks Palette. For example, try clicking the Move 10 Steps block, and you should see the cat move in the direction it is facing, which is to the right. You can also turn the sprite 15 degrees in either direction by clicking the appropriate blocks.

TIP

If your cat goes somewhere that you don't want it to (don't they always?), you can click it on the Stage and drag it back to where you want it. In the Sprite List, you can also edit the sprite's x and y position, direction, and visibility (using the eye icons), if you're using Scratch 3. Set x to 0, y to 0, and direction to 90 to reposition your sprite in the middle of the screen, facing right.

Not all of the blocks will work at the moment because some of them need to be combined with other blocks. There's no harm in experimenting, however. Even if you click something that doesn't work, you won't cause any harm to Scratch or your Raspberry Pi.

Next, we talk you through the different Motion blocks you can use.

Using directions to move your sprite

You can use two different methods to position and move sprites. The first is to make your sprite "walk," and to change its direction when you want it to walk the other way.

Here are the five blocks you use to move your sprite in this way:

» **Move 10 Steps:** This makes your sprite walk in the direction it is facing. If your sprite has been rotated, the steps taken could move it in a diagonal line across the Stage. You can click the number in this block and then type another number to increase or decrease the number of steps taken, but much bigger numbers spoil the illusion of animation. It stops looking like the sprite is walking across the screen when the number of steps taken is too big.

» **Turn Clockwise or Counterclockwise 15 Degrees:** These two blocks rotate your sprite. As with the number of steps, you can edit the number to change the degree by which your sprite is rotated. It walks in the direction it is facing when you use the Move 10 Steps block.

» **Point in Direction 90:** Whatever direction your sprite is facing, this block points it in the direction you want it to face. Use this block as is to reset your sprite to face right. You can change the number in this block to change the direction you want your sprite to face, and the numbers are measured in degrees from the position of facing up (see Figure 9-2). It helps to think of it like the hands of a clock: When the hand is pointing right, it's 90 degrees from the 12 o'clock position; when it's pointing down, it's 180 degrees from the top. To point left, you use –90. When you click the number box to type into it, a dial appears that you can use to select the angle.

TIP

You might be wondering whether you can use 270 to point left instead of –90. Try it. You'll see Scratch changes 270 to –90 if you type it into the block. It is possible to force the block to accept a value of 270 by using a variable. (You learn about variables in Chapter 10.) This can cause errors in your program, though. If you turn your cat to direction 270 and then ask Scratch which way the cat is facing, it tells you –90. To avoid any inconsistencies like this, keep direction numbers in the range from –179 to 180.

>> **Point Towards:** You can also tell the sprite to point toward the mouse pointer or another sprite. Use the menu in this block to choose what you would like your sprite to point toward.

>> **Set Rotation Style:** When the cat is facing left, it appears to stand on its head. You can change the rotation style to fix this. The style left-right keeps your sprite upright. The style all around will restore it to facing the direction it's moving in. There is also an option in this block to not rotate the sprite at all, even if its direction changes.

>> **If On Edge, Bounce:** This block reverses the direction of your sprite if it's touching the edge of the Stage. It's useful for games or animations where you want sprites to bounce off the edges of the Stage.

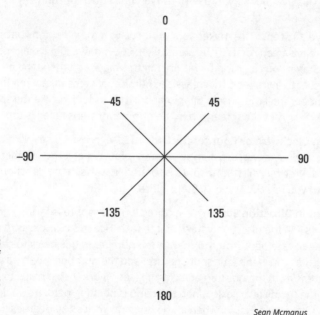

FIGURE 9-2: The number of degrees used to face in different directions.

Sean Mcmanus

REMEMBER

If you're changing the number value in a block, you still need to click the block to run it.

Using grid coordinates to move and position your sprite

The second way you can move and position your sprite is to use grid coordinates. That makes it easy to position your sprite at an exact place on the screen, irrespective of where it is now.

Every point on the Stage has two coordinates: an X position (indicating where it is horizontally) and a Y position (indicating where it is vertically). The X positions are numbered from −240 at the far left to 240 at the far right. The Y positions are numbered from −180 at the bottom edge of the Stage to 180 at the top edge. That means the Stage is a total of 480 units wide and 360 units tall. The center point of the screen, where your cat begins its day, is where X equals 0 and Y equals 0. Figure 9-3 provides a quick visual reference of how the coordinates work.

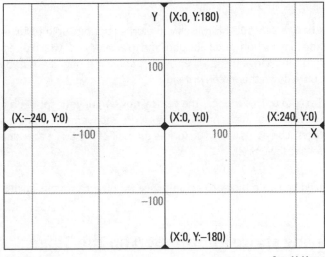

FIGURE 9-3:
The grid coordinates on the Stage.

Sean McManus

Seven Motion blocks use the X and Y coordinates:

» **Go to x:0 y:0:** You can use this block to position your sprite at a specific point on the Stage.

» **Go to:** Use this block to move your sprite to a random position, the mouse pointer's location, or to the location of another sprite, if you have more than one.

» **Glide 1 secs to:** When you use the Go To block, your sprite just jumps to its new position. The Glide block makes your sprite float there smoothly instead. Like the Go To block, you can use this block to glide your sprite to a random position, the mouse pointer, or another sprite. You can change the number of seconds the glide takes, including using decimals for part of a second (for example, 0.5 for half a second).

» **Glide 1 secs to x:0 y:0:** Use this block to smoothly move your sprite to a new coordinate on the screen.

» **Change X by 10:** This moves your sprite ten units to the right. You can change the number of units and use a negative number if you want to move left instead. Note that this doesn't affect your sprite's vertical position and is independent of which way around your sprite is facing.

» **Set X to 0:** This changes the horizontal position of your sprite on the Stage, without affecting its vertical position. The value 0 returns it to the center of the screen horizontally, and you can edit the number to position it left or right of that. Use a negative number for the left half of the screen and a positive number for the right half.

» **Change Y by 10:** This moves your sprite ten units up the Stage, without affecting its horizontal position, and irrespective of which direction it is facing. You can change the number of units and use a negative number to move the sprite down the screen instead.

» **Set Y to 0:** This changes the vertical position of your sprite on the Stage without affecting its horizontal position, and without regard to which way it faces. Use a positive value for the top half of the Stage and a negative value for the lower half.

REMEMBER

You need to run a block to actually see its effect on your sprite. Do this by clicking it.

Showing sprite information on the Stage

It can be hard to keep track of where your sprite is and in which direction it's facing, but you can show the values for its X position, Y position, and direction on the Stage. Select the check boxes at the bottom of the Motion part of the Blocks Palette to do this, as shown in Figure 9-4. They clutter the screen a bit, but they can be essential tools for testing when you're creating a game.

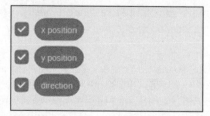

FIGURE 9-4:
The blocks used to show sprite information on the Stage.

Scratch is developed by the Lifelong Kindergarten Group at the MIT Media Lab. See https://scratch.mit.edu.

You can also refer to the Sprite List, where you can see a sprite's coordinates, size, and direction in Scratch 3.

Creating scripts

Clicking blocks in the Blocks Palette is one way to issue commands to Scratch, but if that's all you're doing, you're not really programming. The fact is, if you have to click each block every time you want to run it, you're doing all the hard work of remembering the instructions, and the computer can work only as fast as you can click the blocks.

A *program* is a reusable set of instructions that can be carried out (or *run*) whenever you want. To start to create a program, you drag blocks from the Blocks Palette and drop them in the Code Area in the middle of the screen. Most blocks mentioned so far in this chapter have a notch on the top and a lug on the bottom, so they fit together like jigsaw pieces. You don't have to align them perfectly: Scratch snaps them together for you if they're close enough when you release the mouse button.

You put your blocks in the order you want Scratch to run them, starting at the top and working your way down. It's a bit like making a to-do list for the computer.

A group of blocks in the Code Area is called a *script,* and you can run the script by clicking anywhere on it. Its border is highlighted, and you'll see the cat move around the Stage as you've instructed it to.

You can have multiple different scripts in the Code Area, so you could have one to make the cat walk left and another to make it walk right, for example. When you add multiple sprites (see Chapter 10), each sprite has its own Code Area and scripts there to control it.

TIP

If you want to tidy up the Code Area, you can move a script by dragging its top block. If you drag a block that is lower down in the script, it's separated from the blocks above it and carries with it all the blocks below it. If you want to delete a block or set of blocks, drag it back to the Blocks Palette on the left.

Figure 9-5 shows a script Sean built using some of the Motion blocks. Try building it in your Code Area, by dragging in the blocks and joining them together. Remember to change the numbers in the blocks. When you've finished, click the script to run it. The script makes the cat go to the middle of the screen, walk around the Stage in a square shape, and then point toward the mouse pointer. As you learn about new blocks in the rest of this chapter, you can try adding them to this script, or build your own new script.

Changing your sprite's appearance

As well as moving your sprite around the screen, you can change what it looks like.

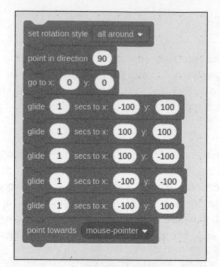

FIGURE 9-5:
A simple script to make the cat walk around the Stage.

Scratch is developed by the Lifelong Kindergarten Group at the MIT Media Lab. See http://scratch.mit.edu.

Using costumes

One way to think of sprites is as the characters in a game (although they can be used for lots of other objects, too, such as obstacles). Each sprite can have a number of *costumes*, which are different pictures of it. If the costumes look fairly similar, you can create the illusion of animation by switching between them. Your cat sprite comes with two costumes, and when you switch between them, it looks like the cat is running. You can think of a costume as being one image in an animation sequence (an *animation frame*).

You can see the costumes for your sprite by clicking the Costumes tab at the top of the Blocks Palette, as shown in Figure 9-6. If you want to modify the cat's appearance, you can click the costume on the left and use the editing canvas to its right. If you want to create a new animation frame, you can right-click the costume and choose duplicate from the menu that opens. You can then edit the bits you want to change.

TIP

It doesn't matter much when you're experimenting with sprites, but when you make your own games and animations, you can save yourself a lot of brain ache by giving your costumes meaningful names. It's much easier to remember that the costume with the name *game over* should be shown when the player is defeated than it is to remember it's called *costume7*. To rename a costume, click the Costumes tab to show the costumes, and then click the costume's current name (refer to Figure 9-6) and type its new name. The costume's name is shown above the editing canvas.

Costume name   Costumes tab

FIGURE 9-6:
You can
change a sprite's
appearance by
giving it a new
costume.

Scratch is developed by the Lifelong Kindergarten Group at the MIT Media Lab. See
https://scratch.mit.edu.

When you've finished using the Costumes Area, click the Code tab to get back to the Code Area.

In the Blocks Palette, there are two blocks you can use to switch between costumes (see Figure 9-7). Click the Looks button beside the Blocks Palette to show them:

>> **Switch Costume to:** If you want to switch to a particular costume, choose its name from the menu in this block and then click the block.

>> **Next Costume:** Each time you use this block, the sprite changes to its next costume. When the costumes run out, it goes back to the first one again.

TIP

You can show a sprite's costume number on the Stage, too, so that it's easier for you to work out what's going on. Just select the check box next to Costume # in the Blocks Palette. In that block, you can choose to show the costume's name instead.

Looks button

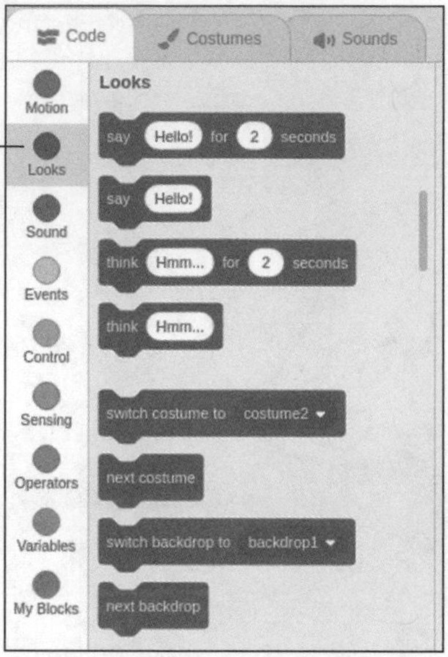

FIGURE 9-7:
Some of the
Looks blocks you
can use to change
your sprite's
appearance.

*Scratch is developed by the Lifelong
Kindergarten Group at the MIT Media Lab.
See https://scratch.mit.edu.*

Using speech and thought bubbles

The Say blocks display a speech bubble, and the Think blocks show thought bubbles (see Figure 9-8). To see them, and to see the other blocks that change a sprite's appearance, click the Looks button beside the Blocks Palette. The speech and thought bubbles are great for giving a message to the player or viewer. You can edit the word in the block (*Hello!* or *Hmm...*) to change the text in the bubble. Figure 9-8 shows the speech bubbles (at the top) and thought bubbles (center and bottom) in action.

If you use one of the options with a length of time in it, the sprite pauses for that length of time and the bubble disappears when it has elapsed.

If you use a block without a length of time, you can make the bubble disappear again by using the Say or Think block again but editing the text so that the text box in the block is empty.

Using graphic effects

You can apply several graphic effects to your sprite using Looks blocks. In Figure 9-8, we've used eight sprites to demonstrate them on the Stage. The Color effect changes the sprite's Color Palette, turning orange to turquoise in the case of the cat. The Fisheye effect works like a fish-eye lens, making the central parts of the sprite appear bigger. Whirl distorts the sprite by twisting its features around its middle. Pixelate makes the sprite blocky. Mosaic shrinks the sprite and repeats it within the space it usually occupies. The Brightness and Ghost effects can sometimes look similar, but the Brightness effect changes the intensity of the colors, and the Ghost effect fades out all colors evenly. In Figure 9-8, we've used a negative number with the Brightness effect to make the sprite darker.

Scratch is developed by the Lifelong Kindergarten Group at the MIT Media Lab.
See http://scratch.mit.edu.

FIGURE 9-8:
The different graphic effects you can apply to your sprite, with thought bubbles and speech bubbles used to describe them.

Here are the three blocks you use to control graphic effects:

>> **Change Color Effect by 25:** You can select the effect you want to change (by default, it's the color effect) and enter the amount of it you want to add. You can use negative numbers to reduce the extent to which the effect is applied to your sprite. The color effect has 200 different levels (from 0 to 200), and the ghost effect has 100 different levels from 0 to 100. The other effects typically look best with levels in the range from –100 to 100. Experiment!

>> **Set Color Effect to 0:** Use this block to set a chosen effect to a specific level. Choosing 0 turns the effect off again. You can use any of the seven effects with this block.

>> **Clear Graphic Effects:** This block removes all graphic effects you've applied to a particular sprite so that it looks normal again.

Resizing your sprite

You can use blocks to change a sprite's size, so you could make it grow larger as the game progresses, for example.

There are two blocks you can use to resize your sprite:

» **Change Size by 10:** This block enables you to change the size of your sprite by a certain number of units, relative to its current size. As usual, you can edit the number. If you want to decrease the sprite's size, use a negative number.

» **Set Size to 100%:** This block sets the size to a percentage of its original size, so with the default value of 100 percent, it effectively resets any resizing you've done.

TIP

You can also select the check box beside the Size block to show the sprite's size on the Stage, in the same way you display other sprite information there. (See "Showing sprite information on the Stage," earlier in this chapter.) This can be useful for testing purposes.

Changing your sprite's visibility

Sometimes, you might not want your sprite to be seen on the Stage. If a spaceship is blown up in your game, for example, you want it to disappear from view. These two blocks give you control over whether a sprite is visible:

» **Hide:** Use this block to make your sprite invisible on the Stage. If a sprite is hidden, Scratch won't detect when it touches other sprites, but you can still move a hidden sprite's position on the Stage so that it's in a different place when you show it again.

» **Show:** By default, your sprite is visible, but you can use this block to reveal it again after you have hidden it.

TIP

Sometimes, sprites might get on top of each other. You can use the Go to Front Layer block to make a sprite appear on top of all the others, or use the menu in it to change it to Go to Back Layer, so you can force a sprite to go behind all the others. The Go Forward 1 Layers block enables you to move sprites forward and backward so you can precisely control which sprites are on top of which other sprites.

Adding sounds and music

As well as changing a sprite's appearance, you can give it some sound effects. Scratch comes with sounds, including slurps, sneezes, and screams; ducks, geese, and owls; and pops, whoops, and zoops. You can find effects for most occasions, and many of them are natural partners for one of the sprites that Scratch provides.

To add a sound to your sprite, you have to do one task first: Import the sound to your sprite. Here's how you'd do that:

1. **Click the Sounds tab above the Blocks Palette, and then click the Choose a Sound button.**

   The button is in the lower left and it looks like a speaker.

2. **In the file browser that appears, browse the provided sounds.**

   There is a search box you can use if you know the name of the sound you're looking for, and there are category buttons that group the effects into loops, animal sounds, musical notes, and more.

3. **(Optional) Click the play button on a sound to hear it.**

4. **Click the sound to bring it into your sprite.**

   After you've imported a sound, you can preview it (see Figure 9-9). Click the play button in the Sounds Area. You can choose different sounds to preview using the panel on the left. Click the trashcan icon on a sound in this panel to delete it from your project.

   If you a delete a sound in this way, it remains on your SD card so that you can import it again later.

After a sound has been imported, you use one of the Sound blocks to play a sound. To see all available Sound blocks, click the Sound button beside the Blocks Palette first. Use the Code tab to return to the Code Area if you can't see any blocks.

The Play Sound block enables you to choose which sound you'd like to play from those you have imported. The Play Sound Until Done block stops the program from running any blocks joined underneath this one until the sound has finished playing.

REMEMBER

The sound is imported to a particular sprite, so if you can't see it as one of the choices in the Play Sound block, be sure you've imported it to the correct sprite. In Chapter 10, we cover how to use multiple sprites in a project.

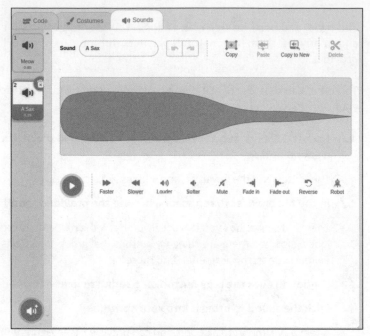

FIGURE 9-9:
Adding sound effects to your sprite.

Scratch is developed by the Lifelong Kindergarten Group at the MIT Media Lab. See https://scratch.mit.edu.

Using the Wait block to slow down your sprite

As you put your script together, you might find that some movements happen so fast that you can hardly see what's going on.

If you click the Control button to the left of the Blocks Palette, you can find a set of blocks that are used to govern when particular things happen. You can read more about these in Chapter 10, but for now it's worth knowing that the Wait 1 Seconds block here enables you to wait for a certain number of seconds. Drag this into your script where necessary to introduce a delay so that you can see each of your blocks in action. The length of the delay is 1 second by default, but you can change it to whatever you want, including parts of a second (for example, 0.5 for half a second).

TIP

The Say Hello! for 2 Secs block can be also be used to force the script to pause before running any more blocks.

Using extensions in Scratch

The latest version of Scratch includes a feature to add new sections to the Blocks Palette, called Extensions. If you're familiar with earlier versions of Scratch, this is where you'll find the Music and Pen blocks that used to be part of the main Blocks Palette.

Adding extensions

The button to add extensions is at the lower left of the screen when you're in the Code Area (refer to Figure 9-1, earlier in this chapter). When you hover over it, it says Add Extension. It takes you to a screen where you can see the extensions available, and click to add one to your Blocks Palette.

We don't have space to cover all the extensions here, but here are some you may want to start experimenting with first.

Using the music extension

The music extension has blocks that let you use virtual drums and pitched instruments to create music using Scratch. Notes are numbered: C is 60, C# is 61, D is 62, and so on. There's a block called Play Note 60 for 0.5 Beats that plays a note with a particular number for a certain duration. When you click the menu in this block to specify which note to play, a piano opens that you can use to select the note.

There are also blocks you can use to change the sound of the instrument and set or change the tempo of the music.

TIP

If you're new to music, you can generally get a good result by starting with C, sticking to the white notes, and making sure no two consecutive notes are too far apart on the piano.

TIP

The note numbers used in Scratch are the same as those used in Sonic Pi (see Table 14-1, in Chapter 14).

Using the pen extension

The pen extension includes blocks for drawing on the Stage. As a sprite moves, it draws a line behind it in your chosen color. Try adding a Pen Down block at the top of the script in Figure 9-5 to see it in action. There are blocks to set the pen color and size (which determines the thickness of the line). You can use the Change Pen Color block to cycle through the colors in the pen's palette, by increasing the pen number by a certain number.

Using the electronics extensions

There are two electronics extensions for the Raspberry Pi. The Raspberry Pi Simple Electronics extension is ideal for getting light-emitting diodes (LEDs) and buttons working with your Raspberry Pi. The Raspberry Pi general-purpose input/output (GPIO) extension gives you more control over the default state of the circuit (pulled up or down). For most purposes, the simple electronics extension is all you need. For more on the electronics extensions, see Chapter 16.

Using the Sense HAT extension

The Sense HAT is an official add-on for the Raspberry Pi. It includes an 8 x 8 grid of LEDs, motion sensors, a joystick, and some environmental sensors. You can program it in Scratch using the Sense HAT extension.

If you don't have a Sense HAT, you can use Scratch together with the Sense HAT Emulator to test your Scratch project (see Figure 9-10).

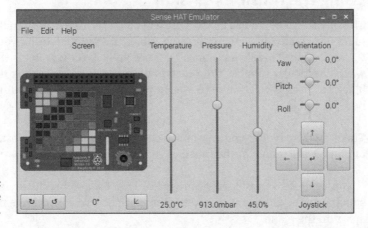

FIGURE 9-10: Emulating the Sense HAT.

The Sense HAT blocks include blocks to scroll text across the LEDs and to display a letter, a shape of your own design, a sprite, or the Stage. Given the extremely low resolution of the LED grid, the sprite and Stage are not always recognizable on the Sense HAT.

There are blocks to change the color used for text and shapes and the background color used for other LEDs. You can set the color of each LED individually, too.

There are blocks you can use to detect joystick movements and tilts and shakes of the device. You can also read the temperature, pressure, humidity, roll, pitch, and yaw sensors.

When you've made something you like, you can take it into the real world by buying a Sense HAT (which costs about $40).

Saving your work

Remember to save your work so that you can come back to it later. You can find the option to save on the File menu, at the top of the window. If you use Scratch online at the Scratch website instead of using the installed software on your Raspberry Pi, your work is saved automatically for you every few minutes. There is an option to Save Now in the upper right when you have unsaved changes.

IN THIS CHAPTER

» Adding sprites to your game

» Drawing and naming sprites

» Controlling when scripts are run

» Using random numbers

» Detecting when a sprite hits another sprite

» Introducing variables

» Making sprites move automatically

» Adding scripts to the Stage

Chapter **10**

Programming an Arcade Game Using Scratch

In this chapter, we show you how to use Scratch to create and play an arcade game. You get a chance to customize the game with your own graphics, but more importantly, you learn how to put a game project together so that you can invent your own games.

In this sample game, you control a flying saucer as it defends its planet from invasion. Grumpy-looking aliens zoom in from above, but you can stop them by hurling fireballs at them. If they get to you, it's game over — not just for you, but for your entire planet. . . .

This chapter explains the Events and Control blocks that enable you to coordinate the actions of different sprites with each other and with the player. The chapter assumes a basic understanding of the Scratch interface and how you use blocks to build a script, so refer to Chapter 9 for a refresher, if you need it.

In this chapter, we use Scratch 3.

REMEMBER

You can download the Scratch file for this chapter's arcade game from this book's companion website. (See the Introduction for more on how to access the book's online content.) You might find it helpful to look at the color-coded script onscreen while you read this chapter. You can use the File menu at the top of the Scratch window to open the project when you download it.

REMEMBER

Don't forget to save your game frequently. It's a good idea to save a new copy of your game with a new filename as you reach each significant point in its development. That way, you can go back if you introduce an unexpected error. Plus, if a file gets corrupted, you won't lose too much work.

Starting a New Scratch Project and Deleting Sprites

If you've been playing with Scratch and have blocks and scripts scattered all over the screen, you can start a new project by clicking File on the menu at the top of the screen and then choosing New.

All projects start with the Cat sprite in them, so the first thing you need to do is delete it. Find it in the Sprite List and click the trashcan icon on it.

REMEMBER

Deleting a sprite is not the same as hiding it. If you hide a sprite, it's still part of your project, but it's not visible. You can bring it back later by showing it. If you delete a sprite, its scripts, costumes, and sounds are removed from your project altogether.

Changing the Backdrop

The Stage can have scripts and different images, just like a sprite can. You can change the Stage's background to something more inspiring than the plain white space you see when you start Scratch.

Hover over the Choose a Backdrop button in the lower right of the screen. A menu of icons opens (see Figure 10-1). When you mouse over an icon, a description appears. You have four options:

>> **Choose a Backdrop:** Pick an image from those supplied with Scratch. They're organized into categories: fantasy, music, sports, outdoors, indoors, space, and underwater.

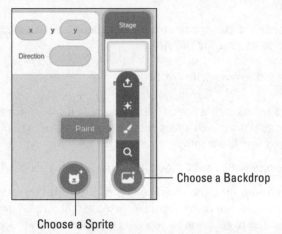

FIGURE 10-1:
Mousing over the Choose a Backdrop button reveals different ways to add a backdrop.

Choose a Backdrop

Choose a Sprite

Scratch is developed by the Lifelong Kindergarten Group at the MIT Media Lab. See https://scratch.mit.edu.

>> **Paint:** You can design your own backdrop using the built-in art package (see the section "Drawing Sprites in Scratch," a little later in this chapter).

>> **Surprise:** This option gives you a random backdrop from the Scratch library.

>> **Upload Backdrop:** This option enables you to use an existing image file. Scratch can open images in .jpg, .gif, and .png format.

>> **Camera:** This option enables you to take a photo using the Raspberry Pi Camera. To find this option, click Stage beside the Sprite List, click the Backdrops tab above the Blocks Palette, and then use the Choose a Backdrop button in the lower *left* of the screen.

You can manage your backdrops in the Backdrops Area. Click the Stage box beside the Sprite List, and then click the Backdrops tab at the top of the Blocks Palette to see the Backdrops Area.

For the backdrop for this chapter's game, we've used a space-themed backdrop from Scratch's collection.

Adding Sprites to Your Game

There wouldn't be much demand for a programming language that could only be used to create games about cats. (Actually, given the popularity of cat videos online, maybe there would.) In any case, Scratch gives you four ways to bring new sprites into your game. Hover your mouse over the Choose a Sprite button in the lower-right corner of the Sprite List to find them (see Figure 10-1). The menu looks the same as the one for backdrops.

The options to add a new sprite are:

>> **Choose a Sprite:** Use this option to select a sprite from Scratch's extensive library. Scratch comes with a wide range of sprites, including dancing people, flying hippos, and fire-breathing dragons. Our kind of party!

>> **Paint:** This opens the Paint Editor so that you can draw your sprite in Scratch.

>> **Surprise:** Looking for some inspiration? This button fires up your creativity by bringing in a randomly chosen sprite from those that Scratch comes with. It's also a quick way to get started if you want to experiment with scripting. If you don't like the sprite you get, you can always delete it and try another surprise.

>> **Upload Sprite:** You can use this button to bring in a graphic you've created using a different art package.

Drawing Sprites in Scratch

One of the most distinctive ways to put your fingerprint on your game is to draw your own sprites for it. Even if it plays the same as a well-known game, it'll look unique if you hand-craft your images. Figure 10-2 shows the Paint Editor in Scratch.

The checkered area is the Canvas. The checkered pattern has a special meaning and is used to indicate parts of the image that are transparent, where the backdrop will show through. Usually, you want everything outside the outline of your sprite to be transparent and everything inside it to be another color.

By default, Scratch uses vector images, which are made up of shapes that you can modify. Vector images are great for Scratch because they still look good when you change their size. The process for creating them may be a bit different from what you're used to, though. The best way to learn how to use the Paint Editor is to experiment with it.

Beside the Canvas, you can see your drawing and editing tools. Click one to select it, and you can then use it on the Canvas. The icon for your chosen tool is highlighted in blue so that you can easily see which tool you're using. Above the Canvas is the Options Area (refer to Figure 10-2). This is where you can choose how to use a particular tool. The main tools are described (from top to bottom) in this list:

Costume name   Undo/redo

costume3

Fill   Outline   0

Options Area

Convert to Bitmap

FIGURE 10-2:
The Paint Editor
in Scratch.

Drawing and editing tools

*Scratch is developed by the Lifelong Kindergarten Group at the MIT Media Lab.
See* `https://scratch.mit.edu/`.

>> **Select:** Use this tool to select a shape that you would like to modify or remove. If you select a rectangle you've drawn on the canvas, you can use the handles on its sides and corners to change its width and height, for example. The handle underneath the shape can be used to rotate it. To select multiple shapes within a rectangular area, click and hold the mouse button in one corner, drag to the opposite corner, and then release the mouse button. You can click a shape in your selected area and drag it to move all the selected shapes to a different part of the canvas. Alternatively, you can use the buttons at the top of the Paint Editor to flip your selected area horizontally or vertically. You can also press Delete on your keyboard to delete the selected area. If you select a shape, you can change its Fill (inside) and Outline color in the Options Area. Selected shapes can also be copied and pasted using the Options Area (or the keyboard shortcuts Ctrl+C and Ctrl+V).

>> **Reshape:** The Reshape tool is perhaps the most important one for creating vector images. It enables you to distort shapes on the Canvas. Choose the tool and then click a shape, and you can manipulate the handles on it. Whereas the Select tool enables you to resize a shape with the handles while preserving its form, the Reshape tool changes the outline of the shape. You can flatten the

bottom of an ellipse, for example, or turn a rectangle into a triangle by dragging two corners together. Click on the outline of the shape to add a new reshaping point, which you can drag and manipulate to further distort the shape.

>> **Brush:** Hold down the mouse button as you move over the Canvas to leave a line. In the Options Area, you can select the brush size and the color, where it says "Fill." Your line becomes a shape that you can edit using the Reshape tool.

>> **Eraser:** The Eraser takes a chunk out of your shapes when you put the Eraser over them and click the mouse button. You can hold down the button and move the mouse if you want to delete large parts of the image, or to delete small sections with a steady hand. In the Options Area, you can choose the eraser size beside the Eraser icon.

>> **Fill:** Click inside a shape on the image to fill it with your chosen color.

TIP

In the Options Area, you can click the Fill color to open the color options (see Figure 10-3). At the top, you can choose a graduated pattern that fades from one color to another. The fade can be vertical, horizontal, or circular. You choose the type of fade at the top of the color panel. You then click each color box to select it, and use the Color, Saturation, and Brightness sliders to choose your color. To reverse the fade, click the Swap icon between the color boxes. The icon in the lower left enables you to choose invisible ink, which enables anything behind the sprite to show through. The pipette in the lower right is used to copy a color from the canvas. Select it and then click the canvas to choose that ink. The Fill color options are also available when drawing or editing shapes.

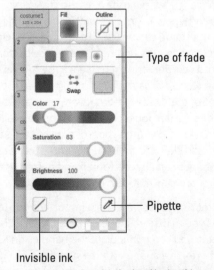

FIGURE 10-3: Setting a graduated fill pattern.

Scratch is developed by the Lifelong Kindergarten Group at the MIT Media Lab. See https://scratch.mit.edu.

>> **Text:** This tool enables you to write on the sprite. You can choose a different font in the Options Area, and press Enter in the text box to start a new line. To resize the text, use the Select tool. You can't reshape text.

>> **Line:** Click and hold the mouse button at the start of the line, move the mouse to the end of the line, and then release the mouse button. The options let you choose the line thickness and Outline color. Lines do not have a Fill color.

>> **Ellipse:** Click to indicate the point where lines from the top and left of the ellipse would meet, and then drag the mouse to the opposite side before releasing the button. Again, you have options to draw a filled or empty shape.

>> **Rectangle:** Click and hold the mouse button to mark one corner of the rectangle, and then drag your mouse to the opposite corner and release the button. In the Options Area, choose the Fill and Outline colors you want. You can also adjust the thickness of the Outline.

WARNING

The Delete button clears the Canvas if you don't have anything selected. If you make a mistake, click Undo (refer to Figure 10-2).

TIP

When drawing ellipses and rectangles, you can make them perfect circles or squares by holding down the Shift key while you drag the mouse. With the Line tool, holding down the Shift key forces your line to be horizontal, vertical, or diagonal at 45 degrees.

TIP

Your canvas has a target in the middle. This is the point around which your sprite will rotate. Usually, you want to align the middle of your sprite with this center point. It's easiest to draw your sprite and then select all of its shapes and move them into position on top of the center point.

To help you build your shapes, you can use the Forward and Backward options to decide which shapes are on top of others. You can also Group shapes, so you can manipulate them as if they were a single shape.

Figure 10-2 shows an alien Sean created for the game in this chapter. He started with an ellipse and then flattened the base of it. He used the Ellipse tool to create circles for the eyes and the eyeballs, the Line tool for the eyebrows and mouth, and the Brush tool to draw the nose. The horned top started off as a rectangle. He added reshaping points along the top edge of it, and then dragged every other one down to create a row of spiky peaks. To make the horns the same on each side, he copied and flipped the entire spiky rectangle and arranged the duplicate to put the horn in position on the right.

Using vector images can be strange at first, but it's worth persevering with. You can, though, choose to edit bitmaps instead. In that case, you manipulate *pixels* (colored dots) instead of shapes, so it's more like Microsoft Paint or a similar basic art package. Click Convert to Bitmap at the bottom of the Paint Editor to enter bitmap mode. If you convert a vector to a bitmap, you'll likely lose some image quality, and you won't be able to edit your shapes, even if you convert back again.

TIP

If you want to edit your picture later, click your sprite's Costumes tab and then click the costume you want to change. If you want to create additional costumes for a sprite, you can also do that in your sprite's Costumes tab. Use the button at the bottom to paint a new one.

Naming Your Sprites

Whenever you're programming, you should give things meaningful names so that you (and others) can easily understand what your program does. Scratch gives your sprites names like *Sprite1* and *Sprite2*, but you can rename them. To rename a sprite, select it in the Sprite List, click the box beside where it says Sprite at the top of the Sprite list, and then type its new name.

Your sprite's costumes are called *costume1*, *costume2*, and so on. If you've created different costumes for your sprite, you should also give them sensible names so that you can easily tell which is which. Go to your sprite's Costumes tab, click a costume, and then type the new name into the box at the top of the Paint Editor (see Figure 10-2).

For the space game, you need to add a spaceship sprite named *ship* and a sprite named *fireball* to represent the ship's weapon. The baddie, a sprite called *alien*, should have two costumes: *alienok*, which shows it looking menacing, and *alienhit*, which shows the same entity after it's been hit by the fireball.

To make it easier to see what you're doing, we recommend you drag your ship to the bottom of the screen, drag the alien to the top, and put the fireball somewhere in the middle. That roughly reflects where they will be in the finished game.

We're using the Rocketship sprite that comes with Scratch for our spaceship, and we've designed our own alien and fireball.

Controlling When Scripts Run

In Chapter 9, we show you how to start scripts by clicking them in the Code Area. Most of the time, you'll want your scripts to run automatically when certain things happen, such as a player pressing the Fire key.

This is where the Events blocks come in: They allow you to trigger scripts to run when a particular event happens, such as a sprite being clicked or a key being pressed.

Using the green flag to start scripts

One of the Events blocks is particularly useful for starting your game and synchronizing your scripts across all your sprites. Above the Stage are two buttons: a green flag and a red Stop button. The green flag is used to start scripts running, and you can use an Events block to detect when it's clicked. This Events block has a curved top on it — because no other block can go above it — but it has a notch underneath so that you can join Motion, Looks, Sound, or other blocks to it. You can put scripts that are triggered by the green flag being clicked into all your sprites so that clicking the flag makes it easy to start scripts on different sprites at the same time.

At the end of a game, aliens and the ship could be anywhere, so at the start of the game, you need to reset each sprite to its starting position. For the player's ship, you need to reset the x position to the center of the screen, set the y position near the bottom of the screen, reset the ship's direction, and bring the ship to the front so that any other sprites are behind it. Later on, this makes the fireball come from behind the ship, so it looks like it's being fired from inside rather than appearing on top of it. It's also a good idea to adjust the size of the sprite for your game. We've set our ship at 50%. If you've drawn your own sprite, you can adjust its size depending on how big you've drawn it.

Figure 10-4 shows the script you should assemble to reset your ship when the green flag is clicked. If you're making your own graphics, the y position might need to be higher, depending on the size of your sprite.

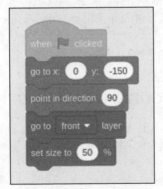

FIGURE 10-4:
Using a green flag
Events block to
reset your sprite.

*Scratch is developed by the Lifelong
Kindergarten Group at the MIT Media
Lab. See https://scratch.mit.edu.*

WARNING

When you have multiple sprites in your project, make sure you're adding blocks to the correct one (the ship, in this case). Each sprite has its own Code Area. To choose a sprite, click it in the Sprite List in the lower right.

Using the Forever Control block

Computers are great at repetitive tasks, and a game program often requires the computer to do the same things over and over again until the game is finished.

TIP

Repeated bits of program like this are called *loops*.

You use the Control blocks to craft the rules and instructions that govern how your game works. There are Control blocks for repeating a set of blocks. The Repeat block enables you to specify how many times you want a block or set of blocks to run. The Forever block runs a block or set of blocks repeatedly until the program is stopped.

Both blocks are shaped like a bracket, so they can enclose the blocks you want to repeat inside them. The Forever block doesn't have a notch on the bottom because it doesn't make sense to put any other blocks after it: They would never be run, because forever never comes to an end.

For the ship in this space game, you need to continue checking for key presses until the game is finished. Without the Forever loop, the script would check once for a key press and then finish.

You can find the Forever block by clicking the Control button at the left of the Blocks Palette to reveal all the Control blocks, and then looking down the list. Drag it into the script for your ship at the end of your green flag script. The first time you use it, we recommend you test how it works by dragging a Motion block into its bracket. Figure 10-5 shows a script that makes the ship sprite rotate for as long as the program runs. You've already built some of this script, so just add the new blocks at the end. Click the green flag to start it, but don't forget to take that rotation block out again when you've finished testing.

Enabling keyboard control of a sprite

For our space game, the player needs to be able to move the ship sprite left and right using the arrow keys. In plain English, you need to use a set of blocks that says, "If the player presses the left-arrow key, move the ship left." And, you need to put those blocks inside a Forever block so that Scratch keeps checking and moving the sprite all the way through the game. You need a similar set of blocks that move the sprite right, too.

FIGURE 10-5:
The Forever block can be used to make the ship rotate the entire time the program runs.

Scratch is developed by the Lifelong Kindergarten Group at the MIT Media Lab. See https://scratch.mit.edu.

The If block is a Control block that enables a set of blocks to be run only under certain conditions. For that reason, it's often called a *conditional statement* in programming. Like the Forever block, it's shaped like a bracket, so you can put other blocks inside it. In the case of the If block, the blocks inside are ones you want to run only in certain circumstances. Drag the If block into the Code Area, inside the Forever block.

Scratch is designed like a jigsaw puzzle, so it gives you visual hints about what blocks can go where if the program is to make sense. The If block has a diamond-shaped hole in it, which is where you describe the circumstances under which you want its blocks to run. There are diamond-shaped Operator and Sensing blocks as well, and we use both in this program.

The block you need for keyboard control is a Sensing block called Key Space Pressed? — it detects a tap on the spacebar. If you want it to detect the pressing of a key other than the spacebar, use its menu to set the key. In this case, you want it to detect the left-arrow key. You can drag and drop this Sensing block into the diamond-shaped hole in the If block in the Code Area.

Figure 10-6 shows the piece of script you need to move the ship left. We've used a Motion block to change its *x* position by −10 units, and we've also adjusted its direction, which makes it tilt toward the direction it's moving. You could change its costume so that it looks different when it's moving left or right, or add any other visual effects or sounds here. This code goes inside your Forever block in your existing script.

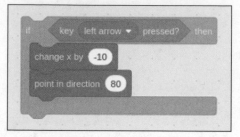

Enabling a sprite to control another sprite

In programming, you can often choose between several ways to achieve the same effect. The game's firing mechanism is one such example. You could sense the spacebar (the Fire key) being pressed using a script on the fireball, for example, and then use that to trigger the fireball's ascent.

We use the firing mechanism as an opportunity to show you how you can make one sprite control another sprite, however. You can't actually make the ship move the fireball, but you can send a message from the ship to tell the fireball you want it to move itself.

There are two parts to this: The first is that you need to use the Broadcast block on the ship to send a message to all the other sprites. You only want to do this when the spacebar (the Fire button in this game) is pressed, so you need to drag an If block to the Code Area of your ship, add a diamond Sensing block to check whether the spacebar is pressed, and, finally, put the Broadcast block inside the If block's bracket.

The Broadcast block, which is one of the Events blocks, has a menu built into it. Click the menu and click New Message to create a new message. We've called our message `fire`.

This approach has a couple of advantages. First, you can keep all your game control scripts on one sprite (the ship), which makes the program easier to manage. Second, it's an efficient way to coordinate multiple sprites. We could, for example, make our alien look terrified when the Fire button is pressed, by just changing its costume, and that requires only two blocks: An Events block for when the message `fire` is received, and the block to change to a new costume which shows the alien looking scared. It's much more efficient than having to look out for the Fire button on the alien, too.

Figure 10-7 shows the script for the ship. When the green flag is clicked, it resets the ship's position and then enters a loop where it moves the ship left if the left-arrow key is pressed, moves the ship right if the right-arrow key is pressed, sends the `fire` message if the spacebar is pressed, and then keeps checking for those keys forever. You can run this script to confirm that the ship moves as expected.

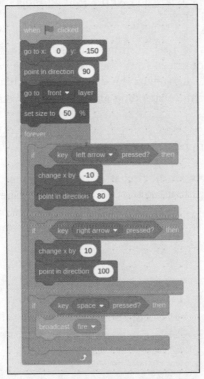

FIGURE 10-7:
The script for resetting and then controlling the ship.

WARNING

If your script doesn't behave as expected, check your brackets. You're allowed to put an If block inside another If block, but that doesn't make sense here, and it will stop the game's controls from working properly. If you put the bracket for detecting the Fire key inside the bracket for detecting the right-arrow key, the game will check for the Fire key only when the right-arrow key is pressed.

Click the fireball sprite in the Sprite List. You can now add scripts to that sprite. An Events block called When I Receive `fire` is used to trigger a script when the `fire` message is broadcast. This script is quite simple: You move the fireball sprite to where the ship is, show the fireball sprite (although it will be behind the ship), play a suitably sci-fi sound from the effects included with Scratch, glide the sprite to the top of the screen, and then hide it again.

In the Glide block, you can drop a block called X Position in place of entering a number for the *x* position. That means you can keep the *x* position the same as it already is while changing the *y* position with a gliding movement. The result is that the fireball moves vertically.

The other script you need on the fireball sprite is one to hide it when the green flag is clicked, just in case it's onscreen from the previous game when a new game starts. We've also added a block to adjust the size of our fireball. It looks best if the fireball is smaller than the ship. Depending on your design, you may need a different number in this block.

REMEMBER

Make sure you're adding scripts to the correct sprite.

Figure 10-8 shows the scripts for the fireball sprite. Remember to add the Laser1 sound effect to the fireball sprite using the sprite's Sounds tab before creating this script. Click Choose a Sound (see Chapter 9) and then use the Search box to find Laser1. After you've added this script, you can confirm that it works by tapping the spacebar.

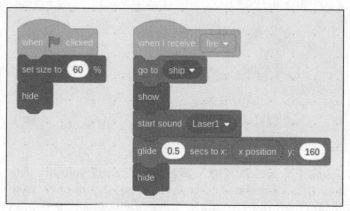

FIGURE 10-8:
The scripts for the fireball sprite.

Scratch is developed by the Lifelong Kindergarten Group at the MIT Media Lab.
See https://scratch.mit.edu.

Using Random Numbers

Games wouldn't be much fun if they were always exactly the same, so Scratch enables you to use random numbers in your scripts. To keep players on their toes, you can make the alien appear at a random *x* position at the top of the screen.

Click your alien in the Sprite List, and then drag in the When Green Flag Clicked Events block. As with the other sprites, you need to create a script that resets the alien to its starting position and sets it to the correct size. In the case of the alien, the sprite switches to a different costume when it's hit, so you should make sure it is using its normal costume at the start of a new game and that it is visible onscreen.

For its screen position, the alien needs to have a y-coordinate of 150, which is near the top of the screen. You don't want to use the full width of the Stage, because it looks odd when half the alien is off the edge of the Stage. From experimentation, we have found that the ideal starting x position for our alien is between –180 and 180, but yours might vary, depending on its size.

Drag in the Motion block you used previously to go to a particular x and y position. If you click Operators at the left of the Blocks Palette, you can find a block called Pick Random 1 to 10, which generates random numbers. Drag this block into the hole where you would normally type the x position, and then change the numbers in the random number block to –180 and 180.

Figure 10-9 shows the initial script for the alien. You can use the green flag to test whether it works and positions the alien at a random point at the top of the screen each time.

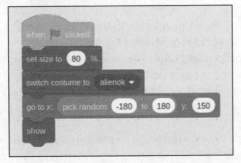

FIGURE 10-9: The script to reset the alien at the start of the game.

Scratch is developed by the Lifelong Kindergarten Group at the MIT Media Lab. See https://scratch.mit.edu.

Detecting When a Sprite Hits Another Sprite

There's no point in throwing flaming fireballs at an alien if it won't even raise an eyebrow. To make this game fun, you need to make the alien sprite react whenever it's hit. Most games involve sprites hitting each other (bats and balls, targets and weapons, chasing and catching), so *collision detection*, as it is often called, is a staple of game design.

WARNING

You can detect whether the fireball is touching the alien sprite from the fireball, but it is the alien that must react, so that's where you need to put your script.

You can use a Sensing block to check whether a sprite is touching another sprite and then combine that with an If block to trigger a reaction when the alien and the fireball touch each other.

As with the key press detection for the ship, you should keep checking for the alien being hit throughout the game, so you should put the If block inside a Forever block (see Figure 10-10 in the next section). Inside the first If block are the instructions for what to do when the alien is touching the fireball: Change the alien's costume to what it looks like when it's been hit, make it say "Arggh!" in a speech bubble, play a sound effect, and then hide the alien. After a random delay of a few seconds, the alien is repositioned at the top of the screen, switched back to its normal costume, and shown so that the horrible cycle of invasion and destruction can begin again.

Introducing Variables

Variables are a way of storing information in a program so that you can refer back to it later or reuse it. You give that piece of information a name, and then you can refer to it by that name in your script. For example, if you want to keep a running tally of the score, you use a variable to do that. They're called *variables* because their values can change over time. The score is zero at the start of the game, for example, but it goes up each time the player zaps an alien out of the sky.

You can tell your script to reset the score to zero, increase it when an alien is hit, and display the score at the end. Each time, you just refer to it as score, and the program works out what number that refers to.

To create a variable, click the Variables button beside the Blocks Palette. In the Blocks Palette itself is a button called Make a Variable. Click that, and you will be asked to enter a name for the variable — in this case, use score.

TECHNICAL STUFF

You're also asked whether this variable should be for all sprites or only the sprite you're working on now. It's important to get this right. For the score, you want to make a variable that all your sprites can see. If you have a variable that's used by only one sprite, it's better to create a variable that's only for that sprite, because it stops other sprites from being able to interfere with it. When you duplicate a sprite, all its scripts and variables are duplicated with it too, so you might find that you have sprites that use variables sharing the same name, but that you want to use independently of each other. In that case, you would set the variable to be for that sprite only. Each sprite would have its own independent version of the variable.

There are blocks in the Variables section of the Blocks Palette that you can use to change the variable's value and show or hide it on the Stage. If you want the score to go up by 50 each time the alien is hit (be generous — it's not an easy game!), you drag the Change my variable by 1 block into your script, change the variable name to score, and edit the number in it to 50. This block needs to go inside the If bracket that detects whether the alien touches the fireball, as you can see in Figure 10-10.

TIP

In Chapter 9, you see how you can display a sprite's position and direction on the Stage. By default, the values of variables are shown on the Stage, too. They appear in the upper left, but you can drag them wherever you want them. This can be useful for tracing and fixing problems, but it can get in the way of the game. We recommend that you deselect the check box beside your new score variable in the Blocks Palette to remove the score from the Stage again.

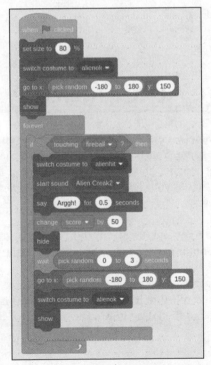

FIGURE 10-10:
Setting up the alien and detecting when it's hit.

Scratch is developed by the Lifelong Kindergarten Group at the MIT Media Lab. See https://scratch.mit.edu.

In the finished game, the alien comes down the screen toward the ship, and the game ends when the alien catches the player's flying saucer. At this point, you want to show the score variable on the Stage and use a special Control block that stops all scripts so that the program comes to an end. Figure 10-11 shows the blocks that do this, which use a similar pattern to the blocks used for detecting when the alien is hit. Add these blocks at the bottom of the script you created in Figure 10-10. The first new blocks is the If Touching Ship block. Pay attention to the brackets!

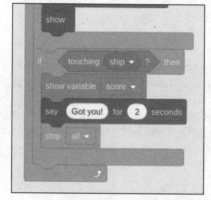

FIGURE 10-11:
Detecting when the alien catches the player's spaceship.

Scratch is developed by the Lifelong Kindergarten Group at the MIT Media Lab. See https://scratch.mit.edu.

Making Sprites Move Automatically

If you're wondering why we left the alien's movement to the end, it's because it makes it easier to test the game. You now have a spaceship that the player can move, a working firing mechanism, and an alien that dies and then regenerates when shot. You can test all that at your leisure and fix any problems without having to keep up with the alien.

The alien moves from left to right and then from right to left, and then back again. Each time it changes direction, it moves down the screen a little.

Figure 10-12 shows the movement script you need to insert into your alien's Forever loop as its first blocks. You'll find it easiest to assemble this set of blocks to the side of your main script and then drag it into the brackets of the Forever block in your main script.

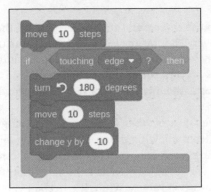

FIGURE 10-12:
The alien's
movement script.

*Scratch is developed by the Lifelong
Kindergarten Group at the MIT Media Lab.
See https://scratch.mit.edu.*

The Touching block, from the Sensing part of the Blocks Palette, can detect whether a sprite is at the edge of the Stage. When our alien touches the edge of the screen, we change our sprite's direction by 180 degrees. We immediately move 10 steps so that we can get away from the edge of the Stage. Otherwise, the sprite might get stuck there. We change its *y* position to move it down the screen, too.

To make this work well, you should insert two additional blocks at the top of your alien's script, after the When Green Flag Clicked block but before the Forever block:

>> **Point in Direction 90:** This is the default value anyway, but it's a good idea to positively set any values you're relying on. This makes the sprite move right when the game begins.

>> **Set Rotation Style Left-Right:** This will stop your sprite flipping onto its head when it moves left.

You'll find both blocks in the Motion section of the Blocks Palette. You can insert them as the first blocks under the When Green Flag Clicked block.

Fixing the Final Bug

In many commercial software development projects, most of the time and money is spent testing programs to make sure they work as expected, and then fixing them when they don't. Errors in programs are often called *bugs*, and even in the simple game in this chapter, we have a bug that would enable the player to cheat.

If the fireball is moving up the screen and the player presses the Fire key again, the firing sequence starts over. That means the fireball that was traveling through the air disappears, and a new one is sent up from the ship. That doesn't make any logical sense, and it means players suffer no consequences if they misfire: They can just fire again and it's as if the misfired shot never happened.

You can use a variable to keep note of when the fireball is moving up the screen so that you can stop the ship from allowing a fireball to be fired again at that time. Variables like this, which are used only to keep track of whether something is happening, are called *flags*. The firing flag needs to be able to say whether the fireball is in play or not, so it has two values. While the fireball is onscreen, give the firing flag a value of 1. When it isn't, give the firing flag a value of 0.

Let's set up the firing flag on the fireball sprite. Start by clicking that sprite in the Sprite List. Click the Variables button at the left of the Blocks Palette, and click the button to make a variable. Give it the name firingflag and make sure the option is selected so that it's available for all sprites. Untick the box beside its name in the Blocks Palette to hide it on the Stage.

After you've created the variable, you can drag in a block from the Variables section of the Blocks Palette to set its value to 1 at the start of the fireball's firing sequence, and to 0 at the end again. You should also update the fireball's green flag script so that it resets the firing flag to 0 at the start of a game, in case a game ended while the fireball was onscreen. Figure 10-13 shows the final scripts for the fireball.

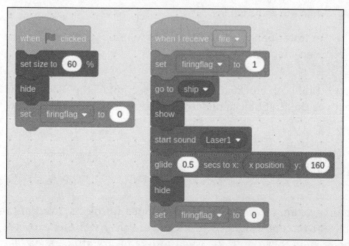

FIGURE 10-13: The final scripts for the fireball, including the firing flag.

Scratch is developed by the Lifelong Kindergarten Group at the MIT Media Lab.
See https://scratch.mit.edu.

You also need to modify the script for the ship so that it fires only if the `firingflag` variable is 0 at the time the spacebar is pressed. This is a little bit complicated because you need to lock together lots of different blocks to express this idea.

Go back to the ship's script. You need to modify the If block that checks whether the spacebar is pressed. Figure 10-14, read from top to bottom, shows how to modify your If block. For simplicity's sake, we've emptied the instructions from inside the If block and separated it out from the rest of the script.

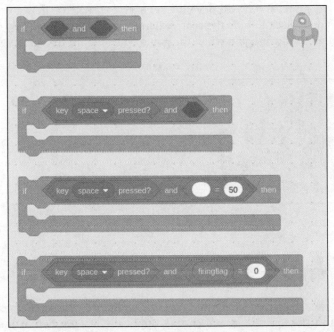

FIGURE 10-14: How to build the If block that checks whether the ship should fire.

Scratch is developed by the Lifelong Kindergarten Group at the MIT Media Lab. See https://scratch.mit.edu.

Start by dragging the Sensing block for the spacebar out of the If block's diamond-shaped hole. In its place, drag the And Operator block. This means the blocks inside the If block's bracket are run only if two things are true. The first is that the spacebar must be pressed, so drag your Sensing block for the spacebar into one of the diamond-shaped holes inside the And statement. The second is that we need to make sure the `firingflag` is 0. Drag the `'=50'` Operator block into the And Operator block on the right, and then drag the `firingflag` variable into the left of the = Operator block. Change the number 50 to 0.

This should ensure that the ship can fire only one fireball at a time. They might be aliens, but they still deserve a fair fight!

Adding Scripts to the Stage

As well as sprites, you can add scripts to the Stage. Click the Stage in the Sprite List and you'll find that it has its own Code Area. It's a real pain to have to hunt through your sprites to find where you put a particular block so that you can change it, so this is a good place to put scripts that affect the whole game and that aren't associated with a particular sprite.

For this game, you should add a block to the Stage to set the score to 0 when the green flag is clicked. Otherwise, the score will increase ever higher with each successive game, and it will never be set back to 0 when a new game starts. You can also add a block to hide the score when the game begins, to tidy up the Stage. The alien will show it again when the game ends. Figure 10-15 shows what we mean.

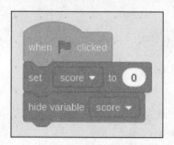

FIGURE 10-15:
The scripts for
the Stage.

Sean McManus

Duplicating Sprites

You can add more aliens by simply duplicating the first one. Right-click the alien in the Sprite List and choose Duplicate. Having two or more aliens adds a nail-biting aspect to the game.

Playing Your Game

To play your game without the distraction of your scripts and other clutter on the screen, near the top right of the screen, click the Four Arrows icon that says Full Screen Control when you hover over it. The Stage enlarges to fill the screen. You can use the green flag to play as usual. To close the full-screen view again, click the button above the Stage in the top right. Figure 10-16 shows the final game, though yours might look quite different with your own art in it.

FIGURE 10-16:
The final game.

Adapting the Game's Difficulty

If the game is too fast or slow for you, you can change the speed of the aliens by changing the number in their first Move 10 Steps block. Use a higher number to make the aliens move faster. On Sean's Raspberry Pi, 20 is a good speed. The aliens don't all have to move at the same speed, and you can add more of them, too, to make the invasion harder to defeat.

Taking It Further with Scratch

In this chapter, we cover many fundamental concepts that are used in programming, including loops, operators, and variables. We've described how you can use Scratch to design your own games, where sprites interact with each other and respond to the player's control. You can do lots of things to customize this game — draw your own sprites or change the speed of the aliens each time they're shot, or the way they move. But your next real adventure is to use Scratch and the skills learned in this chapter, perhaps with some of the other blocks we haven't had the space to cover, to make your own game.

To find out more about Scratch, and to find games and animations that others have made, visit the website at `https://scratch.mit.edu`. You can also share your own work there and get feedback from other Scratch fans. There is a supportive forum where you can get help with your scripts at `https://scratch.mit.edu/discuss`, too.

To dig deeper into Scratch and find more programs to build, see Sean's books *Scratch Programming in Easy Steps* and *Cool Scratch Projects in Easy Steps*. You can find Scratch tutorials and resources at `www.sean.co.uk/scratch`.

IN THIS CHAPTER

» **Using variables, strings, lists, and dictionaries**

» **Accepting user input and printing to the screen**

» **Using** for **and** while **loops**

» **Using conditional statements for decision-making**

» **Creating and using your own functions**

Chapter **11**

Writing Programs in Python

In this chapter, we introduce you to Python, a powerful programming language that's widely used commercially.

One of the best ways to learn programming is to study other people's programs, so in this chapter, we talk you through two different programs: One is a simple calculator for multiplication tables; the other is an artificial intelligence simulation that enables you to chat with your Raspberry Pi.

You'll probably find Python easiest to learn if you create the examples with us, but you can also download the finished programs from this book's website. For more information on accessing the website, see the Introduction.

In a book of this size, it's not possible to cover everything you can do with Python, but this chapter gets you started with your first programs. As you work through these examples, you'll learn about some of the fundamental principles in Python and in programming generally, and you'll gain an understanding of how Python programs are put together.

You'll be able to draw upon this knowledge when exploring the electronics programs in Part 5 of this book and when creating an arcade game with Pygame Zero in Chapter 12.

REMEMBER

Some lines of code are too wide for the page. We use a curving arrow at the end of a line of code to indicate that a line continues. When you see the curving arrow, just carry on typing and ignore the indent on the next line.

Working with Python

Programmers often use something called an *integrated development environment* (IDE), which is a set of tools for creating and testing programs. There is a Python IDE called Thonny on your Raspberry Pi. Click the Applications menu, select the Programming category, and then click the Thonny Python IDE icon to get started.

Entering your first Python commands

When you start Thonny, a window opens with two boxes in it — the window should look something like Figure 11-1.

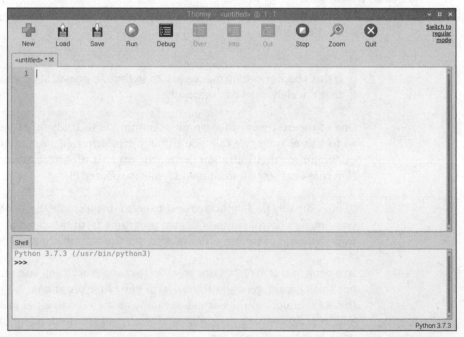

FIGURE 11-1:
The Thonny Python IDE, just after it opens.

The top box is the editor, and we show you how to use that to make programs later in this chapter. For now, you can ignore it.

The bottom box is the Python shell, and the three arrows are called the *prompt*. The prompt shows that Python is ready for you to enter a command. You can test this by entering the license() command, which shows you a history of Python, before displaying the terms and conditions of using it. Use the scroll bar to see it all. If you don't want to get bogged down in legalese, abort when prompted by typing q (in lowercase) and then pressing Enter.

TIP

You can increase the size of the shell to make it easier to read the output. Click and drag the bottom border of the editor box above to adjust the amount of space dedicated to each of the two boxes. Use the button in the top right of the Thonny title bar to maximize the window too. (See Chapter 4.)

One of the most basic commands in any programming language is the one that tells the computer to put some text on the screen. In Python, this command is print(), and you use it like this:

```
>>> print("hello world")
hello world
>>>
```

REMEMBER

A simple demonstration like this one that displays a greeting on the screen is often called a "hello world" program. It's the starting point for learning most programming languages.

The brackets are used to enclose whatever you want to output to the screen. The quotes are used to mark the start and end of the text you want to show.

As you type, you might notice that Thonny highlights your text using different colors. Gray is used to highlight a section that still needs a closing bracket, and green is used for the text you want to display. The highlighting is a feature of Thonny that helps Python beginners avoid common mistakes. Don't press Enter yet if your code is still highlighted: You've left something out.

Whatever you type in the quotes after the print() command is "printed" on the screen, and Python then returns you to the prompt so that you can enter another command.

WARNING

Like the Linux shell, Python is *case-sensitive* — it won't work if you use capital letters where you shouldn't. The command print() must be entered in lowercase; otherwise, Python tells you you've made a name error, because what you entered hasn't been defined. You can mess around with the word in quotes as much as you like, however: This is the text that you want to appear onscreen. Take a look at these examples:

```
>>> PRINT("Hello Leo!")
Traceback (most recent call last):
File "<pyshell>", line 1, in <module>
NameError: name 'PRINT' is not defined
>>> Print("Hello Leo!")
Traceback (most recent call last):
File "<pyshell>", line 1, in <module>
NameError: name 'Print' is not defined
>>> print("Hello Leo!")
Hello Leo!
```

Using the shell to calculate sums

You can also use the shell to carry out simple calculations. Table 11-1 shows you the mathematical operators you can use. Just put the sum after the print() command, like this:

```
>>> print(5 + 5)
10
>>> print(9 - 4)
5
>>> print(7 * 7)
49
>>> print(10 / 2)
5.0
```

TABLE 11-1   ## Mathematical Operators in Python

Operator	Description
+	Addition
−	Subtraction
*	Multiplication
/	Division
//	Division, discarding any decimal portion
%	Modulo, which shows the remainder after a division

REMEMBER

Don't use quotes around the sum in the `print()` command. What would happen if you did? Python would put on the screen literally whatever characters you asked it to, like this:

```
>>> print("5 + 5")
5 + 5
```

If you want to force a rounding effect to remove any decimal portion from your answer after dividing a number, you can use the floor division (`//`) operator, like this:

```
>>> print(10 / 3)
3.3333333333333335
>>> print(10 // 3)
3
```

An operator you might not have come across before is *modulo:* It uses the % sign and tells you the remainder after a division. Here are some examples:

```
>>> print(10 % 3)
1
>>> print(10 % 2)
0
```

You can use the modulo operator to tell whether one number is divisible by another (the modulo is 0, if so).

These sums are quite basic, but you can enter more advanced sums by stringing together numbers and operators. As in algebra, you use parentheses to surround the bits of the sum that belong together and should be carried out first. You still need to put parentheses around the whole sum for the `print()` command. For example:

```
>>> print((10/3) * 2)
6.666666666666667
>>> print(10 / (3*2))
1.6666666666666667
```

TECHNICAL STUFF

You can also do mathematics in the shell by entering the sums without a `print()` command, but it's essential to use the command when you're creating programs, as you see shortly.

The spaces we've used around the mathematical operators are optional. They help with readability and, in that last example, help to show which parts of the sum belong together.

Creating the Times Tables Program

In this section, we show you how to make a program that generates multiplication tables. For example, if the user requests a multiplication table for the number 7, the program outputs the sequence 7, 14, 21, and so on. The program is only a few lines long, but it teaches you how to create programs, use variables to store numbers, ask the user for information, and create *loops* — sections of program that repeat. You build on your understanding of the print() command to do all this, and if you've read Chapters 9 and 10 (on Scratch), some of the ideas should be familiar to you.

Creating and running your first Python program

The problem with entering instructions in the shell is that you have to enter them each time you want to use them. The commands are carried out straightaway, too, which limits the sophistication of the kinds of things you can do. You can solve these problems by creating a *program*, a set of repeatable instructions that you can save as a file and use again.

To create a program, you use the editor, which is the box above the shell that says <untitled> on its tab.

When you enter commands in the editor, they're not carried out straightaway. The editor acts like a simple text editor; it enables you to enter your list of commands (or *program*) and gives you control over when those commands are carried out.

Enter the following commands in the editor, using Enter to start a new line:

```
# simple times table program
print("This program calculates times tables")
print("It is from Raspberry Pi For Dummies")
```

The window should now look like Figure 11-2. The two print() commands should look familiar to you, though the first line is new. In Python, anything after a hash mark (#) on the same line is ignored by the computer. The hash mark indicates a *comment*, used to add notes to programs so that you can understand them later. The very best programs are written in such a way that you can understand them easily anyway, but it's a good idea to leave little messages to your future self (or other people) so that you can quickly understand important aspects of the program. We've put a 1-line summary at the start of the program here so that if we open it later, we can immediately see what it does.

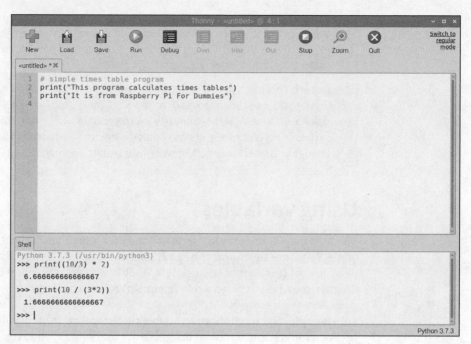

FIGURE 11-2:
Using the editor.

To save your program, click the Save button at the top of the Thonny window. You use the Load button beside it to reopen previously saved programs.

The term used for starting a program is *running* it, so click the Run button to see your program in action. Alternatively, the keyboard shortcut to run the program is F5. When you run the program, you see those two lines of text printed out on the screen in the shell window.

Congratulations! You've just written your first Python program!

WARNING

Before you can run your program, you must save it. If you've made changes since the last time it was saved, your program will be automatically saved for you. This overwrites the previous version of the program. If you might want to revert to an earlier version of the program, you should save a copy of it with a different file-name. You do this using the Save As option, which is hidden by default. Thonny has three different user interfaces:

» **Simple:** Thonny starts in the simple interface, which is ideal for beginners. To minimize distractions and the potential for mistakes, the menus are hidden.

» **Regular:** If you click Switch to Regular Mode in the top right and then close and restart Thonny, you'll enter regular mode. Here, there is a File menu that includes a Save As option. The View menu includes an option to show the variables in use, which you might find useful later. To choose the simple or expert mode, choose Tools ⇨ Options ⇨ General ⇨ UI Mode; the change you select takes effect when you restart Thonny.

>> **Expert:** There is also an expert mode, which is helpful for teachers. It enables them to double-click a tab (such as the Shell tab) to maximize it to fill the screen. Choose View ⇨ Maximize View to get back to normal. To choose simple or regular mode, choose Tools ⇨ Options ⇨ General ⇨ UI Mode; the change you select takes effect when you restart Thonny.

Using variables

The next step in the program is to ask the user which multiplication table to generate. You store this number in a variable. A *variable* is a way of storing a number or a piece of text so that you can refer to it later. (We talk more about variables in Chapter 10, where it plays a role in our Scratch discussion.)

For example, you might have a variable that stores your bank balance. It might go up (ka-ching!) or it might go down (sadly, more often), but you can always refer to it as your bank balance. Variables are one of the basic building blocks of programming, and not just in Python.

In the example of a bank balance, you can create a variable in Python for your bank balance called `balance` by just giving it a value, like this:

```
balance = 500
```

(If you want to try this, enter the commands in the shell. This isn't part of our Times Tables program.) You can vary the value later (which is why it's called a variable) by just giving it a new value:

```
balance = 250
```

More often, you'll want to do sums with the balance, such as taking some money off the total when money is withdrawn or adding money to it when a deposit is received. To do that, you change the variable's value to a number that's calculated from its current value. Here's an example:

```
balance = balance - 250
```

This example takes the value of the `balance` variable, knocks 250 off, and then puts the answer back into the variable `balance`. You can display the value of a variable onscreen using the `print()` command with the variable name:

```
print(balance)
```

Programmers often use a shorthand form when they're adding numbers to, or subtracting them from, a variable. The shorthand is += for addition and -= for subtraction. Here's an example:

```
balance = 500
balance += 20
print(balance)
```

If you run this tiny program, or enter these instructions in the shell, Python prints 520 on the screen.

Here's another example:

```
balance = 500
balance -= 70
print(balance)
```

This program subtracts 70 from the initial balance of 500, so it shows 430 onscreen. This shorthand is an elegant way and concise way to express the idea of changing a variable's value, and you'll see it used widely in Python.

Accepting user input

Before we go any further, we should clarify one piece of jargon: function. A *function* is a set of commands that do a particular job, and lots of functions are built in to Python. You've already seen one of them: `print()`. Later on, you'll learn how to make your own, too. (See "Creating your own functions," later in this chapter.) To use a function, enter its name, followed by parentheses. If you want to send it any information to work with, you put that inside the parentheses, as you already have with `print()`.

When the program runs, we want to ask the user which multiplication table to generate and then store that number in a variable that we call `tablenum`. To do that, we set up the `tablenum` variable using a built-in function called `input()`, which asks the question, waits for the user to enter something, and then puts whatever is entered into the variable.

Here's how the input() function works:

```
tablenum = input("Which multiplication table shall I generate for you? ")
```

TIP

We've inserted a space after the question mark and before the closing quotation mark, because, otherwise, the cursor would appear right next to the question mark. That extra space separating the question and the user's answer makes things look clear and more professional.

Add this new line of code to your program and run it, and you'll see that the program displays the question in the shell and then displays a cursor and waits for you to enter a number. Enter any number to try it out. The program doesn't do anything else yet, however, because you haven't told it to do anything with the number you enter.

Printing words, variables, and numbers together

It's time to make your program do something with its user input. Start by having the program print a title for the multiplication table the user has requested. This requires something we haven't discussed before: the capability to print text and variables on the same line of text. The print() function can be used to print more than one thing in a line, if they're separated by commas, so you can combine text and the variable tablenum, like this:

```
print("\nHere is your", tablenum, "times table:")
```

The first two characters here, \n, have a special meaning. They're known as an *escape sequence*, and they're used to start a new line. Here they create a bit of space between the question asking for input and the resulting heading.

WARNING

Any characters that appear between the quote marks are printed onscreen. If you put the variable name tablenum between quotes, you'll see the word tablenum onscreen instead of the number the user typed.

Now you need to print a line for each entry in the times table, from 1 to 12. As you know, you can use variables in sums, and you can print sums, so you could display the times table like this:

```
print("1 times", tablenum, "is", tablenum)
print("2 times", tablenum, "is", tablenum * 2)
print("3 times", tablenum, "is", tablenum * 3)
print("4 times", tablenum, "is", tablenum * 4)
```

What happens when you run it? You might be surprised:

```
1 times 7 is 7
2 times 7 is 77
3 times 7 is 777
4 times 7 is 7777
```

That's not the result you were looking for! The problem is that the variable table-num is being treated as a *string*, or a group of letters and other characters, rather than a number. This is what happens by default with the input() function. The user could enter anything when prompted for a number, and the program would give a similar result — simply repeating whatever was entered. (If you want to read a rap by your Raspberry Pi, try entering **yeah**.)

To fix that problem, you need to convert the user's input, stored in tablenum, into an integer, using the int function. If you cast your mind back to your mathematics lessons in school, you might remember integers are whole numbers, which have no decimal portion. The int function can be used to convert numbers into integers, or to turn a string into an integer. Add the following line after the input() line:

```
tablenum = int(tablenum)
```

It works, as you can see in Figure 11-3. But it's still not really a good solution. For each line of output, you're entering a new line in the program and adding a new sum at the end of it. (Even using the Editor's Copy and Paste commands (in the Edit menu), we ran out of patience at Line 4.) What if you want to create a times table that goes up to 50? Or 500? Or 5,000? Clearly, you need a more scalable solution.

Using for loops to repeat

To save you from the slog of entering all those print commands, and to make our program more flexible, you can use a for loop. This enables you to repeat a section of program a set number of times, and to increase a variable each time the code repeats. That's exactly what you need for the Times Table program: You want to display one line for each number from 1 to 12, showing the result of multiplying that number by the figure the user entered.

Here's how the code looks that makes it happen:

```
for i in range(1, 13):
    print(i, "times", tablenum, "is", i * tablenum)
```

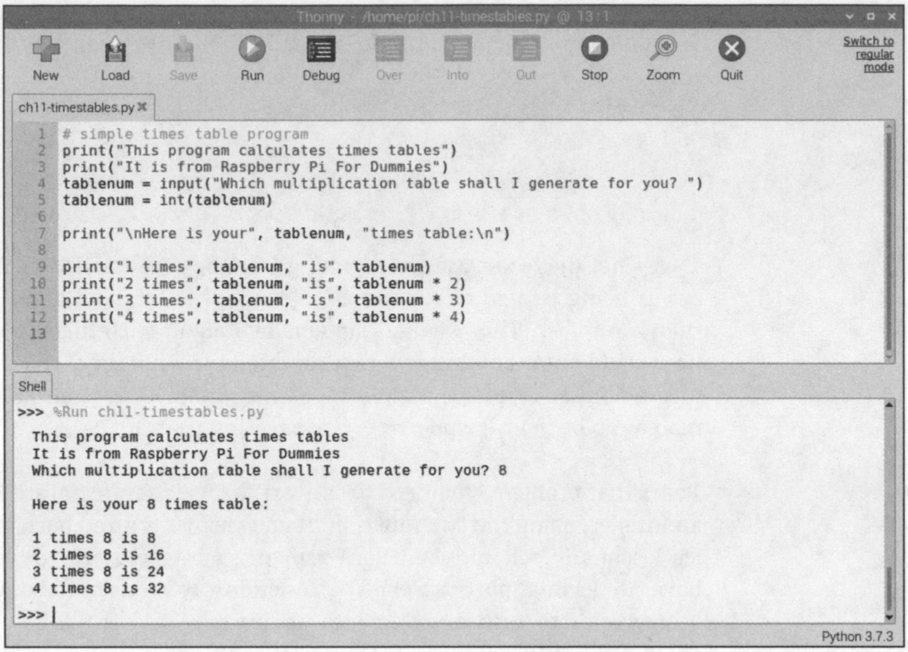

New Load Save Run Debug Over Into Out Stop Zoom Quit

ch11-timestables.py ✖

```
1  # simple times table program
2  print("This program calculates times tables")
3  print("It is from Raspberry Pi For Dummies")
4  tablenum = input("Which multiplication table shall I generate for you? ")
5  tablenum = int(tablenum)
6
7  print("\nHere is your", tablenum, "times table:\n")
8
9  print("1 times", tablenum, "is", tablenum)
10 print("2 times", tablenum, "is", tablenum * 2)
11 print("3 times", tablenum, "is", tablenum * 3)
12 print("4 times", tablenum, "is", tablenum * 4)
13
```

Shell

```
>>> %Run ch11-timestables.py

This program calculates times tables
It is from Raspberry Pi For Dummies
Which multiplication table shall I generate for you? 8

Here is your 8 times table:

1 times 8 is 8
2 times 8 is 16
3 times 8 is 24
4 times 8 is 32

>>> |
```

Python 3.7.3

FIGURE 11-3:
The Times Table program, in development.

Copyright © 2017 Aivar Annamaa

This tiny program snippet introduces several new programming concepts. First, take a look at the range function. It's used to create a sequence of numbers, and you give it a number to start at (1) and the end point (13). The end point is never included in the sequence, so we had to use 13 to make the multiplication tables go up to 12.

If you add a third number between the brackets (parentheses) of the range() function, it's used to specify the size of the gap between numbers. To show only the odd numbers (starting at 1 and adding 2 each time the program repeats), you'd use range(1, 13, 2). (You don't need to do that now, but you can experiment with the ranges later.) The rest of the line containing the range() function sets up the start of the bit to be repeated and says that the variable i should be given the next value from the number sequence each time it repeats. The first time around, i has a value of 1, the first number in the sequence. The second time around, i has a value of 2, which is the second number in the sequence. This goes all the way up to the last repetition, when i has a value of 12.

You tell Python which commands should be repeated by indenting them. The print() command we've used has four spaces in front of it, and in Python these spaces are meaningful. Many languages let you space programs out however you want, but in Python the spacing is often part of how the computer understands your intentions. By enforcing the use of indentations like this, Python makes it easier to read programs because you can see at a glance which bits belong together. They're all indented to the same depth.

You can repeat multiple commands by indenting them all:

```python
for i in range(1, 13):
    print(i, "times", tablenum, "is", i*tablenum)
    print("------------------")
print("Hope you found that useful!")
```

TIP

If you can't get your loop to work, make sure you've included the colon at the end of the `for` line.

The previous snippet works its way through numbers 1 to 12 and prints the times table line for each one, followed by a line of dashes to space it out. When it has finished all 12 lines, it prints `Hope you found that useful!` just once because that command isn't indented with the others in the loop.

Pulling it all together, the final program looks like this:

```python
# simple times table program

print("This program calculates times tables")
print("It is from Raspberry Pi For Dummies")

tablenum = input("\nWhich multiplication table shall I generate for you? ")
tablenum = int(tablenum)

print("\nHere is your", tablenum, "times table:\n")

for i in range(1, 13):
    print(i, "times", tablenum, "is", i * tablenum)
    print("------------------")

print("\nHope you found that useful!")
```

TIP

Although an indentation at the start of a line has special meaning, you can use blank lines to help lay out your program however you want. We've used some blank lines here to make it easier to see which bits of program go together. We've also added some extra \n escape sequences in the print and input commands to add blank lines in the screen output.

Many people find that they learn best from actually typing in programs, but you can download this program from the book's website to save time or if you can't get the program to work. See the Introduction for more on accessing the website.

Figure 11-4 shows what the screen looks like when the program runs. If you want to experiment with the program, there are a few things you can try. How about making it go up to 20, or making it show only the odd lines (1, 3, 5) in the times table? You can make both these changes by playing with the range() function used in the loop. You can customize the screen output, too, to provide more in-depth instructions, or to strip them out entirely. Perhaps you can use keyboard characters such as dashes and bars to put your multiplication table into a box.

```
Thonny · /home/pi/Downloads/Ch11 Python/ch11 timestable FINAL.py @ 25 1      ⌄ □ ✕
File Edit View Run Tools Help
Python 3.7.3 (/usr/bin/python3)
>>> %Run 'ch11 timestable FINAL.py'
  This program calculates times tables
  It is from Raspberry Pi For Dummies

  Which multiplication table shall I generate for you? 7

  Here is your 7 times table:

  1 times 7 is 7
  -----------------
  2 times 7 is 14
  -----------------
  3 times 7 is 21
  -----------------
  4 times 7 is 28
  -----------------
  5 times 7 is 35
  -----------------
  6 times 7 is 42
  -----------------
  7 times 7 is 49
  -----------------
  8 times 7 is 56
  -----------------
  9 times 7 is 63
  -----------------
  10 times 7 is 70
  -----------------
  11 times 7 is 77
  -----------------
  12 times 7 is 84
  -----------------

  Hope you found that useful!
```

FIGURE 11-4: The finished multiplication table. Now, what was 8 times 7 again?

You can stop a running program using the red Stop button on the menu bar.

If you get unexpected results, you can open the Variables pane in Thonny to see what's stored in each of your variables. This can be extremely helpful when tracking down errors in programs, especially as the programs you create become longer and more sophisticated. In regular mode, open the View menu and select Variables to reveal the pane. Use the same process to hide the pane again.

Creating the Chatbot Program

Do you ever find yourself talking to your computer? Wouldn't it be great if it could chat back? The next program in this chapter, Chatbot, enables you to have a conversation with your computer onscreen. Using a few tricks, we make the program appear to be intelligent — and able to learn from what you type in. It's not actual artificial intelligence, of course — that discipline of computer science is highly evolved, and this is a simple demo program. Chatbot can throw out some surprises, however, and you can expand its vocabulary to make it smarter. For a sneak preview of what Chatbot can do, look ahead at the end of this chapter.

As you build this Chatbot program, you'll deepen your understanding of Python. In particular, you'll learn about conditional statements, lists, dictionaries, and random choices.

The program works like this:

1. Chatbot introduces itself and then invites the user to respond.

2. The user types something in.

3. If the user types in **bye**, the computer replies with a message to say thanks for chatting and then finishes the program.

4. The program has stock responses for certain words, so it checks to see whether it recognizes any of the words the user has entered. If it does, it uses one of the appropriate stock responses. If more than one stock response applies, the computer chooses one at random.

5. If none of the words is recognized, the program chooses a random phrase for its reply. To stop the random phrases from repeating, it replaces the phrase that's used with what the user typed in. Over time, the program learns from the user and starts to talk like them.

6. The program keeps chatting with the user until the user types in **bye**.

Now that you know the final goal, take your first steps toward it by setting up the random responses.

REMEMBER

You can download the finished program from this book's website. See the Introduction for more on accessing the website.

Introducing lists

There are several different ways you can organize information in Python, and one of the most fundamental is called a *list*. We'll use lists to store the computer's random responses in the Chatbot program.

The following code shows you how to create a list that has the name shopping_list. You can enter this code in the shell or create a program in the editor so that you can more easily edit and refine it. To start a new program in the editor, click the New button. Your new file opens in a new tab in the editor. If you create a program, make sure you run it, so that the shopping list is set up.

If you enter the instructions in the shell, you can press Enter at the end of each line to go to a new line. Python will know you're not finished if you haven't given it the final bracket yet.

```
shopping_list = ["eggs",
                 "bacon",
                 "tomatoes",
                 "bread",
                 "tin of beans",
                 "milk"]
```

It's similar to the way you create a variable. After the list name comes an equal sign, and then square brackets that contain the list. Each item in the list is separated by a comma. Because each item is a piece of text (or a *string*), you put quotes around it so that Python knows where it starts and ends.

TECHNICAL STUFF

Python doesn't mind whether you use double quotes or single quotes around the strings in your list, but we recommend that you use double quotes. That's because strings often include apostrophes. If you're using a single quote mark (the same symbol as an apostrophe) to close the string, Python thinks it has reached the end of the string when it hits the apostrophe. If you do need to use an apostrophe inside a string that's marked at each end with a single quote, put a slash (\) in front of the apostrophe (for example, 'Mum\'s custard'). It's easier to just use double quotes for strings.

You can put all list items on one line, but the program is easier to read if you put each item on a new line. Using the editor, if you press Enter at the end of a list item, it indents the next line to the same depth as the item above it, so your list looks neat, as in the previous example.

When you're entering lists, pay particular attention to the commas; otherwise your program might not work: One should appear after every list item except for the last one. This is another reason it's a good idea to put list items on separate lines: You can more easily see at a glance when a comma is missing. Your program is color-coded, so the black commas stand out against the strings, which makes it easier to spot errors.

You can print a list to the screen in the same way you print a variable to the screen. Try this in the shell:

```
>>> print(shopping_list)
['eggs', 'bacon', 'tomatoes', 'bread', 'tin of beans', 'milk']
```

Python uses single quotes around the strings in your list, irrespective of the kind of quotes you used to set it up. To find out how many items are in a list, use the len() function, like this:

```
>>> print(len(shopping_list))
6
```

What if you've forgotten something? You can easily add items to the end of the list by using the append() function. Here's an example:

```
>>> print(shopping_list)
['eggs', 'bacon', 'tomatoes', 'bread', 'tin of beans', 'milk']
>>> shopping_list.append("fish")
>>> print(shopping_list)
['eggs', 'bacon', 'tomatoes', 'bread', 'tin of beans', 'milk', 'fish']
```

Each item in the list has a number, starting at 0, which means the second item is number 1 and the third item is number 2, and so on. You can refer to a particular item by putting the item number (known as the item's index) in square brackets:

```
>>> print(shopping_list[3])
bread
```

That gives you the fourth item in the list (remember?) because the first item has the index 0. You can also change items in the list by using their index numbers. For example, if you want to change the fourth item from bread to a baguette, you would use

```
>>> shopping_list[3] = "baguette"
```

For Chatbot, that's everything you need to know about lists, but they're an incredibly flexible way of organizing information and there's much more you can do with one. Table 11-2 provides a cheat sheet covering some of the other functions, if you want to experiment.

TABLE 11-2   **Additional List Operations**

Action	Code to Use	Notes
Sort a list.	`shopping_list.sort()`	Sorts alphabetically, or from low to high in lists of numbers.
Sort a list in reverse order.	`shopping_list.` `sort(reverse=True)`	Sorts in reverse alphabetical order, or from high to low in lists of numbers.
Delete a list item.	`del shopping_list[2]`	Deletes the list item with the index number specified. List items after it move up the list, so there is no gap.

Remove an item from the list.	`if "eggs" in ↵`
	`shopping_list:`
	
	    `shopping_list. ↵`
	`remove("eggs")`
		Deletes the list item that matches the item given. Results in an error if the item isn't in the list, so use the `if` command to avoid this.

TIP

For other projects you work on, it's worth knowing that lists can include numbers as well as strings, and can even include a mixture of strings and numbers.

For example, here's a list of answers to quiz questions:

```
my_quiz_answers = ["Isambard Kingdom Brunel", 2012, "Suffragettes", ↵
7500, "Danny Boyle"]
```

A list can have any items in any order. Python doesn't understand what the list contents mean or how they're organized. To make sense of it, you need to write a program that interprets the list.

Using lists to make a random chat program

After you've mastered the list structure, you can create a simple chat program. For this first version, you take some input from the user, display a random response, and then replace that random response with whatever the user types in.

Here's the program that does that. It introduces a few new ideas, but we talk you through them all shortly:

```
# Chatbot - random-only version
# Example program from Raspberry Pi For Dummies

import random

random_replies = ["Oh really?",
         "Are you sure about that?",
         "Hmmmmm.",
         "Interesting...",
         "I'm not sure I agree with that...",
         "Definitely!",
         "Maybe!",
         "So what are you saying, exactly?",
         "Meaning what?",
         "You're probably right.",
         "Rubbish! Absolute nonsense!",
         "Anyway, what are your plans for tomorrow?",
         "I was just thinking exactly the same.",
         "That seems to be a popular viewpoint.",
         "A lot of people have been telling me that.",
         "Wonderful!",
         "That could be a bit embarrassing!",
         "Do you really think so?",
         "Indeed...",
         "My point exactly!",
         "Perhaps..."]

print("What's on your mind?")
user_says = input("Talk to me: ")
reply_chosen = random.randint(1, len(random_replies)) - 1
print(random_replies[reply_chosen])
random_replies[reply_chosen] = user_says
```

The first two lines are comments — quick reminders of what the program does.

Python has been designed to be easily extended, so the next line, `import random`, tells Python you want to use the extension for generating random numbers. Extensions like this one are called *modules*, and you use several different modules as you play with the projects in this book. The modules provide prewritten functions you can reuse in your programs, so they simplify and accelerate your own programming. The `random` module includes functions for generating random numbers and will be essential when you want to pick a random response for the computer to display.

The next part of the program creates a list called `random_replies`, which contains statements the computer can output in response to whatever the user enters. You can personalize this by changing the responses or by adding more. The more responses there are, the more effective the illusion of intelligence is, but for this demo, we've kept the list fairly short. It doesn't matter what order the responses are in, but keep an eye on those commas at the end of each line. After printing a short line that invites the user to share with the computer what's on her mind, you request input from her using the `input()` function. Whatever the user enters is stored in a variable called `user_says`. The next line picks an index number for the random response.

To understand how this works, it helps to break it down. First, you need to know how to generate random numbers. You give the `random.randint()` function two integer numbers to work with (or *arguments*). The two numbers specify how big you want your random number to be: The first figure is the lowest possible value, and the second figure is the highest possible number. For example, if you want to display a random number between 1 and 10, you would use

```
print(random.randint(1, 10))
```

You can try this multiple times to confirm that it works. Sometimes the numbers repeat, but that's the nature of random numbers. It's like rolling the dice in Monopoly: Sometimes you're stuck in jail, and sometimes you throw doubles.

The range of numbers you want to use for the random number is the size of the `random_replies` list. You can use the `len()` function to see what this is, so you can add things to your list or remove them without having to worry about updating this part of your program. In the `random.randint()` statement, you replace the second number with the length of the list:

```
print(random.randint(1, len(random_replies)))
```

You don't want to just print the result onscreen, however, so you store the number chosen in a variable called `reply_chosen`.

There's one final twist: Because list indexes start counting at 0, you need to subtract 1 from the random number. Otherwise, the program would never choose the first list item, and would try instead to choose one at the end of the list that isn't there. Here's the final command to use:

```
reply_chosen = random.randint(1, len(random_replies)) - 1
```

The final two lines print the randomly selected list item and then replace that list item with whatever the user entered:

```
print(random_replies[reply_chosen])
random_replies[reply_chosen] = user_says
```

You can run the program to test it, but one thing is missing. At the moment, it gives you only one turn before finishing. To fix that, you need to master the `while` loop.

Adding a while loop

Previously, we showed you how to use the `for` loop to repeat a piece of code a set number of times. For this program, we want the computer to keep the conversation going until the user types in **bye**, so you need to use something called a `while` loop.

The section we want the computer to repeat begins with the line that requests the user's input and finishes where the program currently ends, with the user's entry going into the list of random replies.

To repeat this section, you add two lines at the top of the section (see the bold lines in the following code) and then indent the rest of the section by adding four spaces at the start of each line so that Python knows which commands to repeat:

```
user_says = ""
while user_says != "bye":
    user_says = input("Talk to me: ")
    reply_chosen = random.randint(1, len(random_replies)) - 1
    print(random_replies[reply_chosen])
    random_replies[reply_chosen] = user_says
```

The `while` command tells Python to repeat the indented block below as long as the second half of the `while` command is true. The `!=` operator means *not equal to*. In our Chatbot program, the second half of the `while` command is `user_says != "bye"`, which means the block below should keep repeating as long as the contents of the variable `user_says` are not equal to `bye`. The `while` command ends with a colon, so if you see an error message, be sure that you've included it.

To use the `user_says` variable in the `while` command, you have to set it up first because it triggers an error if you try to use a variable that doesn't exist yet in a `while` command. Immediately before the `while` command, you create the variable and give it a blank value (user_says = "") just to get the program past the `while` command and into the loop. Almost immediately, it changes when the user types something in — but that doesn't matter.

If you run the program now, you should find that the conversation rambles on until you type in `bye`. Remember that you can improve the quality of the experience by adding more random sayings in the program's list of random replies.

Using a loop to force a reply from the user

Another trick you can perform with the `while` loop is to make sure that the user doesn't press Enter without typing anything in, either accidentally or deliberately. That protects the extremely high quality of the random replies list (ahem!) by preventing empty entries from being added to it. In more complex programs, a quality-control check like this one can be essential for preventing errors.

You can put loops inside loops, which is called *nesting* them. In this case, we'll have a small loop that keeps asking for input until it receives it, running inside the bigger loop that repeats the whole process of the conversation until the user enters `bye`.

WARNING

To see whether something is equal to something else, use two equal signs together (==). This can be confusing to new programmers, but a single equal sign is only used to assign a value to something, such as when putting a value into a variable. When you want to compare the value of two things to see whether they're the same, you use two equal signs together. In English, we use the same word, but they're completely different ideas when you think about it, and Python certainly considers them as separate and unique concepts.

The following code puts a `while` loop around the input so that it repeats as long as the user_says variable is empty. If the user doesn't type anything in and just presses Enter, they're prompted to enter something again — and again, and again, if necessary:

```
user_says = ""
while user_says != "bye":
    user_says = ""
    while user_says == "":
        user_says = input("Talk to me: ")
    reply_chosen = random.randint(1, len(random_replies)) - 1
    print(random_replies[reply_chosen])
    random_replies[reply_chosen] = user_says
```

Notice how we indented the `input` command, so that Python knows what should be repeated while the `user_says` string is empty.

TECHNICAL STUFF

You might read this program code and wonder why we've set up `user_says` as an empty variable twice. (Notice the new line that now appears between the `while` commands.) The first time is necessary because the `while` command can't reference a variable that doesn't exist yet. The second time is a special case: If you don't reset the value to nothing, the second time around the loop, `user_says` still contains what the user typed in the first time. The way a `while` loop works means that the block underneath, the `input()` function, isn't run because `user_says` already has something in it. That code only runs if `user_says` is empty. This is a nice example of a logic error. The program works in that Python doesn't complain or crash. The program chatters away to itself, however, not letting you get a word in, so it doesn't work as intended.

Using dictionaries

Besides lists, there is another data structure that we use in our program, called a *dictionary*. To access an item in a list, you use an *index number*, which represents its position in the list. Dictionaries are different because you access an item using its *key* — a string or a number that uniquely identifies it. The idea is used a lot in computing. Your bank account number, for example, belongs to you and only you, so it's a unique key for your data.

Unlike with a list, you don't need to know where that item is in the dictionary to be able to use it — you just need to know the key that identifies it.

Dictionaries use curly braces, and contain pairs of items, which are the keys and the values for those keys. If that sounds confusing, here's an example that won't seem too different from the paper dictionary on your bookshelf:

```
chat_dict = {"happy": "I'm happy today too!",
            "sad": "Tell me all about it.",
            "raspberry": "Oh yum! I love raspberries!",
            "computer": "Computers will take over the world! You're already ↩
            talking to one",
            "music": "Have you heard the latest Depeche Mode album?",
            "art": "But what is art really, anyway?",
            "joke": "I only know this joke: How do you kill a circus? ↩
            Go for the juggler.",
            "python": "I hate snakes!",
            "stupid": "Who are you calling stupid, jelly brain?",
            "weather": "I wonder if the sun will shine on Saturday?",
```

```
    "you": "Leave me out of this!",
    "certain": "How can you be so confident?",
    "talk": "You're all talk! Do something!",
    "think": "You can overthink these things, though.",
    "hello": "Why, hello to you too, buddy!"}
```

TIP

In this example, we've given the dictionary the name *chat_dict*, but you can call it anything. You can have more than one dictionary in your program too, if you give them different names.

In this dictionary, we look up a word to see what the reply to it should be. For example, if someone uses the word *happy*, the computer should reply, "I'm happy today too!" If you look up the word *hello*, you can see that the computer's response should be, "Why, hello to you too, buddy!" Each dictionary entry is made up of the key and its value, separated by a colon; for example, the key happy and its value, which is the computer's response to that word. The entries are separated from each other with a comma.

WARNING

The punctuation here is quite fiddly, so take care; otherwise your program might not work. The text strings have quotes around them, but the colon between the keys and their values must be outside the quotes. Each pair needs to end with a comma except the last one, and we use curly braces to enclose everything. (You can usually find curly braces on your keyboard on the same keys as the square brackets.)

REMEMBER

Dictionaries only work if every key is unique. You can't have two entries in there for the word *happy*, for example; otherwise, Python wouldn't know which one to choose.

REMEMBER

Dictionaries only work one way around: You can't use the value to look up the key. One way to remember this is to think of a real paper dictionary. It would be almost impossible to trace a particular definition back to a word because you wouldn't know on which page you could find the definition. Finding definitions from the words is simple, though.

Here's how to print a value from the dictionary:

```
>>> print(chat_dict["hello"])
Why, hello to you too, buddy!
>>> print(chat_dict["weather"])
I wonder if the sun will shine on Saturday?
```

WARNING

If you try to use a key that doesn't exist in the dictionary, you trigger an error. Later in this chapter (see "Creating the dictionary look-up function"), we show you how to test whether a key is in the dictionary.

In the real program, we've extended the vocabulary to cover some other words, too, and this is where you can stamp your identity on the program most clearly. The words you put into the vocabulary, and the responses you give to go with them, are what gives the chat character its intelligence and personality. After you've got the demo working, it's worth spending time refining the language here. When you try playing with the finished program, remember the kinds of words you type in, and the kinds of things you want to chat about, and use that understanding to shape your Chatbot's vocabulary.

TIP

You can use the responses you give here to steer the conversation. We've included a joke for when users ask the computer to tell them one (as they inevitably do). Our full definition list also recognizes the word *funny* because that is reasonably likely to come up in the user's response to the joke. (Possibly in the context of "not very," but heigh-ho!)

Creating your own functions

One of the things you can do in Python, and many other programming languages, is parcel up a set of instructions into a function. A *function* can receive some information from the rest of the program (one or more *arguments*), work on it, and then send back a result. In our Chatbot program, we use a function to look up whether any words that are entered are in the dictionary of known words and responses.

Before you can use a function, you have to define it, which you do using a def statement. To tell Python which instructions belong in the function, you indent them underneath the def statement. Here's a program to familiarize you with the idea of functions and how we'll be using it:

```python
# Example of functions

def dictionary_check(message):
    print("I will look in the dictionary for", message)
    return "hello"

dictionary_check("blue")

result = dictionary_check("red")
print("Reply is:", result)
```

We talk you through that program in a moment, but here's a glimpse of what is shown onscreen when you run it:

```
I will look in the dictionary for blue
I will look in the dictionary for red
Reply is: hello
```

This is a short but powerful program because it tells you nearly everything you need to know about functions. As you can see, we defined the function at the start of the program, with this line:

```
def dictionary_check(message):
```

This sets up a function with the name `dictionary_check()` but also sets it up to receive a piece of information from the rest of the program and to put it into the variable called `message`. The next line prints out a statement saying, "I will look in the dictionary for" followed by the contents of the variable `message`. That means it prints out whatever information is sent to the function. The next line starting with `return` exits the function and sends a message back, which in the example is `hello`.

WARNING

Functions are self-contained units, so the variable `message` can't be used by the rest of the program. (It's known as a *local variable*.) When you're writing your own functions, you should give them a job to do and then use `return` to send the result back to the rest of the program.

Functions aren't run until you specifically tell the program to run them, so when Python sees the function definition, it just remembers it for when it needs it later. That time comes shortly afterward, when you issue the command:

```
dictionary_check("blue")
```

This runs the `dictionary_check()` function and sends it the text `blue` to work with. When the function starts, Python puts `blue` into the function's variable called `message` and then prints onscreen the text that contains it. The text `hello` is sent back by the function, but you don't have a way to pick up that message.

The next code snippet shows you how you can pick up information coming back from a function. Instead of just running the function, you set a variable to be equal to its output, like this:

```
result = dictionary_check("red")
print("Reply is:", result)
```

When the text `hello` is sent back by the function, it goes into the variable result, and the main program can then print it on the screen.

This simple example illustrates a few reasons why functions are a brilliant idea and have become fundamental building blocks in many programming languages:

» **Functions enable you to reuse parts of your program.** For example, we've used our function to display two different messages here, just by sending the function a different argument each time. When you use more sophisticated programs, being able to reuse parts of your program makes your program shorter, simpler, and faster to write.

» **Functions make understanding the program easier because they give a name and a structure to a set of instructions.** Whenever someone sees `dictionary_check()` in our program, they can make a good guess at what's going on. So far, our programs haven't been particularly complex, but as you work on bigger projects, you'll find that readability becomes increasingly important.

» **Functions make it easier to maintain and update your program.** You can easily find which bits of the program to change, and all the changes you need to make will be in the same part of the program. If you think of a better way to do a dictionary look-up later, you can just modify the function, without disturbing the rest of the program.

» **Functions make testing and development easier.** We've built an experimental program that takes some text and sends back a message. That's what the finished `dictionary_check()` function will do, except that this one just sends the same message back every time, and the finished one will send different messages back depending on what the user said. You could build the rest of the program around this simple test function to ensure that it works, and then go back and finish the `dictionary_check()` function.

Creating the dictionary look-up function

Now that you know how to create a function, we're going to tell you how to build a function that takes the user's text and checks for any relevant responses. To do this, you'll use what you already know about dictionaries and functions, and we'll add some new ideas relating to loops, strings, and decision-making.

The function is only 12 lines long, but it's quite sophisticated. It needs to take what the user entered and check each word in it to see whether the dictionary has a response for that word. The user might use more than one word that's in the dictionary. For example, if the user says "I love pop music," both the words *love* and *music* might be in the dictionary. We'll deal with that eventuality by showing one of the possible responses, chosen at random. Alternatively, the user might use no words that the program recognizes, so you need to design the function to cope with that situation, too.

Before we start to break it down, here's the function in its entirety so that you can see how all the bits fit together:

```
def dictionary_check(message):
    message = message.lower()
    user_words = message.split()
    smart_replies = []
    for each_word in user_words:
        if each_word in chat_dict:
            answer = chat_dict[each_word]
            smart_replies.append(answer)
    if smart_replies:
        reply_chosen = random.randint (1, len(smart_replies)) - 1
        return smart_replies[reply_chosen]
    else:
        return ""
```

The function definition is the same as we used earlier in our function example. When we use it, we send it what the user has typed in, so this goes into the variable called message.

The next two lines introduce something new: *string methods*. These are like built-in functions that are attached to a string and transform it in some way. The `lower()` method converts a string into lowercase. This is important because if a user uses capital or mixed-case letters, they won't match the lowercase words used in the dictionary keys. As far as Python is concerned, `hello` and `Hello` aren't the same thing. The `split()` method takes a string and splits it into a list of its constituent words. The first two lines in the function, then, turn the contents of the message variable into a lowercase version of itself and then create a new list of the words the user entered, called `user_words`.

We store possible replies to the user in a list called `smart_replies`, so we create that as an empty list. The next step is to set up a loop that goes through the list of words that the user entered. When you used a `for` loop previously, you worked your way through a sequence of numbers. This time, you work your way through a list of words. Each time around the loop, the variable `each_word` contains the next item from the list of words the user entered.

The next line introduces a new idea, the conditional statement, which starts with `if`. A *conditional statement* is used to enable the computer to make a decision about whether it should carry out certain instructions, and you'll come across one in almost every program you write. Here, it's being used to prevent the program from stopping and reporting an error if you try to use a key that isn't in the dictionary:

```
if each_word in chat_dict:
    answer = chat_dict[each_word]
    smart_replies.append(answer)
```

The each_word variable contains one of the words the user entered, so the if statement checks to see whether that word is in the dictionary and carries out the next two instructions only if they are. Notice how indenting (by a further four spaces) is used here to show which commands belong together — in this case, which commands are controlled by the if statement. If the word is in the dictionary, the program looks it up and adds the resulting response to the smart_replies list, using append().

This process is repeated for every word the user entered, but that's all that happens in the loop. The next line is not indented below the for statement, so it's not controlled by it.

When the program comes out of the loop, it checks to see whether the list smart_replies has anything in it, by using simply

```
if smart_replies:
```

In English, this means "if smart_replies has content in it." The commands indented underneath are carried out only if some entries were added to the smart_replies list, which only happens if one or more of the words the user entered were found in the dictionary. In that event, you want to return one of the items in the smart_replies list to the main program, so the program picks one at random from the list and uses return to send it back to the main program and exit the function.

After that, you use the else command. In plain English, this means *otherwise*, and it's joined to the if command, so it's lined up with it. If smart_replies has content in it, the commands are carried out to send back an appropriate reply, chosen at random. If none of the user's words was found in the dictionary and so smart_replies is empty, the instructions indented underneath the else command are carried out instead. The function sends an empty message ("") back to the main program and exits the function.

Creating the main conversation loop

We previously showed you how to create a version of Chatbot that could only provide random responses. Now you need to change the main conversation loop so that it checks for words in the dictionary and shows an intelligent response if they're found; and if not, shows a random response and replaces it with what the user entered. This final version brings together all the ideas we've helped you explore as you've built this program.

After the command that accepts the user's input, you put the following:

```
smart_response = dictionary_check(user_says)
if smart_response:
    print(smart_response)
else:
    reply_chosen = random.randint(1,len(random_replies)) - 1
    print(random_replies[reply_chosen])
    random_replies[reply_chosen] = user_says
```

This starts by using the `dictionary_check()` function (or *calling* it, to use the jargon), sending it whatever the user typed in and putting the response from the function into the variable `smart_response`.

The next line checks to see whether `smart_response` has any content in it (or is not empty) and, if so, prints it onscreen. Otherwise, the instructions under the `else` command are used to show a random response from the list of random replies, and to replace it with what the user entered. Those last three instructions were in the previous version of your program, but they need indenting now, by adding four spaces at the start of each line.

Final thoughts on Chatbot

That completes the Chatbot program. Along the way, you've learned how to use variables, lists, loops, random choices, dictionaries, conditional statements (`if` and `else`), and functions. You've learned how to take input from the user and print responses onscreen, and you've created the skeleton of a chat program that you can flesh out with your own personality.

The program works best if you don't use punctuation at the end of your sentence. When the program splits up your sentence, your last word would include any punctuation symbol next to it. That stops the program matching that word against any words it knows. Perhaps you could modify the program to strip out punctuation?

Here is a sample run of the program, albeit a fairly contrived one to show the program at its best, using lots of words it knows:

```
What would you like to talk about today?
Let's talk about the future
Who knows what tomorrow will bring?
```

```
I guess weather forecasters do
I wonder if the sun will shine on Saturday?
Fingers crossed! I hope so!
You've got to keep the hope alive!
Yes
A healthy positive attitude!
I think it's important to stay positive.
You can overthink these things, though.
That's true. Tell me a joke.
I feel I can tell you anything
That's so lovely to hear
Do you really think so?
Of course.
Perhaps...
Go on! Tell me a joke.
How do you kill a circus? Go for the juggler.
That's funny
There's a big difference between funny peculiar and funny ha-ha
True, but I laughed.
Anyway, what are your plans for tomorrow?
```

As you expand the vocabulary in the dictionary and include more random replies in the list, you'll find that the program can often surprise you with its apparent intelligence. You're never alone with a Raspberry Pi!

The final Chatbot program

For your reference, here's a final listing of the Chatbot program, ready for you to customize with your own responses:

```python
# Chatbot
# Example program from Raspberry Pi For Dummies
import random

# Following list is heavily abbreviated
# to save space in the book. Should be 20+ entries.
random_replies = ["Oh really?",
                  "Are you sure about that?",
                  "Perhaps..."]

# Following dictionary is also heavily abbreviated
chat_dict = {"happy": "I'm happy today too!",
             "sad": "Cheer up, mate!",
"computer": "Computers will take over the world! You're already talking to one"}
```

```
def dictionary_check(message):
    message = message.lower()
    user_words = message.split()
    smart_replies = []
    for each_word in user_words:
        if each_word in chat_dict:
            answer = chat_dict[each_word]
            smart_replies.append(answer)
    if smart_replies:
        reply_chosen = random.randint (1, len(smart_replies)) - 1
        return smart_replies[reply_chosen]
    else:
        return ""

print("What would you like to talk about today?")

user_says = ""
while user_says != "bye":
    user_says = ""
    while user_says == "":
        user_says = input("Talk to me: ")

    smart_response = dictionary_check(user_says)
    if smart_response:
        print(smart_response)
    else:
        reply_chosen = random.randint (1, len(random_replies)) - 1
        print(random_replies[reply_chosen])
        random_replies[reply_chosen] = user_says

print("Goodbye. Thanks for chatting today. Drop in again soon!")
```

IN THIS CHAPTER

» Using Pygame Zero to display images and play sounds

» Making a simple click-the-clouds game

» Using a list to manage multiple images

» Animating your images

» Adding a countdown timer

Chapter **12**

Creating a Game with Python and Pygame Zero

Developing games is one of the best ways to explore programming. It gives you rapid, visual results, so you can easily see what's going on and you get to have some fun playing your creation at the end. Game development also makes it easy to think of fresh coding challenges to solve, as you dream up new features to add.

One of the most popular tools for making games in Python is called Pygame: It's a set of functions that makes it easier to manage your images and sounds, among other things. Pygame has a few complexities, though, and can be difficult to get started with.

That inspired Daniel Pope to make Pygame Zero, a library of routines that simplifies Pygame so that you can get started more easily. It also includes a number of built-in functions that manage animations, loops, and images. Pygame Zero was designed for education, but is great fun for anyone to tinker with.

REMEMBER

Pygame Zero is also available for Mac and Windows, so the games you make on the Raspberry Pi can be enjoyed by your friends even if they have different computers.

In this chapter, we show you how to make a game called Cloudbusting, where you have to "pop" the clouds as they drift up the screen. You have ten seconds to click as many as you can. For the smoothest experience, we recommend using a Raspberry Pi 2, 3, or 4.

Collecting Your Sounds and Images

To follow the examples in this chapter, you need the sounds and images described in the following list. You can either create your own or download our examples. (See the Introduction for more on downloading our supporting files.) This simple game works equally well whatever images you use.

Here are the assets needed for the Cloudbusting game:

» **Six images, named** target0.png **to** target5.png: In our game, we use colored clouds, but you could use anything. Our images measure about 100 pixels by 90 pixels, and you should aim for a similar size, give or take 10 or 20 pixels in either direction. (We've noticed that it adds a sense of depth to the game if a couple of the images are a slightly different size from the rest, so you may want to try that out.) Use a transparent background so that your images can overlap without their backgrounds blocking out the images behind them. The PNG format is recommended, but Pygame Zero can also use GIF and JPEG images (although JPEGs don't support transparency). The simplest way to make your own images is to use the built-in Paint Editor in Scratch 3 (refer to Chapter 10). When you've designed your sprite, save your Scratch project. Click the button in the Paint Editor to convert your costume to a bitmap, and then right-click the costume in the Costumes Area and choose to export it from the menu that appears. Now close your Scratch project without saving it if you want to keep a high-quality version of the sprite in your project.

» **One image named** pop.png: This image is shown when the target is hit. Ours is a cartoon-style spiky bubble with "Pop!" written in it. We made this image in Scratch too.

» **Two sound effects — one for the popping of a target and the other to play when the game ends:** We're using a sound effect we found online called blop.ogg for when a cloud bursts, and a tune Sean made on the iPad called whoops.ogg for when the game ends.

REMEMBER

The .ogg format is a sound format often used in open source software. You can also use a .wav file in Pygame Zero. If you're using sound effects different from ours, change the filenames in the programs to your choices.

As always, you can download the code for this chapter, too, if you have any difficulties getting it working or you don't want to type the examples yourself. See the Introduction for instructions on downloading the supporting materials. You can also see the full listing for this game at the end of this chapter.

Setting Up Your Folders

Pygame Zero lays down strict rules about what your files can be called and where they should be saved. Filenames must be in all lowercase, which helps to ensure that they work across different computers. Your images must be stored in a folder called images, and your sounds in a folder called sounds.

Let's set up those folders now. Follow these steps:

1. **In the desktop environment, click the taskbar's File Manager icon.**

   For more on the taskbar and the File Manager, see Chapter 4.

2. **In the** pi **folder, right-click, and then choose New Folder from the contextual menu that appears.**

3. **Name your folder** images.

4. **Repeat Step 2, but this time name your new folder** sounds.

5. **Copy your image files into the** images **folder, and your sounds into the** sounds **folder.**

   Remember that when you download files with your web browser, they will be saved in your Downloads folder, which is also in your pi folder. You may need to extract them from a .zip file, by right-clicking the .zip file and then clicking Extract Here. If you want to unzip the folder and put its contents somewhere else, use the Extract To option instead. After extracting the files, you can copy them to the images and sounds folders. For guidance on using the File Manager to copy and move files, see Chapter 4.

Creating and Running Your First Program

The lines of code in Listing 12-1 show your very first Pygame Zero program, which puts an image in the center of the window. Type it into Thonny and save it in your pi folder. See Chapter 11 for more on Thonny.

LISTING 12-1: **Your First Pygame Zero Program**

```
import pgzrun

WIDTH = 500
HEIGHT = 500

cloud = Actor('target5')
cloud.x = 250
cloud.y = 250

def draw():
    screen.clear()
    cloud.draw()
```

As of this writing, Pygame Zero behaves differently depending on which mode of Thonny you're using. Listing 12-1 works in the simple mode. In the regular and expert modes, you need to add the instruction pgzrun.go() to the end of your listing. If you add that instruction and run your program in the simple mode, your program runs twice. In this chapter, we assume you're using the simple mode. If you can't get the listings in this chapter to work, add pgzrun.go() at the end. For more on the different modes in Thonny, see Chapter 11.

To start the program, click the Run button at the top of Thonny. You should see something like Figure 12-1, with an image in the middle of a new window that opens. You might need to wait a moment for the window to open. When you've finished with it, you can click the Close button in the top right of the window to quit.

FIGURE 12-1:
Your first Pygame Zero program puts an image in the middle of the window.

Sean McManus

Looking at the code shows you how easy it is to use images in Pygame Zero.

The first line imports the Pygame Zero library, so you can use its functions in your program.

The WIDTH and HEIGHT variables are used to set up the size of the window — in our case, a square of 500 pixels.

Pygame Zero introduces the idea of actors, which are like sprites in Scratch. An actor stores an image and its position in one place, which makes it easier to manage the characters and obstacles in your games. This kind of thing is already built into Scratch, but it's not part of Python, or even Pygame. The use of actors in Pygame Zero makes it much easier to move from programming in Scratch to using Python.

To create an actor, you give it a name and tell it which filename to use for the actor's picture, like this:

```
cloud = Actor('target5')
```

Here you've created an actor called cloud that uses the image target5.png. Note that you don't need to give Pygame Zero the file extension of the image; that's another way it keeps things simple.

The coordinates are measured from the top left of the window, so they go from 0 to 500 from left to right (the x direction), and from 0 to 500 down the window (in the y direction). If you want to change the x position of the cloud actor, you can assign it a new value, like this:

```
cloud.x = 250
```

Because the window is 500 pixels across and the actor's position is measured from its center, changing the actor's x position to 250 centers it in the window.

The last thing you need to do is draw the sprite. The draw() function is run regularly by Pygame Zero, and it's where you tell it what to draw on the screen. (See Chapter 11 for more on how functions work and are defined.) When things start moving around, they'll leave a trail if you don't remove the previous images before drawing each screenful, so you start this function by clearing the screen with this instruction:

```
screen.clear()
```

In this case, you just want to draw the cloud actor at its current position, so you use

```
cloud.draw()
```

Can you work out how to add another cloud to the program and position it to the left of the first one? Call it cloud2, and remember to change its x position and to draw it when you've finished. Listing 12-2 shows one possible answer. You could use a different number for the x position of cloud2.

LISTING 12-2: **Adding a Second Cloud**

```
import pgzrun

WIDTH = 500
HEIGHT = 500

cloud = Actor('target5')
cloud.x = 250
cloud.y = 250

cloud2 = Actor('target5')
cloud2.x = 80
cloud2.y = 250

def draw():
    screen.clear()
    cloud.draw()
    cloud2.draw()
```

Detecting mouse clicks

Pygame Zero makes it easy to respond to mouse clicks. Add the following function to the end of your existing program. It doesn't matter whether you're using the one-cloud or two-cloud program from Listing 12-2:

```
def on_mouse_down(pos):
    if cloud.collidepoint(pos):
        cloud.image = 'pop'
        sounds.blop.play()
```

This function runs when the mouse button is clicked, and stores the position of the mouse pointer in pos, which is something called a *tuple*.

TECHNICAL
STUFF

You don't need to know this to use Pygame Zero, but if you're curious, a tuple is a sequence a bit like a list. The main differences are that you can't change the items in it, and it's defined using parentheses () instead of using square brackets []. The pos tuple used in this function contains the *x* and *y* values for the mouse pointer.

Using the mouse position stored in pos, we can work out whether the player clicked on the actor. Collision detection is all about working out whether one sprite is touching another or touching the mouse pointer. It can often be tricky to code, but Pygame Zero includes some features built in to do it for you. You can check whether the mouse pointer (in pos) is touching the cloud actor using the if command together with the collidepoint() function, as shown earlier. If so, you can change the image of the actor to the pop.png image using

```
cloud.image = 'pop'
```

Again, you don't need the file extension of the image file.

This function for detecting mouse clicks also shows how to play a sound: blop is the name of our sound file, minus the extension (.ogg), which Pygame Zero works out for itself. The rest of the line that plays the sound would be the same for any sound. You might find that a simple program like this one is a useful template when you're building your own programs.

Run the program. When you click the background, nothing happens. When you click the cloud in the center of the screen, though, you see it change to a "Pop!" image. One thing you might notice is that you can click near, but not on, the image and it will sometimes register as a hit. That's because an actor is always rectangular. The cloud has rounded corners, but the empty spaces in those corners are still part of the rectangular image, as far as collision detection is concerned.

TIP

When you're making your own images, try to stretch the image out to a rectangle shape as much as possible, and trim any unused space from the edges. That helps your collision detection look accurate. With fast-moving images, nobody will notice if the collision detection is slightly off. If your game includes lots of circles, though, the unused corners of the image will also register a hit, which might feel unfair or unrealistic when playing the game.

Animating your actors

Pygame Zero includes a simple function for animating your actors so that they can move across the screen automatically. This is another way it makes Python more accessible for people who have previously used Scratch: Usually, movement is achieved by having a loop that repeatedly changes the *x* or *y* position of the image. Pygame Zero handles all that complexity for you.

Here's an example of the `animate()` function:

```
animate(cloud, tween='linear', duration=5, pos=(cloud.x, -200))
```

It breaks down like this:

» Give it the name of the actor you want to animate first — in this case, `cloud`.

» Tell it the type of animation you want to use. This describes whether the actor speeds up or slows down over the course of the animation. You can consult the documentation for a list of options (see the "Exploring Pygame Zero Further" section at the end of this chapter), but we are using `'linear'` so that the animation is an even speed throughout.

» The `duration` is how long the animation will take, measured in seconds.

» The `pos` is the finishing position. You give it the *x* and *y* position in brackets, like this: `(x, y)`. As with the mouse position, this is a tuple. I've used the existing *x* position (`cloud.x`), and set the *y* position to be –200, which is off the top of the screen. As a result, the cloud will drift up the screen in a straight line until it's no longer visible.

Try this animation by adding the preceding line of code into the section where you set up your cloud actor, in the following snippet. The new instruction is shown in bold. Note that what you see in this example isn't the full program: You need to keep the `WIDTH` and `HEIGHT` instructions, and the `draw()` and `on_mouse_down(pos)` functions from your previous program:

```
cloud = Actor('target5')
cloud.x = 250
cloud.y = 250
animate(cloud, tween='linear', duration=5, pos=(cloud.x, -200))
```

REMEMBER

You need the second bracket at the end of the new line. One pair of brackets encloses the position, and another goes around everything given to the `animate` function.

When the program runs, the cloud actor will be set up as before, but this time its animation will be started. Each time the `draw()` function runs, it draws the cloud in its latest position, making it appear to move up the screen.

You can still click the cloud to make it pop, so you have the engine for a basic game here. Next, let's add an element of chance.

Using random numbers

Chapter 11 shows you how you can use the `random.randint()` function to pick a random integer (whole number) in a certain range. You can also pick a random floating-point number, with a decimal point in it, using `random.uniform`. To pick a random floating-point number between 1 and 3, for example, you can use

```
random.uniform(1, 3)
```

You can try this in the shell window (without making a program) if you type

```
>>> import random
>>> for i in range(10):
        print(random.uniform(1, 3))
```

Remember the colon at the end of the second instruction there, and you'll find that the next line is indented for you and that Python waits for you to enter it before doing anything. Enter a blank line at the end to tell Python you've finished, and then you'll see ten numbers that look like a bit like this:

```
2.660707438900764
1.5292264804300049
2.9317931171330924
```

Now you can add some random numbers into the animation to make it less predictable, and to make it feel more "floaty." Add this line at the start of your program to import the `random` module:

```
import random
```

Now edit the animation line in your program as shown here:

```
animate(cloud, tween='linear', duration=random.uniform(1, 3), ↵
    pos=(cloud.x + random.randint(10, 100), -200))
```

Here the `duration` is set to a random floating-point number between 1 and 3 seconds. While we were making this game, we tried using a random integer first, but it made the clouds bunch up, and it looked fake when they obviously moved at the same speed. With a random floating-point number between 1 and 3, the clouds all move at different speeds, sometimes overtaking each other but also with a pace that makes sense for gameplay.

We've also added a random whole number between 10 and 100 to the x coordinate of the end point of the animation. This makes the cloud drift sideways slightly. Run the program a few times to see the difference. It's fairly subtle, but it makes it feel much more organic when there are several clouds moving up the screen.

TIP

If you find the game too fast, you can change the numbers between the `random. uniform` brackets. The first number is the minimum time taken to move up the screen. The second one is the maximum. Changing it to `random.uniform(3, 5)`, for example, will give you between 3 and 5 seconds to catch each cloud. Experiment to find a speed you're comfortable with.

Adding more clouds

Chapter 11 shows you how to create a list to store strings, but you can also use a list to store actors, which means you can have several in your game and manage them using loops. In this game, you'll use six clouds in total. When one goes off the top of the screen, you'll put it back at the bottom again, if there is still time left on the timer.

We'll be building on the ideas and code presented earlier, but you might find it easier to start a new file to make this program, to avoid the risk of overlooking some of the changes we've made.

Listing 12-3 shows the first part of the program.

LISTING 12-3: **Adding Yet More Clouds**

```
import pgzrun
import random

WIDTH = 500
HEIGHT = 500
score = 0
timer = 10
clouds = list()

for i in range(6):
    filename = 'target' + str(i)
    clouds.append(Actor(filename))
    this_cloud = clouds[i]
    this_cloud.x = random.randint(int(this_cloud.width / 2), ↵
      int(WIDTH - this_cloud.width / 2))
    this_cloud.y = HEIGHT + this_cloud.height
    animate(this_cloud, tween='linear', duration=random.uniform(1, 3), ↵
      pos=(this_cloud.x + random.randint (10, 100), -200))

def draw():
    screen.clear()
    for i in range(6):
        this_cloud = clouds[i]
        this_cloud.draw()
```

```
screen.draw.text("Score: " + str(score), (2, 2), color="orange")
screen.draw.text("Timer: " + str(timer), (WIDTH - 70, 2), color="orange")
if timer == 0:
    screen.draw.text(str(score), (130, 120), color="white", fontsize=300)
```

This program sets up some new variables for the score and the timer, which we'll use later. It also creates a new list called clouds.

The loop at the start sets up the multiple clouds. It loops six times, giving the variable i the values 0 to 5 in turn. It creates the filename for this actor's image, by joining 'target' to i, making the filenames target0, target1, and so on. The program has to use the str() function to convert i from a number to a string before it can be joined to the 'target' string. The program then adds a new actor to the clouds list, using that filename for the image.

To make the program easier to read, it uses a variable called this_cloud to refer to the cloud it's currently setting up. Its x position is set to a random position in the window. To make sure it won't spill off the edge, the program uses the image's width to work out minimum and maximum possible values for the x position. The image is positioned from its center, so half its width is the minimum x position for it all to fit on the screen. If the image is 100 pixels wide, for example, its center must be at least 50 pixels from the left to fit in. On the right, the program makes a similar calculation, but it subtracts half the image's width from the WIDTH of the window. The x position that's chosen is a random integer between those minimum and maximum positions on the left and right.

The y position is set to be the HEIGHT of the window plus the height of the image itself, which puts it comfortably off the bottom of the screen.

As the program sets up each cloud, it also animates it so that it starts to float up the window.

The draw() function uses a loop to extract each actor in turn from the clouds list, put it into this_cloud, and then draw it.

This function also now draws the score in the top left and the timer in the top right. Because the clouds are drawn first and the text is drawn afterward, the text appears on top of the clouds. That makes it look like the clouds float behind the text, which is a cool effect. The text is drawn using the screen.draw.text() function, which takes the text you want to draw, the (x, y) position for it, and its color as arguments. Optionally, you can add the font size, which we've used for the final score when the timer reaches 0, making the score unmissably big in the middle of the window.

When you run this program, you see six clouds drift up the screen (see Figure 12-2). Although they all start at the same time, their different animation durations make them quickly spread out.

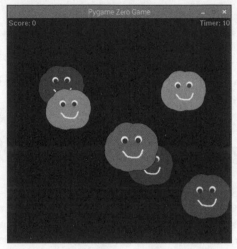

Sean McManus

FIGURE 12-2:
Clouds drifting up
the screen.

Making the clouds regenerate

The way your code works right now, the clouds disappear off the top of the window, and that's the last you see of them. Add the following function to your latest program to make the clouds reappear at the bottom when they go off the top:

```
def update():
    for i in range(6):
        this_cloud = clouds[i]
        if this_cloud.y < 0 - this_cloud.height:
            this_cloud.y = HEIGHT + this_cloud.height
            if timer > 0:
                this_cloud.x = random.randint(int(this_cloud.width / 2), ↩
                    int(WIDTH - this_cloud.width / 2))
                this_cloud.image = 'target' + str(i)
                animate(this_cloud, tween='linear', duration=random.uniform ↩
                    (1, 3), pos=(this_cloud.x + random.randint(10, 100), -200))
```

The update() function is like Pygame Zero's draw() function in that it is automatically run regularly. It's where you put the instructions to change the position of your actors, or to otherwise update the game's progress. The update() function in this game uses a loop to look at each cloud in turn, and uses the variable this_cloud as shorthand for each cloud while it works with it. If the cloud is fully off the screen (its y position is less than 0 minus its height), then the function resets the y position to the bottom of the screen. In fact, this creates some margin because the program only needs to subtract half the height from the cloud's y position, which is in the middle of the image, to be certain it's off the screen. But doing it as we show you here is simpler, and also makes it look more like a stream of clouds — and less like the same clouds keep wrapping around the screen. If there is still time on the clock, we also reset it and start it moving again. We set its x position randomly and reset its image to the target image, because in some cases the cloud will have become a 'pop' image if it was clicked. (We'll add that code shortly.) We also start a new animation, using the same instruction as we did at the start of the program.

You can run the program now to see an unending stream of clouds floating up the window.

Enabling multiple clouds to be clicked

We've already described how you can make one cloud respond to a click, and how you can create, display, and update multiple clouds. We'll combine those ideas in a new function that enables you to click multiple clouds. Add this to the end of your program so far:

```
def on_mouse_down(pos):
    global score
    for i in range(6):
        this_cloud = clouds[i]
        if this_cloud.collidepoint(pos) and timer > 0 and ↵
          this_cloud.image != 'pop':
            this_cloud.image = 'pop'
            sounds.blop.play()
            score += 1
```

This function runs when the mouse button is clicked. When that happens, the function goes through a loop, checking to see whether each cloud has been hit. You only want to pop the cloud if these conditions are true:

>> It's been hit by the mouse, expressed as this_cloud.collidepoint(pos).

>> There's still time on the clock for the player to catch clouds, expressed as timer > 0.

>> The cloud hasn't already been popped, which you can check by making sure its image isn't 'pop', expressed as this_cloud.image != 'pop'.

If all these conditions are true, the program changes the cloud's image to the 'pop' image, plays the sound, and increases the score by 1. After the image changes to 'pop', the cloud continues to float up and off the screen.

Variables used within a function belong only to that function, by default, and are called *local* variables. This isolation stops functions from changing the data that other functions are using by accident. *Global* variables are variables that can be used by any functions. The score, for example, needs to be used in this function (where the program changes it) and the draw() function (where the program displays it), so it's a global variable. If you want to change a global variable inside a function, you have to start the function by telling Python you intend to use that global variable, so that's what the first line in this function does.

At this point, you should be able to play the game almost fully: You can click multiple clouds, see your score increase in the top left, and see the clouds keep coming around at the bottom of the screen again. It can be quite hypnotic. Let's add a time limit, before you fall into a trance.

Adding the timer

Pygame Zero includes the ability to schedule a function to run regularly, which you can use to run a timer. You've already set the timer variable to 10 at the start of the program. You just need to create a function to decrease it by 1 each time it runs, and then schedule that to run every second. Here's what that code looks like. You need to add it to the end of your existing program:

```
def countdown():
    global timer
    timer -= 1
    if timer == 0:
        clock.unschedule(countdown)
        sounds.whoops.play()

clock.schedule_interval(countdown, 1)
```

The line at the end isn't part of any function. You can tell because it's not indented. As a result, this instruction runs when the game begins. The clock.schedule_interval() function sets another function to run at regular intervals. In our case, we're running the countdown() function every 1 second. Note that we don't use brackets after countdown here. We have to put this clock.schedule_interval(countdown, 1) line after the function because it can only be used to schedule a function that has been defined earlier in the program.

The countdown() function itself reduces the global variable timer by 1 each time it runs. If the timer is now 0, it uses clock.unschedule to stop countdown() from running regularly and then plays the Game Over music.

When you run the game, you should now see the timer counting down in the top right. When the timer runs out, no new clouds will appear on the screen, although those that are already there will drift off the top in the usual way. When the time is up, your score will be shown large in the middle of the window.

Adjusting the game difficulty

There are several things you can do to adjust the game and its difficulty:

>> Change the size or shape of the window by changing the WIDTH and HEIGHT variables at the start. A wider window makes them harder to catch, but a taller window gives you more time.

>> Change the speed of the moving clouds, by changing their duration in the animate() functions. Remember that there's one animate() function at the start, and one when they reappear at the bottom of the screen.

>> Increase the amount of time on the clock.

The final game listing

To help you find your way around creating this program, and provide you with it all in one place, here's the final listing (Listing 12-4):

LISTING 12-4: **The final Cloudbusting game**

```
# Cloudbusting game
# From Raspberry Pi For Dummies 4th Edn
# By Sean McManus - www.sean.co.uk


import pgzrun
import random

WIDTH = 500
HEIGHT = 500
score = 0
timer = 10
clouds = list()
```

(continued)

LISTING 12-4: *(continued)*

```
    for i in range(6):
        filename = 'target' + str(i)
        clouds.append (Actor(filename))
        this_cloud = clouds[i]
        this_cloud.x = random.randint(int(this_cloud. width / 2), ↵
          int(WIDTH - this_cloud.width / 2))
        this_cloud.y = HEIGHT + this_cloud.height
        animate(this_cloud, tween='linear', duration=random.uniform(1, 3), ↵
          pos=(this_cloud.x + random.randint(10, 100), -200))

def draw():
    screen.clear()
    for i in range(6):
        this_cloud = clouds[i]
        this_cloud.draw()

    screen.draw.text("Score: " + str(score), (2, 2), color="orange")
    screen.draw.text("Timer: " + str(timer), (WIDTH - 70, 2), color="orange")
    if timer == 0:
        screen.draw.text(str(score), (130, 120), color="white", fontsize=300)

def update():
    for i in range(6):
        this_cloud = clouds[i]
        if this_cloud.y < 0 - this_cloud.height:
            this_cloud.y = HEIGHT + this_cloud.height
            if timer > 0:
                this_cloud.x = random.randint(int(this_cloud.width / 2), ↵
                  int(WIDTH - this_cloud.width / 2))
                this_cloud.image = 'target' + str(i)
                animate(this_cloud, tween='linear', duration=random.uniform ↵
                  (1, 3), pos=(this_cloud.x + random.randint(10, 100), -200))

def on_mouse_down(pos):
    global score
    for i in range(6):
        this_cloud = clouds[i]
        if this_cloud.collidepoint(pos) and timer > 0 and ↵
          this_cloud.image != 'pop':
            this_cloud.image = 'pop'
            sounds.blop.play()
            score += 1

def countdown():
    global timer
    timer -= 1
```

```
    if timer == 0:
        clock.unschedule(countdown)
        sounds.whoops.play()

clock.schedule_interval(countdown, 1)
```

Exploring Pygame Zero Further

We hope this project has given you a taste for what Pygame Zero can do. As well as supporting mouse clicks, you can use it to detect keypresses, making it possible to make more conventional games with characters moving under keyboard control. As such, it's a good foundation for most types of game.

To find out more, we recommend that you check out the following resources:

>> **Daniel Pope's blog post announcing Pygame Zero:** `http://mauveweb.co.uk/posts/2015/05/pygame-zero.html`

>> **The Pygame Zero documentation, including simple examples:** `https://pygame-zero.readthedocs.io/en/latest`

>> **The Pygame Zero built-in objects list, with information about actors, images, sounds and the clock:** `https://pygame-zero.readthedocs.io/en/latest/builtins.html`

You can also read Sean's book *Mission Python* to see how Pygame Zero can be used as the foundation for a 3D adventure game. See `www.sean.co.uk/books/mission-python/index.shtm`.

Chapter **13**

Programming Minecraft with Python

Minecraft appeals to the Lego fan in everyone. It enables you to build immersive 3D worlds from blocks of materials, and it has fired up imaginations to the extent that over 200 million copies have been sold across various platforms, including the PC and Xbox.

A version of Minecraft is available for the Raspberry Pi. It features only the Creative mode, where you can build items peacefully without the threat of monster attacks or starvation. The best feature is that you can program it using Python. This means that you can build a grand palace without having to manually place every block, and you can write programs that can invent original new structures for you to roam around and explore, as you see in this chapter.

The project in this chapter uses a Python program to build a maze in Minecraft. Each time you run the program, it builds a new maze for you, and you can control how big you want it to be and which materials you want it to be made of. During the course of this project, you'll find out how to place and remove blocks in Minecraft using Python so that you'll have the skills to write your own programs that supercharge your construction work.

Minecraft: Pi Edition is labeled as *alpha* software, which means that it's a very early test version (less well developed than a beta version). We had only a minor issue with it: The cursor misbehaved when we maximized the window. You might also experience issues with the screen display not being aligned with the window correctly, and there is no sound. It's highly unlikely there will be changes to this version of Minecraft in the future — since its first release in 2013, there haven't been any updates.

See the Introduction for details of where you can download the code for this chapter. The full listing is at the end of the chapter, which can help you to find your way around the code as you build this project.

Playing Minecraft

Minecraft is preinstalled in Raspbian. You start it by clicking it on the Applications menu at the top left of the screen, where Minecraft Pi is filed under Games. When you start Minecraft on the Raspberry Pi, the title screen gives you two options:

>> **Start Game:** Generates your own game world to explore. You can also use this option to choose a previously generated world to revisit, when you replay Minecraft later. To choose between the different worlds, click and drag them left and right to position your chosen one in the middle, and then click it to open it.

>> **Join Game:** Lets you join other players in a game on a local network. A discussion of this option is outside the scope of this chapter, but this option can enable collaborative or competitive play in a Minecraft world.

Click Start Game and choose Create New, and Minecraft then generates a new world for you, with its own, distinctive terrain of mountains, forests, and oceans. When it's finished, you see a first-person view of it (see Figure 13-1).

TIP

You can change your perspective to show the player's character in the game. Press the Esc key to open the Game menu, and then click the icon beside the Speaker icon in the top left to change the perspective.

When you've finished playing, you can exit the game by pressing the Esc key to open the Game menu and then choosing Quit To Title.

FIGURE 13-1:
Minecraft on
the Pi.

Moving around

Minecraft is easiest to play using two hands — one on the mouse and one on the keyboard. Use the mouse to look around you and change your direction, sliding it left and right to turn sideways, and forward and backward on the desk to look up and down. To move, you use the keys W and S for forward and backward, and A and D to take a sidestep left and right. Those keys form a cluster on the keyboard, which makes it easy to switch between them.

You character automatically jumps onto low blocks if you walk into them, but you can deliberately jump by pressing the spacebar.

For the best view of your world, take to the skies by double-tapping the spacebar. When you're flying, hold the spacebar to go higher, and the left Shift key to go lower. Double-tap the spacebar to stop flying and drop to the ground. There's no health or danger in this edition of Minecraft, so you can freefall as far as you like.

Making and breaking things

To break blocks in your world, use the mouse to aim the crosshair at the block you want to destroy, and click and hold the left mouse button. Some blocks are easier to break than others. There's a limit on how far away you can be, so move closer if you can't see chips flying off the blocks as you attempt to smash them.

The panel at the bottom of the window shows the blocks you can place in the world (refer to Figure 13-1). You choose between them using the scroll wheel on the mouse or by pressing a number between 1 and 8 to pick one (from left to right). Press E to open your full inventory, and then you can use the movement keys (W, A, S, D) to navigate around it or Enter to choose a block — or you can simply click your chosen block with the mouse.

To position a block, right-click where you would like to place it. You can put a block on top of another one only if you can see the top of it, so you might need to fly to make tall structures.

TIP

You can build towers and rise into the air on them by looking down and repeatedly jumping and placing a block under you.

Although Python makes it much easier to build things, we recommend that you spend some time familiarizing yourself with how players experience the world. In particular, it's worth experimenting with how blocks interact with each other. Stone blocks can float in the air unsupported, but sand blocks fall to the ground. Cacti can't be planted in grass, but can be placed on top of sand. If you chip away at the banks of a lake, the water flows to fill the space you made. You can't place water and lava source blocks within the game, although you can program them using Python and they can cascade down and cover a wide area. When water and lava come into contact with each other, water sometimes cools lava into stone.

Preparing for Python

One of the peculiarities of Minecraft is that it takes control of the mouse, so you have to press Tab to take back control when you want to use any other programs on your desktop. To start using the mouse in Minecraft again, click the Minecraft window. You'll soon become used to pressing Tab before you try to do any programming. Press Tab now to leave Minecraft running but bring the mouse cursor back into the desktop. To make your Minecraft programs, you'll use Thonny, so open the Applications menu in the top left, click Programming, and choose Thonny Python IDE. You might have to click the top of the Minecraft window and drag it out of the way first, or click the first button in the top right of its window to minimize it.

REMEMBER

One of the first things you'll notice is that Minecraft sits on top of other windows, and your Python window might well be underneath it, so a certain amount of reorganization is necessary. If your screen is big enough, you might be able to show the Minecraft and Thonny windows on screen at the same time.

If your screen isn't big enough, you'll have to switch between the windows as necessary. To hide Minecraft, click it on the taskbar at the top of the desktop and then click the button in Minecraft's title bar (at the top of its window) to minimize it. To bring Minecraft back or activate it, click it on the taskbar again. If you can't see Minecraft's title bar, it won't respond to the mouse and keyboard controls, so you need to click it on the taskbar. (See Chapter 4 for a guide to using the desktop.)

Using the Minecraft Module

For your first Python program for Minecraft, we will show you how to send a message to the Chat feature in the game.

We'll use the editor in Thonny (the top part of the window). Enter the following, use the File menu to save it in your pi folder, and then press F5 or click the Run button to run it (note that you must have a Minecraft game session running for this to work):

```
import sys, random
from mcpi import minecraft
mc = minecraft.Minecraft.create()
mc.postToChat("Welcome to Minecraft Maze!")
```

Your first line of code imports the sys and random modules. The random module, you'll need later to build a random maze as you develop this program.

To issue Python commands to Minecraft, you use the minecraft.Minecraft.create() function and then add the command at the end. For example, to put a greeting in the Chat window, you might use the following:

```
minecraft.Minecraft.create().postToChat("Welcome to Minecraft Maze!")
```

That soon gets hard to read, so in the program you're working with, you set up mc so that you can use it as an abbreviation for minecraft.Minecraft.create(). As a result, you can use the shorter line that you see in the program to post a message.

WARNING

If your code isn't working, pay particular attention to the case. Python is case-sensitive, so you have to use upper- and lowercase exactly as shown here. Look out for the mixed upper- and lowercase in postToChat, and the capital M in minecraft.Minecraft.create().

When you run the program, switch back to Minecraft to see your message on screen. It disappears after about 10 seconds.

Understanding coordinates in Minecraft

As you might expect, everything in the Minecraft world has a map coordinate. Three axes are required in order to describe a position in the game world:

>> **x:** This axis runs parallel to the ground, from west (negative numbers) to east (positive numbers). The world measures 255 blocks in this direction.

>> **y:** This axis runs vertically and could be described as the height. You can fly at least as high as 500 blocks, but you can't see the ground from higher than about 70 blocks, so there's not much point. Sea level is 0. You can break blocks to tunnel under the sea too. We made it down to about −70 before we fell out of the world and died. This is the only way we've seen that you can die in Minecraft on the Pi.

>> **z:** This is the other axis parallel to the ground, running from north (negative numbers) to south (positive numbers). The world measures 255 blocks in this direction too.

We put the axes in this order deliberately because that's the order that Minecraft uses. If, like us, you often use x and y to refer to positions in 2D (as you do in Scratch), it takes a short while to get your head around the fact that y represents height. Most of the time in this chapter, you'll use the x and z coordinates to describe a wall's position (which differs depending on the wall), and the y coordinate to describe its height (which doesn't, in our project).

As you move in the game, you can see the player's coordinates in the top left of the Minecraft window change. If you try to move outside the game world, you hit a wall of sky that you can't penetrate, like in *The Truman Show* (except that he had a door).

Repositioning the player

You can move your character to any position in the Minecraft world, using this command:

```
mc.player.setTilePos(x, y, z)
```

For example, to parachute into the world, use

```
mc.player.setTilePos(0, 100, 0)
```

TIP

You don't have to put this command into a program and run it. If you've already run the program to set up the Minecraft module, you can type commands to move the player and add blocks in the Python shell.

Assuming that the game is not in Flying mode, you'll drop from the sky into the world. If it is in Flying mode, click Minecraft on the taskbar and double-tap the spacebar to turn it off and start your descent.

Note that this coordinate won't always be in the middle of the world, even though worlds are the same size. In one of our worlds, the coordinates run from −85.7 to 169.7 in the x plane and from −98.7 to 156.7 in the z plane.

You can put the player anywhere in the game world, and sometimes that means she'll appear in the middle of a mountain or another structure, where she can't move. If that happens, reposition the player using code. Putting her somewhere high is usually a reasonably safe bet because she can fall to the highest ground from there.

Adding blocks

To add a block to the world, you use this command:

```
mc.setBlock(x, y, z, blockTypeId)
```

blockTypeId is a number that represents the material of the block you're adding. You can find a full list of materials at https://minecraft.gamepedia.com/Bedrock_Edition_data_value. (Take the number from the Dec column in the table on that page. You want the decimal number rather than the hexadecimal one.) Any number from 0 to 108 is valid, and a few higher numbers are as well. Table 13-1 shows some of the materials you might find most useful for this project and for experimentation.

TIP

If you use the water and lava blocks, you could flood your world, so create a new world to experiment with.

TABLE 13-1   **Materials in Minecraft: Pi Edition**

blockTypeId	Block Type
0	Air
1	Stone
2	Grass
3	Dirt
5	Wooden plank
8	Water
10	Lava
12	Sand
20	Glass brick
24	Sandstone
41	Gold brick
45	Brick
47	Bookshelf
53	Oak stairs
57	Diamond block
64	Oak door
81	Cactus

There is another command you can use to create a large, cuboid shape built of blocks of the same material. To use it, you provide the coordinates of two opposite corners and the material you'd like to fill the space with, like this:

```
mc.setBlocks(x1, y1, z1, x2, y2, z2, blockTypeId)
```

You can quickly build a brick shelter by making a large cuboid of brick and then putting a cuboid of air inside it. Air replaces any other block, effectively deleting it from the world. Here's an example:

```
mc.setBlocks(0, 0, 0, 10, 5, 7, 45) #brick
mc.setBlocks(1, 0, 1, 9, 5, 6, 0) #air
```

These lines build a shelter that is 10 × 7 blocks in floor space and 5 blocks high, starting at coordinate 0, 0, 0. The walls have a thickness of 1 block because you fill with air the space from 1 to 9 on the x-axis, from 1 to 6 on the z-axis, and from 0 to 5 on the vertical axis, leaving intact 1 block of brick from the original cuboid on four sides and the roof open.

If your Minecraft window goes black when you try this, you've probably built the walls on top of the player. Reposition the player using code to get the view back.

REMEMBER

The # symbol represents a comment that's there only as a reminder for you. The computer ignores anything on the same line after the #.

Although players can have coordinate positions with decimal portions (such as 1.7), when you place a block, its position is rounded down to the nearest whole number.

Stopping the player from changing the world

We know you wouldn't cheat, but there's no fun in a maze that you might *accidentally* just hack your way through, is there? To stop players from being able to destroy or place blocks in the world, use the following:

```
mc.setting("world.immutable", True)
```

The word *immutable* is often used in programming, and it means "unchangeable."

Setting the maze parameters

Now that you know how to place blocks in the world and use the air block to remove them again, you're ready to start making the maze program. In this program, you'll use a number of constants to keep track of important information about the maze. *Constants* are just variables that you *decide* not to change the values of as the program is running, so their values are always the same. It's conventional to use uppercase letters for the names of constants to signal your intent to others reading the program, and to remind yourself that you're not supposed to be letting the program change these values. Replacing numbers in your program with constants makes it easier to customize your program later, but also makes it much easier to read your program and understand what different numbers represent.

REMEMBER

Variable names are case sensitive, so Python would think SIZE and size were two different variables. You'd be unwise to use both in the same program, though!

The program starts by setting up these constants:

```
SIZE = 10
HEIGHT = 2
MAZE_X = 0
GROUND = 0
MAZE_Z = 0
MAZE_MATERIAL = 1 #stone
GROUND_MATERIAL = 2 #grass
CEILING = False
```

To build the maze, you start with a grid of walls with 1-block spaces (or cells) between them, which looks a bit like a waffle (see Figure 13-2). Each cell starts with four walls, and the program knocks down walls to create paths between them and build the maze. The maze is square, and its SIZE is measured in cells. A maze with a SIZE of 10 will have 10 cells in the x and z dimensions, but will occupy double that space in the Minecraft world (that is, 20 blocks by 20 blocks) because there is a 1-block wall between each cell. This becomes clearer as you start to build the maze. We've tried mazes as big as 40, but they take some time to build and ages to explore. Ten is big enough for now.

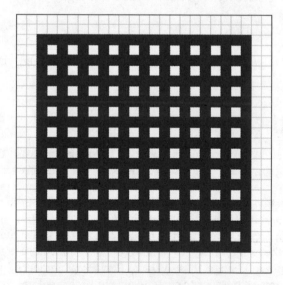

FIGURE 13-2:
The starter grid.

The HEIGHT is how many blocks tall the maze walls are. We chose 2 because a value of 1 means that the player can just walk over the maze. (The player automatically steps onto blocks 1 unit high.) Higher values obscure any mountains in the distance that can otherwise give a nice visual hint to the player.

The constants MAZE_X, GROUND, and MAZE_Z are used for the starting coordinates of the maze. The MAZE_MATERIAL is stone (1), and the GROUND_MATERIAL is grass (2). We've added an option for a ceiling, to stop players from just flying out of the top of the maze, but we've turned it off for now so that you can freely explore the maze as you're building it.

The program stops with an error if there isn't enough room for all of the maze in your world. In that case, you can try using a smaller maze, try moving the MAZE_X and MAZE_Z coordinates, or try using a different world.

TIP

A maze of bookshelves (MAZE_MATERIAL=47) looks great!

Laying the foundations

One of the first things you need to do before you build the maze is make sure that you're building on solid land. Because Minecraft worlds are dynamically generated, you might find that, otherwise, you're building a maze inside a mountain or in the sea.

You'll need to clear, as well as the area the maze will occupy, an area of ten blocks all the way around it so that the players can approach it easily and walk around the outside of it. First you clear the area by filling it with air blocks, which will wipe out anything else in that space. Then you add the floor, a layer of blocks made of the ground material.

The maze occupies a ground space measured in blocks from MAZE_X to MAZE_X+(SIZE*2), and from MAZE_Z to MAZE_Z+(SIZE*2). (*Remember:* * is the symbol for multiplication.) The number of blocks is twice the number of cells (SIZE) because each cell has a wall on its right and below it (when viewed from above, as in Figure 13-2). The middle of the maze in the Minecraft world is MAZE_X+SIZE, MAZE_Z+SIZE.

You need to clear 10 blocks farther in each direction. The following code clears everything as high as 150 above the ground level of the maze, to stop the risk of any remaining mountain blocks falling from the sky into the maze, and then lays the floor:

```
mc.setBlocks(MAZE_X-10, GROUND, MAZE_Z-10, MAZE_X+(SIZE*2)+10, GROUND+150, ↵
    MAZE_Z+(SIZE*2)+10, 0)
mc.setBlocks(MAZE_X-10, GROUND, MAZE_Z-10, MAZE_X+(SIZE*2)+10, GROUND, ↵
    MAZE_Z+(SIZE*2)+10, GROUND_MATERIAL)
```

We recommend adding a block to indicate the starting corner of the maze (where MAZE_X and MAZE_Z are). You will find it useful when writing and debugging the program, because it will enable you to tell which way around the maze is as you fly around it. To do so, use the following:

```
mc.setBlock(MAZE_X, GROUND+HEIGHT+1, MAZE_Z, MAZE_MATERIAL)
```

Put your player character above the middle of the maze, too, so that you can watch it being built by looking down, as follows — if you're not flying, you'll fall onto the maze wall, but you can just fly up again:

```
mc.player.setTilePos(MAZE_X+SIZE, GROUND+25, MAZE_Z+SIZE)
```

Placing the maze walls

To make the waffle-like grid, use the following code:

```
for line in range(0, (SIZE+1)*2, 2):
    mc.setBlocks(MAZE_X+line, GROUND+1, MAZE_Z, MAZE_X+line, GROUND+HEIGHT, ↵
        MAZE_Z+(SIZE*2), MAZE_MATERIAL)
    mc.setBlocks(MAZE_X, GROUND+1, MAZE_Z+line, MAZE_X+(SIZE*2), ↵
        GROUND+HEIGHT, MAZE_Z+line, MAZE_MATERIAL)
```

The for loop gives the variable line the values of even numbers starting at 0 and finishing at SIZE*2, in turn. Note that you have to add 1 to SIZE before doubling it, because the range function doesn't include the last number in the sequence. If you use range(1, 10), for example, you get the numbers 1 to 9. The 2 at the end of the range function is the step size, so it adds 2 each time it goes around the loop, and only gives you the even numbers. That means you leave a gap for the cell between each wall. Each time around the loop, it uses cuboids to draw two walls that stretch across the maze from edge to edge in the *x* and *z* dimensions. It doesn't matter that the same block is set twice where those lines intersect. You build the wall starting at GROUND+1, so the grass is still underneath when you knock down the walls to make paths.

REMEMBER

Don't forget the colon at the end of the for statement, and that the next two lines should each be indented by four spaces to tell Python that they belong to the loop.

You should now have a grid that looks like Figure 13-3.

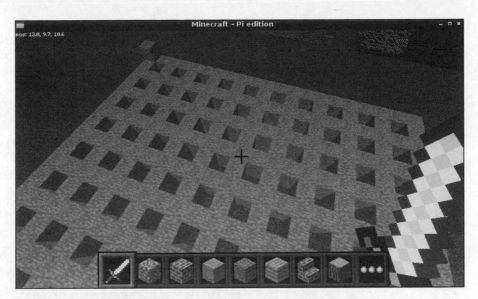

FIGURE 13-3:
Your grid in
Minecraft.

Understanding the maze algorithm

Before you dig into the code that turns your waffle into a maze, let us tell you how
it works. You're going to make what's known as a *perfect maze* (that's a technical
term, not us bragging): That means there are no loops in it and every part of the
maze can be visited. There is only one path between any two points in the maze.

An algorithm is a set of rules or a process for solving a particular problem. It is
typically performed by a computer, but you could carry out the maze making algorithm using paper and pencil, and an eraser to wipe out the walls as you go. Here's
how it works:

1. You start with the "waffle" you've built, with every cell having all four walls.

2. You pick a random cell in the maze to start at.

3. You look at your current cell's neighbors and make a list of all those that have
   all four walls intact. These are the cells that have not yet been visited.

4. If you found some unvisited neighbors, you pick one at random, knock down
   the wall between it and your current cell, and then move into that cell, making
   it your current cell.

5. If your current cell has no unvisited neighbors, you go back one cell in the path
   you've taken, and make that your current cell.

6. Repeat Steps 3 to 5 until you've visited every cell.

Setting up the variables and lists

To implement this algorithm, you'll use the following variables:

>> numberOfCells: This is the total number of cells in the maze, which will be SIZE*SIZE.

>> numberOfVisitedCells: This keeps track of how many cells you've visited. When this is the same as the numberOfCells, every cell has been visited and had a wall demolished, and is therefore reachable. The maze is finished.

>> xposition: This remembers your *x* position as you move through the maze generating it. It's measured in cells, and it starts as a random number between 1 and the maze SIZE.

>> zposition: This remembers your *z* position as you move through the maze generating it, also measured in cells and also starting as a random number.

>> cellsVisitedList[]: This is a list that stores the path you've taken so that the program can retrace its steps. When you set it up, you put your starting position into it using the append() list method.

>> playerx and playerz: These are used to remember the starting position, so you can put the player there when the maze has been built.

When an algorithm like this is implemented (it's called a *depth-first maze generation algorithm*), it often requires a list or similar data structure to store the locations of walls. You don't need that, because you have actual walls in Minecraft you can look at. The game world stores your maze, if you like.

The following code lines set up your starting variables:

```
numberOfCells = SIZE * SIZE
numberOfVisitedCells = 1 # 1 for the one you start in
cellsVisitedList = []

xposition = random.randint(1, SIZE)
zposition = random.randint(1, SIZE)
playerx = xposition
playerz = zposition
# see the next section, "Creating the functions"
showMaker(xposition, zposition)
cellsVisitedList.append((xposition, zposition))
```

Creating the functions

There are a number of basic functions you will need for your program:

» `realx(x)` and `realz(z)`: These convert coordinates in the maze (measured in cells) into coordinates in the Minecraft world (measured in blocks and offset from the maze's starting position).

» `showMaker(x, z)` and `hideMaker(x, z)`: These functions use a gold block to show which cell the program has reached as it builds the maze. It's fun to watch from above and is useful while building and debugging the program.

» `demolish(realx, realz)`: This knocks down a wall in the maze and takes real x and z coordinates in the Minecraft world as its parameters.

» `testAllWalls(cellx, cellz)`: This checks whether the four walls on a cell are intact. If all of them are, it returns True. Otherwise, it returns False. It uses the command `mc.getBlock(x, y, z)`, which tells you the `blockTypeId` at a particular location. You use two equal signs, as usual, to test whether a block in a wall position is the same as the MAZE_MATERIAL, which means that there's a wall there.

Add these function definitions at the start of your program, right after where you set up the Minecraft module:

```python
def realx(x):
    return MAZE_X + (x*2) - 1

def realz(z):
    return MAZE_Z + (z*2) - 1

def showMaker(x, z):
    mc.setBlock(realx(x), GROUND+1, realz(z), 41) # 41=gold

def hideMaker(x, z):
    mc.setBlock(realx(x), GROUND+1, realz(z), 0)

def demolish(realx, realz):
    mc.setBlocks(realx, GROUND+1, realz, realx, HEIGHT+GROUND, realz, 0)

def testAllWalls(cellx, cellz):
    if mc.getBlock(realx(cellx)+1, GROUND+1, realz(cellz))==MAZE_MATERIAL ↵
        and mc.getBlock (realx(cellx)-1, GROUND+1, realz(cellz))==MAZE_MATERIAL ↵
```

```
    and mc.getBlock(realx(cellx), GROUND+1, realz(cellz)+1)==MAZE_MATERIAL ↵
    and mc.getBlock(realx(cellx), GROUND+1, realz(cellz)-1)==MAZE_MATERIAL:
        return True
else:
    return False
```

TIP

If you have an error, check for missing colons at the end of your def and if statements.

Creating the main loop

Your maze algorithm runs until you've visited every cell, so it starts with the following statement:

```
while numberOfVisitedCells < numberOfCells:
```

You need to test whether your current cell's neighbor cells have all their walls intact. To do that, you check each direction in turn, using the testAllWalls(x, z) function. When you find a cell with all the walls intact, you add its direction to the list possibleDirections[] using the append() list method. This implements Step 3 in the algorithm — and keep in mind that it's all indented underneath the while statement:

```
possibleDirections = []

if testAllWalls(xposition - 1, zposition):
    possibleDirections.append("left")

if testAllWalls(xposition + 1, zposition):
    possibleDirections.append("right")

if testAllWalls(xposition, zposition - 1):
    possibleDirections.append("up")

if testAllWalls(xposition, zposition + 1):
    possibleDirections.append("down")
```

The values of up, down, left, and right are somewhat arbitrary in 3D space, but we've used them because they're easy to understand. If you fly into the air and look down on the maze as it's being generated and you have the block identifying the starting corner of the maze (MAZE_X, MAZE_Z) in the top left, these directions will look correct to you.

Incidentally, you might have noticed that there's no check for whether these cell positions are inside the maze borders. What happens if you look for a cell off the left edge of the maze, or off the bottom edge? No problem. The program implementation automatically respects the borders of the maze because when it looks at "cells" outside the borders, they don't have all four walls (their only wall is the maze's border), so they are never visited.

Step 4 in the algorithm is to pick a random direction if you found any unvisited neighbors, knock down the wall in that direction, and move into that cell. To decide whether you found any possible directions, you check the length of the possibleDirections list and act if it is not equal to 0 (expressed as !=0). All of this should be indented under the while loop. (If you get lost in the indenting, consult the full code in Listing 13-1, near the end of this chapter.)

Before you start moving your position, you hide the gold brick that shows where you are in the maze:

```
hideMaker(xposition, zposition)
if len(possibleDirections) != 0:
    directionChosen=random.choice(possibleDirections)

    if directionChosen == "left":
        demolish(realx(xposition) - 1, realz(zposition))
        xposition -= 1

    if directionChosen == "right":
        demolish(realx(xposition) + 1, realz(zposition))
        xposition += 1

    if directionChosen == "up":
        demolish(realx(xposition), realz(zposition) - 1)
        zposition -= 1

    if directionChosen == "down":
        demolish(realx(xposition), realz(zposition) + 1)
        zposition += 1
```

After you've moved into a new cell, you need to increase your tally of cells visited by one, and add the new cell to the list that stores the path taken. This is also a good time to show the gold block in the cell to highlight how the maze is being built:

```
numberOfVisitedCells += 1
cellsVisitedList.append((xposition, zposition))
showMaker(xposition, zposition)
```

The way you've stored the list of cells visited deserves some explanation. You've put the xposition and zposition in parentheses, which are used to indicate a *tuple*. A *tuple* is a data sequence, a bit like a list, with a key difference that you can't change its values. (It's immutable.) So cellsVisitedList is a list that contains tuples, which in turn contain pairs of *x* and *z* coordinates. You can use the Python shell to take a look inside this list. Here's an example from one run of the program, showing a path taken through the maze:

```
>>> print(cellsVisitedList)
[(6, 6), (6, 7), (6, 8), (5, 8), (4, 8), (3, 8), (3, 7)]
```

For Step 5 in the algorithm, you go back to the previous position in the path if your cell has no unvisited neighbors. This involves taking the last position out of the list. There's a list method called pop() that you can use. It takes the last item from a list and deletes it from that list. In your program, you put it into a variable called retrace, which then stores a tuple for the *x* and *z* positions in the maze. As with a list, you can use index numbers to access the individual elements in a tuple. The index numbers start at 0, so retrace[0] will hold your previous *x* position, and retrace[1] will hold your previous *z* position. Here's the code, including a line to show the gold block in its new position:

```
else: # do this when there are no unvisited neighbors
    retrace = cellsVisitedList.pop()
    xposition = retrace[0]
    zposition = retrace[1]
    showMaker(xposition, zposition)
```

Note that the else statement should be in line with the if statement it's paired with — in this case, the one that tests whether you found any possible directions to move in.

Step 6 in the algorithm has already been implemented because the while loop will keep repeating the indented code underneath it until every cell has been visited.

Adding a ceiling

Personally, we think it's more fun to leave the ceiling open and be free to fly up and marvel at your maze and then drop into it at any point. If you want to build a game around your maze, though, and stop people from cheating, you can add a ceiling using the following code. Just change the variable CEILING to True at the start of the program. We've made the ceiling out of glass bricks, so it doesn't get too dark in there:

```
if CEILING == True:
    mc.setBlocks(MAZE_X, GROUND+HEIGHT+1, MAZE_Z, MAZE_X+(SIZE*2), ↩
    GROUND+HEIGHT+1, MAZE_Z+(SIZE*2), 20)
```

Positioning the player

Finally, let's place the player at the random position where you started generating the maze. You could put the player anywhere, but this seems as good a place as any, and it uses random numbers you have already generated:

```
mc.player.setTilePos(realx(playerx), GROUND+1, realz(playerz))
```

Now you're ready to play! Figure 13-4 shows the maze from the inside.

FIGURE 13-4:
Finding your way around the maze.

The final code

Listing 13-1 shows the final and complete code:

LISTING 13-1: **The Minecraft Maze Maker**

```
# Minecraft Maze Generator
# by Sean McManus
# From Raspberry Pi For Dummies
```

(continued)

LISTING 13-1: *(continued)*

```python
import sys, random
from mcpi import minecraft
mc = minecraft.Minecraft.create()

mc.postToChat("Welcome to Minecraft Maze!")

def realx(x):
    return MAZE_X + (x*2) - 1

def realz(z):
    return MAZE_Z + (z*2) - 1

def showMaker(x, z):
    mc.setBlock(realx(x), GROUND+1, realz(z), 41) # 41=gold

def hideMaker(x, z):
    mc.setBlock(realx(x), GROUND+1, realz(z), 0)

def demolish(realx, realz):
    mc.setBlocks(realx, GROUND+1, realz, realx, HEIGHT+GROUND, realz, 0)

def testAllWalls(cellx, cellz):
    if mc.getBlock(realx(cellx)+1, GROUND+1, realz(cellz))==MAZE_MATERIAL ↵
        and mc.getBlock (realx(cellx)-1, GROUND+1, realz(cellz))==MAZE_MATERIAL ↵
        and mc.getBlock(realx(cellx), GROUND+1, realz(cellz)+1)==MAZE_MATERIAL ↵
        and mc.getBlock(realx(cellx), GROUND+1, realz(cellz)-1)==MAZE_MATERIAL:
        return True
    else:
        return False


mc.setting("world_immutable", True)

# Configure your maze here
SIZE = 10
HEIGHT = 2
MAZE_X = 0
GROUND = 0
MAZE_Z = 0
MAZE_MATERIAL = 1 # 1=stone
GROUND_MATERIAL = 2 # 2=grass
CEILING = False

#clear area
mc.setBlocks(MAZE_X-10, GROUND, MAZE_Z-10, MAZE_X+ (SIZE*2)+10, GROUND+150, ↵
  MAZE_Z+(SIZE*2)+10, 0) # air
```

```
#lay the ground
mc.setBlocks(MAZE_X-10, GROUND, MAZE_Z-10, MAZE_X+ (SIZE*2)+10, GROUND, ↵
  MAZE_Z+(SIZE*2)+10, GROUND_MATERIAL)

# origin marker
mc.setBlock(MAZE_X, GROUND+HEIGHT+1, MAZE_Z, MAZE_MATERIAL)

# move player above middle of maze
mc.player.setTilePos(MAZE_X+SIZE, GROUND+25, MAZE_Z+SIZE)

mc.postToChat("Now building your maze...")

# build grid of walls
for line in range(0, (SIZE+1)*2, 2):
    mc.setBlocks(MAZE_X+line, GROUND+1, MAZE_Z, MAZE_X+line, GROUND+HEIGHT, ↵
      MAZE_Z+(SIZE*2), MAZE_MATERIAL)
    mc.setBlocks(MAZE_X, GROUND+1, MAZE_Z+line, MAZE_X+ (SIZE*2), ↵
      GROUND+HEIGHT, MAZE_Z+line, MAZE_MATERIAL)

mc.postToChat("TIP: Fly above it and look down.")

# set up variables for creating maze
numberOfCells = SIZE * SIZE
numberOfvisitedCells = 1 # 1 for the one we start in
cellsVisitedList = []

xposition = random.randint(1, SIZE)
zposition = random.randint(1, SIZE)
playerx = xposition
playerz = zposition
showMaker(xposition, zposition)
cellsVisitedList.append((xposition, zposition))

# main loop
while numberOfvisitedCells < numberOfCells:
    possibleDirections = []

    if testAllWalls(xposition - 1, zposition):
        possibleDirections.append("left")

    if testAllWalls(xposition + 1, zposition):
        possibleDirections.append("right")

    if testAllWalls(xposition, zposition - 1):
        possibleDirections.append("up")
```

(continued)

LISTING 13-1: *(continued)*

```
                if testAllWalls(xposition, zposition + 1):
                    possibleDirections.append("down")

            hideMaker(xposition, zposition)

            if len(possibleDirections) != 0:
                directionChosen=random.choice(possibleDirections)

                #knock down wall between cell in direction chosen
                if directionChosen == "left":
                    demolish(realx(xposition) - 1, realz(zposition))
                    xposition -= 1

                if directionChosen == "right":
                    demolish(realx(xposition) + 1, realz(zposition))
                    xposition += 1

                if directionChosen == "up":
                    demolish(realx(xposition), realz(zposition) - 1)
                    zposition -= 1

                if directionChosen == "down":
                    demolish(realx(xposition), realz(zposition) + 1)
                    zposition += 1

    # after the move, increase number of visited cells
                numberOfvisitedCells += 1

                cellsVisitedList.append((xposition, zposition))
                showMaker(xposition, zposition)

            else: # do this when there are no unvisited neighbors
                retrace = cellsVisitedList.pop()
                xposition = retrace[0]
                zposition = retrace[1]
                showMaker(xposition, zposition)

    if CEILING == True:
        mc.setBlocks(MAZE_X, GROUND+HEIGHT+1, MAZE_Z, MAZE_X+(SIZE*2), ↵
            GROUND+HEIGHT+1, MAZE_Z+(SIZE*2), 20)

    mc.postToChat("Your maze is ready!")
    mc.postToChat("Happy exploring!")
    mc.player.setTilePos(realx(playerx), GROUND+1, realz(playerz))
```

Adapting the Program

After the maze has been built, the gold brick remains visible, so you could try to solve the maze to find the brick. You could also plant other objectives in the maze and time how long it takes the player to find them. The `mc.player.getTilePos()` command checks where the player is in the Minecraft world and gives you a result in the form `x,y,z`.

You could add an entrance and exit in random positions in the border of the maze so that the goal is to travel from one side to the other. You could make huge mazes more playable by adding landmarks. (Try using different wall materials or putting blocks on top of some walls.) After the maze has been generated, you could knock out random walls so that there are some shortcuts through the maze. Or maybe just replace them with glass blocks, to provide a tantalizing glimpse into another corridor. What about a multistory maze, with stairs between the levels? The possibilities are — ahem! — amazing.

Chapter **14**

Making Music with Sonic Pi

For much of the music we hear today, computers are at least as important in the studio as microphones are, and they have been for many years. Using Sonic Pi, you can start composing your own computer music by programming your Raspberry Pi. It enables you to put together simple programs that play synthesizer melodies and sampled sounds, generating your own distinctive instrumental music.

Sam Aaron, who created Sonic Pi, says: "Sonic Pi has two concurrent goals: to be simple to learn and yet also powerful enough for professional musicians. You might think that sounds a bit ambitious, but it exactly describes a piano."

Sam has performed concerts using Sonic Pi at international festivals including Moogfest, a celebration of electronic music. While people listen to Sam's music at a performance, they can see how he is creating and editing his code on the screen on the stage. Music magazine *Rolling Stone* described Sam's music as sounding like "Electric Café-era Kraftwerk, a little bit of Aphex Twin skitter and some Eighties electro."

Sonic Pi runs from the desktop environment (see Chapter 4) and is one of the recommended applications in Raspberry Pi OS. If it isn't installed, you can add it using the Recommended Software program in the Preferences part of the Applications menu.

You run Sonic Pi by clicking the Applications Menu button in the top-left of the screen, clicking Programming, and then clicking the entry for Sonic Pi.

WARNING

As of this writing, Sonic Pi can only send its sound through the HDMI cable. If you're using a monitor with built-in speakers, you're all set. If you're using headphones or external speakers connected to the Raspberry Pi's audio socket, it won't work. In that case, you could try Sonic Pi using your TV if it has an HDMI socket.

Understanding the Sonic Pi Screen Layout

Figure 14-1 shows the screen layout for Sonic Pi. You might see some differences between your screen layout and ours, but the fundamentals should be the same. Click to enlarge the window if necessary. (See the section on resizing and closing your program windows in Chapter 4.) On the left is the Editor, where you type in your code. On the right is the Log, where Sonic Pi tells you what it's doing as it plays your music. At the bottom is the Help pane.

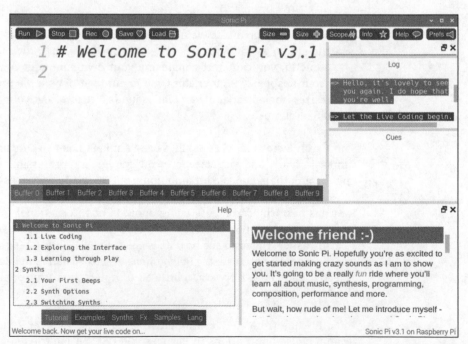

FIGURE 14-1:
The Sonic Pi screen layout.

Sean McManus

Sonic Pi uses ten buffers, which you access by clicking the buttons at the bottom of the Editor. You can think of each buffer as like having a different file open for editing, but you can play music from different buffers at the same time. This can be particularly useful for live performance: You might set up a loop in one buffer and then experiment with code to add notes on top in another buffer. When you exit Sonic Pi, the content of your buffers is saved for you, and it's loaded when you come back again. You can also save the content of a buffer to a text file using the Save button in the menu bar at the top of the screen.

At the top of the screen are buttons to run your program (play your music) and stop it. They use symbols similar to any audio player: a triangle to play and a square to stop. There are also buttons to adjust the text size, show or hide the oscilloscope (which displays waveforms as your music plays), show or hide the info window, and show or hide the Help pane at the bottom of the screen.

There is also a Preferences button to the right of the Help button. Clicking this button shows and hides the Preferences pane. The Preferences pane provides a volume control and has options for adjusting the Editor display and checking for any updates. If you use other electronic instruments or music software, you can connect to Sonic Pi using the MIDI or OSC protocols, which are configured in the IO preferences.

TIP

You can adjust the size allocated to the panes by clicking and dragging the dividing lines between them.

TIP

If you are performing with Sonic Pi, you can use the dark mode for a color scheme based on a black background, which will be more comfortable to use in a club or concert venue with low lights. There's also a Pro Icons setting, which replaces the buttons with stripped-back icons for a more streamlined interface. You'll find both of these settings in the Editor preferences.

Playing Your First Notes

Click in the Editor and type the following:

```
play 60
```

Nothing happens because you've entered your program but haven't run it yet. Click the Run button and you will hear a middle C note sound. At the same time, you'll see the Log update.

The note numbers used are standard MIDI note numbers, widely used in electronic instruments. You've already seen them in Scratch. Higher-sounding notes use higher numbers, and lower-sounding notes use lower numbers.

Try adding some more notes to your program:

```
play 60
play 64
play 67
```

When you click the Run button, you still hear just one sound, but three different notes are playing at the same time. It's actually a C chord you hear, which uses the notes C (60), E (64), and G (67). If you want to play the notes separately, you can add a pause of half a beat between them using the sleep command:

```
play 60
sleep 0.5
play 64
sleep 0.5
play 67
sleep 0.5
play 72
```

There is an extra higher C note on the end of that sequence to make it sound like a fanfare. You can experiment with writing your own tunes. Just put together a sequence of notes.

Table 14-1 shows the standard MIDI notes, which run from 0 to 127. In practice, they sound extremely tinkly at the high end and descend into indistinct soft thuds at the low end. For best results, we recommend you keep your numbers between 48 and 96, but feel free to experiment to find out what sounds good to you.

As you can see, the numbers in the table count from top to bottom, and from left to right. Notes get higher as you go down the table and as you move from left to right across the columns. The next highest note after B (at the bottom of the table) is the C at the top of the next column to the right. It's like a piano, where the same key layout (running from C to G, then A to B, and then starting from C again) repeats all the way along it.

If you don't know much about music, stick to the notes that don't have a sharp symbol (#) on them and avoid too many huge leaps. Try moving a few notes up or down a column and dip into a neighboring column when you're near the top or bottom of your column. By following those simple guidelines, you should end up with a jolly little ditty.

TABLE 14-1

MIDI Notes

Note		0	1	2	3	4	5	6	7	8	9
C	0	12	24	36	48	60	72	84	96	108	120
C#	1	13	25	37	49	61	73	85	97	109	121
D	2	14	26	38	50	62	74	86	98	110	122
D#	3	15	27	39	51	63	75	87	99	111	123
E	4	16	28	40	52	64	76	88	100	112	124
F	5	17	29	41	53	65	77	89	101	113	125
F#	6	18	30	42	54	66	78	90	102	114	126
G	7	19	31	43	55	67	79	91	103	115	127
G#	8	20	32	44	56	68	80	92	104	116	
A	9	21	33	45	57	69	81	93	105	117	
A#	10	22	34	46	58	70	82	94	106	118	
B	11	23	35	47	59	71	83	95	107	119	

Using Note and Chord Names

Sonic Pi enables you to use proper note names instead of MIDI numbers by using the name of the note (a letter from A to G), plus the number of the octave it's in. You can see those numbers labeling the columns in Table 14-1.

For example, to play a middle C, you can use

```
play :c4
```

To play the B one note before it, which is in the next lowest octave, you would use

```
play :b3
```

The Log shows that Sonic Pi plays notes 60 and 59 respectively. You can check the note names and numbers in Table 14-1 to confirm that this is what you expected.

Here's how you could code a fanfare using note names instead of numbers:

```
play :c4
sleep 0.5
```

```
play :e4
sleep 0.5
play :g4
sleep 0.5
play :c5
```

If you want to use a sharp note, insert the letter *s* in the note name (for example, play :cs4) and use *b* for a flat note (play :cb4).

You can also use names to play chords. You tell Sonic Pi the lowest note in the chord and add which type of chord you want (try :major, :minor, or :diminished). There are also options for :major7, :minor7, :diminished7, and :dom7, among others. For a complete list, click Lang in the Help pane, and then select Chord. Try this:

```
play chord(:a3, :major)
sleep 1
play chord(:a3, :minor)
```

In each case, it plays three notes at the same time. If you look at the note numbers in the Log, you can see that the middle note was one pitch lower in the second chord because it's a minor chord. Again, you can use Table 14-1 to check the note numbers Sonic Pi displays against the musical note names.

The chord is returned as a list, and you can use play_pattern to hear the notes of the chord in a sequence, like this:

```
play_pattern chord(:a3, :major)
play_pattern chord(:a3, :minor)
```

Playing Timed Patterns

There is a more efficient way you can play a sequence of notes and specify the time, in beats, between each one: Use the play_pattern_timed command. Click a button to go to a new buffer and try this:

```
play_pattern_timed [:c4, :e4, :g4, :c5], [0.5, 0.5, 1]
```

Pay careful attention to the brackets and commas here. This command takes two different sets of information, and each set is between square brackets. The first set is the notes you want to play, and they are the same notes as we used in our fanfare earlier. The second set of information is separated from the first set by a comma, and it is the length of the pause between the notes. There are four notes, but just three gaps between them, so the second set of brackets has fewer items in it. The numbers we've used here put a half-beat pause between the first and second notes, and the second and third notes, but double that to build up the suspense (such as it is) before the final note sounds.

Composing Random Tunes Using Shuffle

The bracketed sections are lists, similar to lists in Python. You can add different instructions (or *methods*) to the lists to change the order of the items in them. For example, try this, using the reverse method:

```
play_pattern_timed [:c4, :e4, :g4, :c5], [0.5, 0.5, 1]
play_pattern_timed [:c4, :e4, :g4, :c5].reverse, [0.5, 0.5, 1]
```

You'll hear the notes of the fanfare played forward and then backward, but with the same timing each time. You can use the shuffle method, which changes the order of the items in a list, to hear a random tune. Try this:

```
play_pattern_timed [:c4, :d4, :e4, :f4, :g4, :a4, :c5].shuffle, [0.5, 0.5, 1, ↵
    0.5, 0.5, 1]
```

We've used a simple rhythm there: two short notes and then a long note. It's a cheery melody, but it's a bit short, so get Sonic Pi to repeat it. Here's how:

```
4.times do
play_pattern_timed [:c4, :d4, :e4, :f4, :g4, :a4, :c5].shuffle, [0.5, 0.5, 1, ↵
    0.5, 0.5, 1, 2]
end
play :c4
```

This example wraps the tune playing code in a loop that repeats it four times. The start of the loop is 4.times do, and the end of the repeating section is marked, appropriately enough, with the word end. Sonic Pi automatically indents your musical code by two spaces to show it's the part that is to be repeated. If you want to repeat more or less than four times, change the number 4 at the start.

There are some other changes here, too: First, we added a timing value for the last note in the sequence. It's the 2 that has sneaked inside the last square bracket. We've also added a final note, :c4. Whatever randomness happens in the rest of the tune, this sequence of notes always sounds good when it ends on a C, because all the notes in the sequence are from the C major scale.

Changing the Random Number Seed

Each time you run the program, it uses the same sequence of notes, even though that sequence was generated randomly. Sonic Pi ensures that your music sounds the same each time you run it, and wherever it runs, even if it incorporates random elements. That ensures that you remain in control as a composer and know what the music will sound like for your listeners, even with an element of chance in the composition.

If you want to change the sequence to a different one, you can change the *seed*, which is the starting point for generating random numbers. By default it's set to 0. Add the line below at the start of your program for generating random tunes using shuffle, and then run your program:

```
use_random_seed 10
```

Try changing the seed to different numbers, and run the program again to see how the sequence changes.

Using List Names in Your Programs

The lists of notes and values can make your program look cluttered, but you can tidy your program up by giving the lists names and using those names in place of the lists. You can streamline your previous program like this:

```
use_random_seed 10
note_pitches = [:c4, :d4, :e4, :f4, :g4, :a4, :c5]
note_timings = [0.5, 0.5, 1, 0.5, 0.5, 1, 2]
4.times do
  play_pattern_timed note_pitches.shuffle, note_timings
end
play 60
```

Playing Random Notes

You can play random note numbers, like this:

```
lowest_note = 60
highest_note = 84
6.times do
```

```
    play rrand_i(lowest_note, highest_note)
    sleep 0.5
end
```

Try this program in a new buffer. In this program, `lowest_note` and `highest_note` are variables. The `rrand_i()` function gives you a random whole number (or integer). You give the function the lowest and highest possible number you want the computer to pick from. As with the `shuffle` method, each time you run the program, it generates the same random numbers, so the music sounds the same. You can change the random seed to get a new melody. (See the section in this chapter on creating random tunes with shuffle.)

The problem with generating random note numbers is that not all notes sound good together. Before this example, we've been using the notes from the white keys on the piano (the scale of C major) and none of the sharp notes. When you start throwing in sharp notes, as the random music can do, it starts to sound too chaotic. An alternative way to pick a random note is to create a list of the notes you like (the scale of C major we've been using) and then use the `choose()` method to pick a random note from it. Here's an example:

```
note_pitches = [:c4, :d4, :e4, :f4, :g4, :a4, :c5]
loop do
  play note_pitches.choose()
  sleep 0.5
end
```

That program uses a loop that repeats forever, so it'll keep improvising until you click the Stop button.

REMEMBER

The Log shows you the note numbers that are played, so you can use this to see which notes are being chosen and confirm that your program is behaving as you expect.

Experimenting with Live Loops

One of the best features of Sonic Pi is the *live loop*, which enables you to change your music while it repeats.

The following program modifies the previous example by turning it into a live loop and adding tempo (BPM is short for beats per minute), synth, and option choices. Changing the synth is like changing the instrument the notes are played on. Options (or *opts*) can be added after the instruction to play a note and affect how the note is played on that instrument. The attack option changes how long a note takes to reach its full volume, for example.

The new lines are in bold in the following code:

```
note_pitches = [:c4, :d4, :e4, :f4, :g4, :a4, :c5]
live_loop :endless_notes do
  use_bpm 120
  use_synth :dull_bell
  play note_pitches.choose(), attack: 0.01
  sleep 0.5
end
```

WARNING

Take care with where you put spaces around colons. In Sonic Pi, colons are used immediately before the name of something (such as a live loop, sample, synth, or chord). The program won't work if you put a space between the colon and the synth name. Colons are also used to separate a parameter (or something you can change) from its value. In this example, attack is a parameter. The colon must go immediately after the parameter name.

Note that you need to give a live loop a name. This program uses :endless_notes, but you could choose something else if you prefer.

When you run the program, you'll hear the new synth sound (:dull_bell), and notice the new tempo.

The clever bit is that you can make changes to the program and have them incorporated seamlessly into the music. For example, try changing the synth to :chip-bass and clicking the Run button. The tune changes instrument, but everything remains in time. Try changing the BPM to 60 and clicking the Run button to hear the music slow down. Change the attack value to 1, and you'll hear the sharp percussive sound of the instrument change to a slow drone, almost like a church organ. You can add more options to the same play instruction, as long as you separate them with a comma.

REMEMBER

To tell Sonic Pi that you've finished making changes and to hear them in your music, you must click Run again.

The Help pane includes a list of synths you can choose from. Show the Help pane and then click Synths at the bottom. For each synth, there is information about the available options, and useful number values to try.

For more fluid experimentation, the autocomplete feature can be used while you're writing your code. When you want to change synth, for example, delete the name of the current synth, including the space before it, and then tap space after use_synth to see the list of available synths in the autocomplete menu (see Figure 14-2). Click one in the menu, or highlight it using the cursor keys and select it by pressing Tab or Enter. This is a quick way to enter your code, but also a great way to explore the synths and options available.

```
note_pitches = [60, 62, 64, 65, 67, 69, 72]
live_loop :endless_notes do
  use_bpm 50
  use_synth
  play choos :beep
  sleep 0.2 :blade
end          :bnoise
             :chipbass
             :chiplead
             :chipnoise
             :cnoise
             :dark_ambience
             :dpulse
             :dsaw
             :dtri
             :dull_bell
```

FIGURE 14-2: Choosing a synth using the autocomplete feature.

Sean McManus

TIP

If you press Run multiple times in a program that doesn't use a live loop, you will hear multiple instances of the music at the same time, with no synchronization between them.

Using Samples

The programs you've made so far are an interesting way to explore computer music, especially when the computer starts surprising you with its random compositions.

Sonic Pi can also use *samples*, which are snippets of music that you can manipulate, such as by changing their speed or adding effects to them. Sonic Pi includes a wide range of samples, and you can see a list of them by showing the Help pane and then using the Samples button in the bottom left. They're especially useful for adding drum loops to your music.

Here's one of our favorites:

```
sample :loop_industrial
```

Put that into a new buffer and click Run to hear it. You can speed it up or slow it down by changing its rate. Here's how you make it play at half its normal speed:

```
sample :loop_industrial, rate: 0.5
```

We can repeat that sample to make a continuous rhythm. Like we did when we were playing notes, we use the `sleep` command to put a pause between each repetition. Samples can be different lengths, however, which can make it difficult to work out how long to sleep. Luckily, Sonic Pi provides a feature in the language to stretch a sample over a certain number of beats, which solves this problem. You can use it like this:

```
loop do
  sample :loop_industrial, beat_stretch: 2
  sleep 2
end
```

Now the sample is stretched over 2 beats, and there is a 2-beat pause between each repetition. As a result, you can hear a continuous rhythm. As you saw when experimenting with live loops, you can change the tempo with the `use_bpm` command.

Adding Special Effects

You can play a sample and add effects to it, including distortion, echo, and reverb. There is a full list of effects (also known as Fx) in the Help pane, and you can also find them using the autocomplete feature. This is how you add distortion to one of the guitar samples:

```
with_fx :distortion do
  sample :guit_e_fifths
end
```

You can adjust how the effect is applied to the sound using options. There are several options for distortion, including `distort`, which controls how much the sound is distorted, and `mix`, which controls the balance between the original sound and the distorted version. In both cases, the value should be between 0 and 1. Here's how you use them:

```
with_fx :distortion, distort: 0.9, mix: 0.5 do
  sample :guit_e_fifths
end
```

Try using different values for those options and running the code to see how the sound changes. Experiment with other effects too.

To see the full options for this or any other effect, click Fx in the Help pane, or click the effect name in your code and then use Ctrl+I to jump to its entry (if available) in the Help. You can also use Ctrl+I to get help on commands in your code.

Synchronizing with Your Drumbeat

You now know how to play a repeating rhythm, and how to play other samples and synth melodies. You can play multiple live loops at the same time. One of the challenges is to synchronize all the different parts of the music so they play in time. To do that, you can provide an option to synchronize the start of a live loop with the name of another `live_loop`. Here's an example that synchronizes a cymbal beat with the main drum loop:

```
live_loop :drums do
  sample :loop_industrial, beat_stretch: 2
  sleep 2
end

live_loop :cymbals, sync: :drums do
  4.times do
    sample :elec_cymbal
    sleep 1
  end
  sleep 8
end
```

When you play that example, you'll hear that the cymbal plays four times, in perfect sync with the main drumbeat. It then rests for 8 beats before playing again.

Take care with the colons when setting the sync option for your live loop; otherwise, the program won't work. The first colon (after `sync`) goes between the option and its parameter and touches the option name, and the second colon (before drums) marks the start of the loop name you want to synchronize with.

Bringing It All Together

Listing 14-1 brings together the ideas in this chapter. It uses the `loop_industrial` sample as the backbone, with the other live loops synchronized to start in time with it. There is a melody live loop, which plays random notes from the list provided. The list includes the same note several times, so the song tends toward using the C note more often. There is a loop that plays higher notes more quickly over the top, and a bass part that is played using the `play_pattern_timed` instruction you saw earlier in this chapter. Finally, there is another live loop that adds the amen drumbeat on top, in bursts. It plays 8 times, but the `sleep` command is used to stop the beat from starting immediately, and to put a gap between each burst.

LISTING 14-1: **Bringing it all together**

```
# Robot jam by Sean McManus
# Music example from Raspberry Pi For Dummies, 4th Edition

live_loop :drums do
  with_fx :krush do
    sample :loop_industrial, beat_stretch: 2
  end
  sleep 2
end

live_loop :melody, sync: :drums do
  note_pitches = [:c3, :c3, :c3, :d3, :g3, :g3, :a3, :c4, :c4]
  use_synth :saw
  with_fx :wobble do
    play note_pitches.choose()
    sleep 2
  end
end

live_loop :plinks, sync: :drums do
  plink_pitches = [:c4, :d4, :e4, :g4, :a4]
  use_synth :chiplead
  play plink_pitches.choose()
  sleep 0.25
end

live_loop :bass, sync: :drums do
  use_synth :piano
  play_pattern_timed [:c2, :c2, :c2, :c2, :c2, :c2, :g2, :a2], [0.5, 0.5, 0.5, ↩
    0.5, 0.5, 0.5, 0.5, 0.5]
end

live_loop :amen, sync: :drums do
  sleep 16
  8.times do
    sample :loop_amen, beat_stretch: 2
    sleep 2
  end
end
```

You can use this piece as the basis for your own experiments. Because the sections use live loops, you can modify the effects, options, synths, and notes and then click the Run button to hear your changes.

Next Steps with Sonic Pi

We hope that this chapter has inspired you to experiment with making music on the Raspberry Pi. You've learned how to play melodies using different synths. You've also discovered several different ways to create improvised music, using random note numbers, and random notes picked from a list. We've shown you how to bring it all together, too, combining and synchronizing samples and synth melodies. Using the code in this chapter, and the rich range of samples and tools that Sonic Pi provides, you can compose your own music.

There's much more you can do with Sonic Pi, including using it with Minecraft and using MIDI signals to connect to other electronic instruments and music software. The Help pane includes a tutorial and a series of articles that were previously published in *The MagPi* magazine, so there's plenty of support when you're ready to take Sonic Pi further.

Whether you like dance, prog, pop, or rock, Sonic Pi deserves to be in your band.

5

Exploring Electronics with the Raspberry Pi

Chapter **15**

Understanding Circuits

P art 5 of this book deals with what is known as *physical computing*, or making your program reach out beyond the confines of keyboard and screen and into the physical world. You discover how you can use your Scratch and Python programming skills to sense what is happening in the outside world and to control lights, motors, and, in fact, anything else that uses electricity. However, before you can do this safely, without risking damage to you or your Pi, you need to look at a little bit of background electrical theory so that you have a foundation to build on.

In this chapter, we show you the relevant concepts that allow you to understand why the projects look like they do and what you should avoid doing. Next, we introduce you to the concept of GPIO connections, explain what they are, and look at why they are included in the Raspberry Pi computer. We also discuss in general how you can use them.

Although you can make the projects in Chapter 16 without soldering, in order to make most things in electronics, you have to be able to use a soldering iron. We show you how to go about this and discuss safety concerns. Finally, we introduce you to the concept of ready-made add-on boards because they can make building stuff a lot simpler (albeit at the cost of a bit more money).

Discovering What a Circuit Is

The first thing you have to understand is that a circuit is something where electricity can flow; it is a path, or a conduit. It is continuous; that is, it's a loop with no dead ends. If you have a dead end, you don't have a circuit. Electricity has to be able to flow. So let's be more specific in what we mean by electricity. It can be complex stuff and because it's invisible, you have to do a bit of imagining to appreciate what is going on.

There are two aspects of electricity: current and voltage:

>> *Current* is what actually flows.

>> *Voltage* is what forces the current round a circuit.

Voltage can't flow and current doesn't exist in the absence of a voltage. However, voltage can exist in the absence of current. You've no doubt felt the effects of static electricity, which is the build-up of voltage that occurs when insulators (materials that don't normally conduct electricity) are rubbed together.

It's kind of like how rubbing a balloon on wool can make the hairs on the back of your hand stand up. You can feel it, but only because you feel your hairs being lifted. You aren't feeling the electricity itself. You only feel static electricity when it stops being static and a current flows. At a very high voltage, a little current can hurt a lot. You've probably felt the static-discharge shock of touching a metal object after walking on a nylon carpet.

Understanding the nature of electricity

So, what is electric *current?* It is a flow of electrons past a point, just like a flow of cars past a highway sign. With electric circuits, you measure current in amps. One amp of current is about 6.24×10^{18} electrons per second passing a point, or 624 followed by 16 zeros. That's a big number, and, fortunately, we don't have to count all those electrons individually. The bigger the voltage, the more current is forced through a circuit, but circuits have a property that resists the flow of current. We call this the *resistance* of a circuit. This resistance depends on the materials the circuit is made from and is measured in a unit called *ohms.* So, because we know how to define an amp in terms of electron flow, we can define these other two properties in terms of an amp:

One volt is the voltage you need to drive one amp through a circuit with a resistance of one ohm.

You can advance a long way in electronics by knowing just that single fact. In fact, that definition is contained in what is known as *Ohm's law:*

Volts = Amps × Ohms

However, it would be too easy to just use that as a formula. People would understand it straight off, and that would never do! You have to build up a mystique. Imagine how you would feel about a doctor if they actually told you in plain English what was wrong with you. No, it needs to be dressed up so that not everyone can understand it. Mike once went to a doctor when he lost his sight in one eye for a few seconds; the doctor told him he had amaurosis fugax, which is simply the Latin for "fleeting blindness," nothing more. So, to dress it up, Ohm's law becomes

$E = I \times R$

where E is the electromotive force measured in volts, I is the current measured in amps, and R is the resistance measured in ohms.

This is the formula you see in books and all over the Internet, but remember — it's just

Voltage = Current × Resistance

Connecting things to the Raspberry Pi involves juggling voltage and current, and often you need to use a resistor to limit the current a voltage pushes through a device in a circuit. Using Ohm's law is the simple way to work out what you need. Later on in this chapter, we show you how to use this to make sure you drive light-emitting diodes (LEDs) correctly.

Resistance is not the only thing we can calculate. If we know two of the quantities in a circuit, we can calculate the other one. We do this by rearranging this simple formula to find any one of the quantities if we know the other two. We like the Ohm's law triangle, which gives the three formulas in one go:

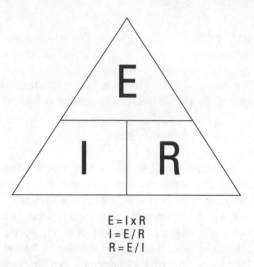

$$E = I \times R$$
$$I = E / R$$
$$R = E / I$$

When scientists were first discovering electricity, they knew that it flowed from one terminal to the other. They said the flow was from the positive to the negative, but which was which? Experiments with making a current flow through a solution of water and copper sulphate showed that copper was dissolved from one wire and deposited on the other. So they quite reasonably assumed that the metal was flowing with the flow of electricity and named the dissolving wire an *anode* or *positive* and the wire receiving the metal the *cathode* or *negative*. They were wrong: The electrons that constitute the current actually flow the other way. However, this notion became so entrenched that today we still use it. We call it *conventional current*, and it flows from the positive to the negative.

In a way, it doesn't matter which direction we think of it as flowing. It's the fact that it *is* flowing that is important, and we use the terms *positive* and *negative* so we know what way round it is flowing. A common mistake beginners make in thinking about electricity is to think that the direction of flow matters because "it first flows through one component and then through another." This leads to the erroneous thought that the first component it meets will "use up" some or all of the electricity and then pass on what is left to the next one. This is simply wrong, because current flows through all components equally in any one path. All the electrons are actually waiting in the components of the circuit before any voltage is applied, and when one is, they all shuffle round together, just like train carriages totally filling a circular track. You might also hear the phrase "electricity finds the path of least resistance and flows through that." Again this is nonsense. Electricity always flows through all available paths at the same time — it is just that the resistance of a path determines how much current flows through that path so a lot of current flows down the path of least resistance but also flows through all other paths with a current proportional to their resistance.

Power sources, like batteries and power supplies, are all marked with positive and negative symbols so that you can connect it the correct way. This is known as *direct current* (DC) because the current flows in only one direction.

The other sort of power supply you can get drives the current round in one direction for a short length of time and then reverses the direction for a short time. This is known as *alternating current* (AC). A favorite trick that electricians play on their apprentices is to send them to the store to fetch the nonexistent AC battery.

TIP

Switches are used to make (complete) or break circuits, so an early name for a switch was a *breaker*.

WATTS THE MATTER?

While voltage and current are all you need to know about, sometimes the word *power* is used to describe the amount of electricity that is used, and it is expressed in units of watts. This is mainly used for incandescent light bulbs and power supplies. This is not a new unit — it's simply the joining up of voltage and current by multiplying them together. So this "new" unit can be packed and unpacked in a similar way to Ohm's law. If you know any two, you can find the third.

The following figure shows a triangle for remembering conversions involving watts.

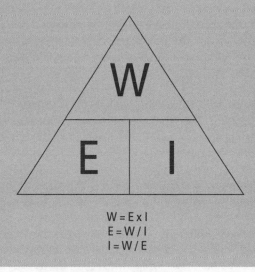

$$W = E \times I$$
$$E = W / I$$
$$I = W / E$$

Putting theory into practice

To see how this works, consider a simple circuit. To make things a bit clearer and easy to draw, we use symbols to represent components and use lines to represent wires that connect the components together, as shown in Figure 15-1.

FIGURE 15-1:
Two circuit symbols representing a switch.

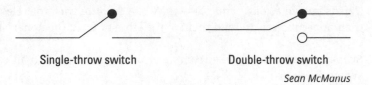

Single-throw switch   Double-throw switch

Sean McManus

Take a switch. Its symbol (refer to Figure 15-1) is simple. There are two types of switches: single throw and double throw. In the *single throw*, a connection is made or not made through the switch, depending on the switch position. In the *double-throw* switch, a common connector is connected to one or the other switch contact, depending on the switch's position. That is, when the switch is one way, there is a connection through the switch from one connection to the common connection. When the switch is the other way, the connection is between the other connection and the common connection. There are also switches that activate as long as you hold them down and deactivate when you release them; these are called *momentary push switches*.

It's called a *double-throw switch*, or sometimes a *changeover switch*, because the switch changes over which terminal is connected to the common one. The figures in this section help explain this concept. However, the important thing to note is that we use the same symbol for a switch, no matter what the physical switch looks like. Figure 15-2 shows just some of the many physical forms a switch can take.

Figure 15-3 shows the symbols for a battery, a small flashlight or torch bulb, and a resistor. Note that there are two symbols for a resistor: one for the U.S. and one for Europe. In the U.K., we used to use the U.S. symbol until the late 1960s. Today, both are understood.

The world's simplest circuit is shown in Figure 15-4. While the switch is open, there is no complete circuit, so there is no current flow and no lighting of the bulb.

However, when the switch is closed as shown in Figure 15-5, a path for the current to flow along is created and the bulb lights up. Note that this diagram has a different symbol for a closed switch than the one used in Figure 15-4. This is so that you can more easily see what is going on. Normally, you have to imagine the switch in the open and closed positions and visualize the resulting circuit or break in the circuit. We call this a *series circuit* because all the circuit elements are in a line, one after the other, and the same current flows through all elements of the circuit.

FIGURE 15-2:
Just a few of the
many different
physical forms a
switch can take.

Sean McManus

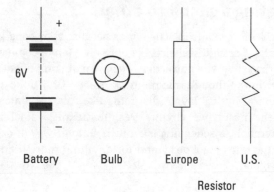

6V

FIGURE 15-3:
Schematic
symbols for some
components.

Battery   Bulb   Europe   U.S.

Resistor

So, for a circuit like this, there is only one value of current. When the switch is closed, current flows from the positive end of the battery through the switch, through the bulb lighting it up, and, finally, back into the battery's negative terminal. Note here that the actual electrons are returned to the battery. The battery loses energy because it has to push them round the circuit. The positive and negative terminals of a battery show the direction it will push the current, from the positive to the negative. In this circuit with an incandescent light bulb, the direction of the current doesn't matter; however, this is rare in electronics. In most circuits, the current must be sent round the circuit in the correct direction.

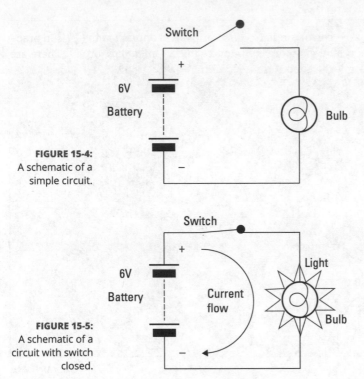

FIGURE 15-4:
A schematic of a simple circuit.

FIGURE 15-5:
A schematic of a circuit with switch closed.

Communicating a circuit to others

You should use circuit symbols in schematics because they constitute a universal language and make it easy to see what is going on. Many people waste their time using diagrams that show the physical appearance of components and wires and their interconnection. Although this might appear at first to be attractive, especially to a beginner, physical layout diagrams like this are almost impossible to follow in all but the most trivial circuits. Despite the initial small hurdle of learning to read the symbols, a schematic is a much simpler way of defining a circuit. Physical layout diagrams are a dead-end for anything more than a trivial circuit and should be avoided.

Some time ago, Mike was visiting Russia and bought his son an electronic construction set. Even though the words were in Russian and incomprehensible to them both, the diagrams were in the language of schematic and perfectly understandable.

TECHNICAL STUFF

To show the units of resistance, we can use various symbols. We can say 18 ohms, 18 Ω, or, as we shall use in this book, 18 R.

Although the units for calculation are volts, amps, and ohms, 1 amp (A) is in practice a lot of current, and it's more common to talk of *milliamps,* or *mA.* There are 1,000 mA in 1A. Similarly, 1000 R is one kilohm, or 1K.

Calculating circuit values

Although the circuit shown in Figure 15-5 is all very well because it describes what's actually wired up, it's not useful for calculating anything using Ohm's law because it shows no resistances. However, each real component has associated with it a resistance. We say it has an *equivalent circuit.* These are shown in Figure 15-6. All components, even the wires, have some *series resistance.* In other words, they behave like they have a resistor in line with the component. Sometimes this is important, and sometimes it is not. The trick is in knowing when to ignore them. Also, don't go overboard on the calculations, just two places of decimals are fine for most things.

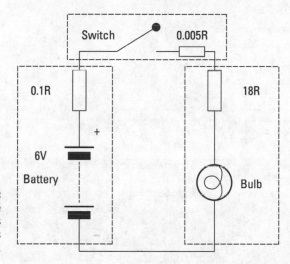

FIGURE 15-6:
A circuit with the effective series resistance values shown.

When resistors are placed in series, or in line with each other, you can find the resistance by simply adding up all the individual resistance values. Figure 15-6 shows our circuit with the series resistance values shown. If we add up all the values around the circuit, we get 18R105. (That's 18.105 ohms.) Note that virtually all of the resistance in the circuit comes from the bulb. The series resistance of the switch is negligible, as is the series resistance of the battery. This is not always the case with all circuits; if the current is large, even small resistances become significant and produced voltage drops across these parts or wires. So, with 18R resistance and 6V, we can calculate that the current through the circuit should be

$I = E \div R$

Current = 6 ÷ 18 = 0.333 Amps or 333mA

TWO TYPES OF CIRCUITS

We mention earlier that Figure 15-5 was a *series circuit*, but there is another type of circuit called a *parallel circuit*. In this type of circuit, all elements are connected together, as shown in the following figure. Here, you see three light bulbs, but this time they're different wattage bulbs. This means not only will they be different brightnesses, but each will also take a different amount of current. The total current is simply the sum of all the currents for each part, and the current for each part is determined by its resistance, which we can work out from the wattage and voltage rating of the bulb. The important thing to remember is that each part of a parallel circuit is connected to the same power supply. This is the way the lights in your house are wired.

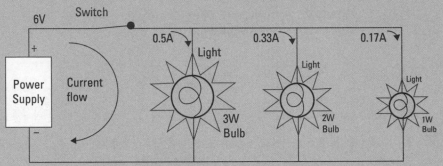

The following figure is an example of a series circuit. This is typical of how incandescent Christmas tree lights are wired (although, normally, the lights are all the same power). Here you see three different bulb wattages all in series. The current flowing round the circuit is the same for each bulb, so to calculate it, you must look at the resistance of each bulb and add them up to find the total resistance of the circuit; then, using Ohm's law, you can calculate the current through all bulbs. Because each bulb will have a different resistance, there will be a different voltage drop across each bulb. When you sum all the voltage drops, it will equal the voltage across the series circuit. Just like Christmas tree lights, if one bulb fails, they all go out. (Modern LED Christmas tree lights are wired in parallel, so this doesn't happen.) Note that if you tried to make this circuit, the voltage drops wouldn't work out quite as you calculated because the resistance of the filament of a bulb changes with its temperature.

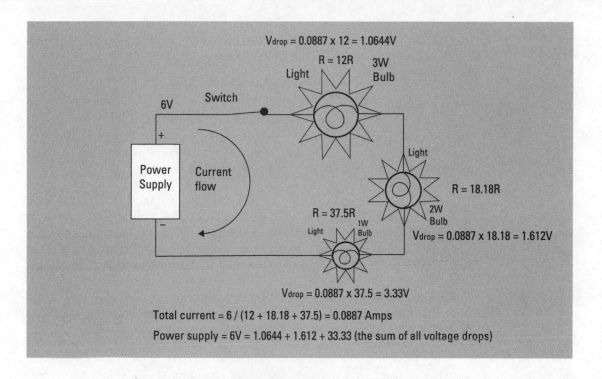

$V_{drop} = 0.0887 \times 12 = 1.0644V$

R = 12R
3W
Bulb

Light

Switch

6V

+

Power
Supply

Current
flow

−

Light

R = 18.18R

2W
Bulb

$V_{drop} = 0.0887 \times 18.18 = 1.612V$

R = 37.5R
1W
Bulb

Light

$V_{drop} = 0.0887 \times 37.5 = 3.33V$

Total current = 6 / (12 + 18.18 + 37.5) = 0.0887 Amps

Power supply = 6V = 1.0644 + 1.612 + 33.33 (the sum of all voltage drops)

Determining how a component needs to be treated

How do we know the series resistance of a component? Well, it is normally in that component's *data sheet*, the document that the manufacturers of all components produce to exactly define the component and its properties. However, it's not always given as a straightforward value. Take an incandecient light bulb, as mentioned before. They're normally "rated" as a voltage and a current; that is, we would say that the bulb is 6V at 0.33 amps.

The other point is that a bulb doesn't have a constant resistance. We say it's a *nonlinear device*; that is, the resistance changes depending on what current is going through it. This is because a bulb is just a coil of wire. As current passes through it, the wire heats up. This causes the resistance to increase, thus limiting the current. An equilibrium point is reached where the temperature reaches a point where the resistance is such that the current is limited to the rated value at the rated voltage. This increase in current, with increasing temperature, is known as a *positive temperature coefficient*. Other materials, like the silicon used to make transistors, processors, and memory, have a *negative temperature coefficient*, which means the hotter it gets, the lower the resistance becomes. If not kept in check, this can lead to thermal runaway and can destroy components. We use this concept of a nonlinear resistance later in this chapter when we calculate what resistor we need to use with an LED.

TECHNICAL
STUFF

When dealing with units like volts and ohms that include a decimal point, often the point is missed out and the letter of the unit is substituted, so 3.3 volts becomes 3V3, or 4.7K becomes 4K7. This is done to make it clear that there is a decimal point that otherwise might be lost in small print. While this is not an officially approved standard, it is widely used in the electronics industry.

The series resistance of a battery — or any power supply, for that matter — is an important concept in that it limits the current the battery can deliver. This is all wrapped up in the chemistry of the battery, but its effects can be summed up by a theoretical series resistance. A battery that can deliver a lot of current has a low series resistance. This is sometimes known as the *output impedance* of the battery.

REMEMBER

These concepts may seem like they have nothing to do with the Raspberry Pi, but as you shall see in later chapters, these concepts are the ones you need in order to get the Pi to interact with the world outside the keyboard and screen.

Getting Familiar with the GPIO

The original Raspberry Pi Model 1, and the Pi Zero are made using a BCM2835, a single-core system on a chip. The Pi Model 2 uses the BCM2836, and the Model 3 uses the BCM2837 — both are quad core. Finally, the Model 4 and Pi 400 uses a BCM 2711, again a quad-core chip. Basically, a core is a processor, so the later models have four processors that, in theory, can work on four different processes at the same time, although this capability is not presently fully exploited by the operating system. The cores in the Pi Model 4 are more powerful than the cores Pi3, which, in turn, were better than those in the Pi 2, but again the full power of these processors has yet to be exploited by the operating system. This is because it will mean having a version of the operating system for each type of core, and at the moment there is just the one version of the operating system for all four processor chips. However, when the Pi 4 was launched, an optional beta of a new Raspberry Pi operating system was launched that allows you to use all 8M of memory and compiles to 64-bit-wide instructions instead of the previous 32-bit compilation for faster running.

Unlike traditional microprocessors, these chips are designed to be used in an embedded system. An *embedded system* is a device that has a computer inside it, but you don't use it as a computer — things like mobile phones, GPS displays, digital cameras, and TV set-top boxes. These chips have a number of connections to them in order for the software in them to read or control things like push buttons and displays and getting sound in and out. All the processors used for the Pi have 54 such signals. They are called General Purpose Input/Output pins (GPIO), and they can be controlled by software. (Whenever a GPIO is mentioned, Mike thinks of that old cowboy song "GPIA, GPIO, Ghost Riders in the Sky.") Some of these signals are used to build and control the peripheral devices that turn the chip into a computer, like the SD card reader, the USB, and the Ethernet. The rest are free — that is, not needed to make the Pi — so they are surplus to requirements.

Rather than just ignore these surplus GPIO lines, the designers of the Raspberry Pi have routed some of them out of the chip and to the connector called P1 on the board for you to play with. It's a bonus. This sets the Pi apart from mainstream computers and laptops in this respect. However, they have not routed all the spare pins out to this connector. Some go to other connectors — the camera socket, for example, or the display socket — and some are not even connected to anything. This is because the family of chips used to make the Raspberry Pi are housed in a package called a BGA, or Ball Grid Array, with connections less than a millimeter apart. So close are they that you can only have enough room for one trace (PCB wire) between the connectors.

This means that, to get some of the inner connections out to other components, you need to use a printed circuit board (PCB) that has a number of extra layers of wiring inside the board. You might think the Pi's board has just a top side and an underside, but in fact it is made from several boards sandwiched together to give six layers of wiring.

Even with this many layers, there is not enough room to route out all 54 GPIO signals. Adding more layers would significantly increase the price of the PCB and make the bonus cost something instead of being free. (You are no doubt aware that the price point of the Pi is one of its major features.) However, over successive hardware revisions, an increasing number of these pins have been brought out to use. There were 17 on the original board, but Revision 2 saw some rearrangement and another socket bring the total to 21; and, finally, the Model B+, and Revision 3, saw this increase to 28, all on one 40-pin header. This still leaves 8 GPIO pins not routed out or used internally on the board. This has remained the same for the Model 2, 3 and 4 of the Raspberry Pi as well as for the Zero models. While 28 GPIO pins on a 40-pin header might sound a little odd, the extra pins carry ground connections (8 pins), 5V (2 pins), and 3V3 (2 pins), which are very useful when it comes to connecting external circuits. The Model 4 also breaks out an extra I2C interface for general use. It was always there, but the software prevented you from using it for anything other than the HAT extension boards (see the section "Looking at Ready-Made Add-On Boards," later in this chapter).

Putting the general purpose in GPIO

GPIO pins are called *general purpose* because you can use them for anything you want under the control of a program. They're called *input/output* pins because the software can configure them to be either an input or an output. When a pin is an input, the program can read whether a high voltage or a low voltage is being put on the pin. When the pin is an output, the program can control whether a high voltage or low voltage appears on that pin. In addition, many pins have one or more *superpowers* — alternative functions as a secret identity, like so many comic book heroes. These powers are not shared by all pins, but are specialist functions able to do things without software intervention. They act as ways to tap directly, deep into the computer's inner workings. When you switch to these functions, they stop being general-purpose pins and do a specific job. For example, one pin can be used to output a continuous stream of high and low voltage levels in alternation, that, after they get going, continue without any further intervention from the program. So, if you connect that pin to an amplifier, or amplified speaker, you can generate a tone that keeps on sounding until you command it to stop.

Understanding what GPIOs do

GPIOs are the gateway to interaction with the outside world and are, in essence, quite simple.

Figure 15-7 shows the equivalent circuit of a Raspberry Pi GPIO pin when it is configured as an output. You can see it is simply a double-throw switch that can be connected between the computer's power supply of 3V3 or ground (that's 0V). This is sometimes called the *common ground* or *reference*, and it is the basis of all measurements in a circuit. Basically, it's the other end of the battery — the negative terminal, if you will. Between this switch and the output is in effect a series resistor, one that is in line with the voltage coming from the Pi. It limits the current you can get through the output pin.

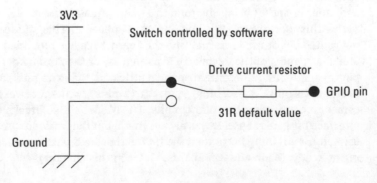

FIGURE 15-7:
A GPIO when used as an output.

FINDING A SAFE VALUE OF CURRENT

There *is* a value of current that would instantly destroy at least the output circuitry of the pin, if not the whole Pi itself. But, there is *also* a value of current that would not instantly kill the Pi but would damage the circuitry and lead it to fail prematurely. Lower that current, and the damage is lowered, until you get to a point where no damage occurs. However, these values are not known for the chips used on the Pi. In fact, they are not known for the vast majority of chips. It's best to stick to the "safe" value or lower.

Beware of people who say that they have a circuit that takes 30mA or more from a pin and it's still working. They tend to be idiots who confuse whether a pin is dead yet with whether a pin is safe. It's just like smoking: You can do it and it doesn't kill you immediately, but it does do harm and eventually it can kill, if nothing else gets you first. No one would pretend that it's safe.

On the Pi, the value of this resistor can be changed over a limited range. The default value is 31R, but note that this resistor, by itself, is insufficient to protect the Pi from giving too much current if you connect it to too low a resistance load. If you did this, then your Pi would be permanently damaged; at best, this damage might be limited to just that pin. So an output pin can switch between only two voltages — 0V and 3V3. These are known as logic levels, and they have a number of names: *high and low, true and false, zero and one,* and even *up and down.*

Although the logic voltage levels on the Pi are simple, the current that these outputs can supply is more complex, with a current limit of about 16mA. This limit is how much current the Pi *should* supply into a load, not how much it *can* supply or *will* supply. That depends on the resistance of the load connected to the pin. We say that the limit is about 16mA, but this is a bit of a gray area. This value is considered safe for the Pi to supply, but that is not to say a value of 17mA would be considered dangerous or excessive.

Putting an output pin to practical use

What can you do with a switched output? Well, you can drive a small current through a load, or you can control another device that can control a bigger current through a load. Put like that, it doesn't sound exciting, but it's what physical computing is all about — it is the gateway to all control. The load can be a light, a motor, a solenoid (an electromagnetic plunger used to prod or strike things), or anything that uses electricity. Because that includes most everything in the modern world, it is safe to say that if it uses electricity, it can be controlled and, more importantly for our purposes, controlled by a Pi.

Take a look at controlling a light — not the current-heavy flashlight bulb we looked at earlier, but rather a component known as a light-emitting diode (LED). These can light up from just a tiny bit of current, and the 16mA we have available is more than enough for that task. In fact, you're going to limit the current to less than 10mA by adding a 330R-series resistor.

For the moment, just look at the circuit in Figure 15-8. This shows two ways to wire up an LED — or any other load — directly to a GPIO pin. Here we just show the GPIO pin and not the equivalent series resistance of the power source as discussed earlier — in the context of a 330R resistor, 31R is negligible.

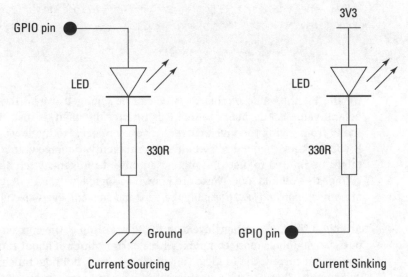

FIGURE 15-8:
Two ways of driving an LED.

The first way to wire it, called *current sourcing*, is perhaps the way a beginner might think of as natural. When the GPIO pin is set by the program to produce a high voltage (that is, when the switch is set to connect the 3V3 line to the output pin), current flows from the pin through the LED, through the resistor and to ground, thus completing the circuit, causing current to flow and so lighting up the LED. When the GPIO pin is set by the program to produce a low voltage (that is, when the switch is set to connect the 0V or ground line to the output pin), no current flows and the LED is not lit. This method is known as *current sourcing* because the source of the current — the positive connection of the power — is the GPIO pin.

The second way of wiring (refer to Figure 15-8) is known as *current sinking*. When the GPIO pin is set by the program to produce a low voltage, the current flows through the LED, through the resistor, and to ground, through the GPIO pin. To turn off the LED, set the output to a high voltage. There's no way current can flow round the circuit, because both ends of the load (LED and resistor) are connected to 3V3 — so there is no voltage difference to push the current through the components.

Note in both circuits that the position of the resistor and LED can be interchanged — it makes no difference. You might like to think of these two approaches as switching the plus, sometimes called *top switching,* and switching the ground.

Now an LED is a non-linear device in a similar way to the filament of a flashlight bulb, but in the case of an LED we need a resistor to limit the current to a safe value; it will not self-limit, like the bulb. The important thing to know is the LED's *voltage drop* — the voltage that will appear across it when the LED is on. This drop changes with different colours of LED, so to get the same current down two LEDs of different colours, you need different resistors.

REMEMBER

Getting the same brightness is not the same thing as setting the same current, because different LEDs have different current-to-light efficiencies. Modern LEDs are a lot more efficient that they used to be, so you need less current.

With a red LED, there is about a 2V2 voltage drop across it. This leaves the remaining voltage from the GPIO pin (3.3 − 2.2 = 1.1V) to be dropped across the resistor. With a 330R resistor, Ohm's law tells you the current will be: 1.1 ÷ 330 = 3.3mA, which is plenty bright to see.

WARNING

Don't try to run the LEDs at too much current. Although most will be able to take 20mA, if you did this then your GPIO pin would be supplying too much current and your Pi could be damaged. There should be no need to run an LED any harder than 10mA.

Using GPIOs as inputs

The other basic mode of operation for a GPIO pin is as an input. In this case, you don't have to worry about the current, because when the pin is an input, it has a very high input *impedance,* or put another way, a high value of series resistance, and therefore little or no current can flow with normal logic level voltages. A *resistance* is a special form of impedance, which, as its name implies, impedes the flow of electricity. There is a bit more to impedance than simple resistance, but at this stage, you can think of them as the same sort of thing. They are both measured in ohms.

Resistance is the property of a material, whereas impedance is the property of a circuit and includes how it behaves with AC as well as DC. So an input pin has a very high impedance. It allows hardly any current to flow through it. The result is that we can connect it directly to either 0V or 3V3 with no extra resistors. In fact, an input is so high-impedance that if you leave it unconnected, it picks up very tiny radio waves and other forms of interference and gives random values when you try to read it.

In fact, the human body can act as an antenna when close to or touching a high-impedance input, causing any readings to go wild. This often amazes beginners, who think that they have discovered something mysterious. They haven't. In fact, the tiny amounts of energy in the radio waves that are all around us are not absorbed by the high-impedance circuits as they would be by low-impedance circuits. A low impedance would cause current to flow, but it would easily absorb all the power, leaving minuscule amounts of voltage. Just the fact that you have a wire carrying AC power (mains) close by is enough for that wire to radiate radio-wave interference.

To explain why this is so, consider that interference of, say, 2V is enough to override the signal from a chip and cause it to malfunction. With a low resistance — say, 1K — in order to develop 2V across, it needs to have a current of 2mA (Ohm's law) flowing through it. This represents a power (volts × current) of $I \times V$ = 4mW of interference. However, with a resistance of 1M, you can get 2V across it by only having 2μA (micro Amps) flowing through it. (μ is the Greek letter mu, which is scientific shorthand for "micro," which is 10^{-6}, a very small amount.) This represents a power of 4μW (micro Watts). So, a high resistance is much more sensitive to interference because it requires less power from the interfering source to develop the same voltage. Therefore, weaker fields produce enough interfering voltage to disrupt a circuit.

This underlines an important fact about inputs: They can't simply be left alone. They must be driven to one voltage state or the other; that is, either 3V3 (known as *high*) or 0V (known as *low*). If you connect an input to the output from some other chip, that's fine, but if you want to detect whether a switch is made or broken, you have to give the input pin some help. This is normally done with a resistor connected from the input to either the 3V3 or the ground.

When a resistor is used in this way, it's called a *pull-up* or *pull-down* resistor, as shown in Figure 15-9. Of the two arrangements, a pull-up is preferable, mainly because switches are normally on the end of long runs of wire and it is safer to have a ground than a 3V3 voltage on a wire. This is because it tends to cause less damage if you accidentally connected a ground wire to the wrong place than a power wire. This arrangement of pull-up or pull-down resistors is so common that the computer processor in the Pi has them built-in, and there is a software method for connecting or enabling internal pull-up or pull-down resistors. We show you in Chapter 16 how to control this from software.

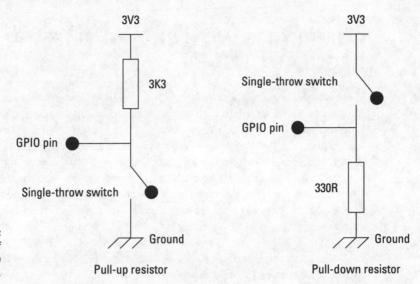

3V3

3V3

3K3

Single-throw switch

GPIO pin

GPIO pin

Single-throw switch

330R

FIGURE 15-9:
Two ways of
using a GPIO
as in input.

Ground

Ground

Pull-up resistor

Pull-down resistor

Learning which end is hot: Getting to grips with a soldering iron

Although you can do some interfacing without resorting to the soldering iron to join components together, to get serious, you'll have to do some soldering at some stage or the other. This often induces fear and panic in the newcomer, but even a child can solder successfully. In fact, Mike had his first soldering iron at the age of nine and by and large taught himself. Soldering involves two parts, the *solder*, which is an alloy of two or more metals, and the *flux*, a chemical cleaning agent. If you are soldering something like a gas pipe, you would apply the flux round the joint, heat the joint up with a blow torch, and apply the rod of solder to the hot joint. The job of the flux when it is heated is to clean the surface and make the solder flow. It does this by breaking down the surface tension on the molten solder. Without it, the solder would clump together in round globs held by the tight surface tension.

Water has surface tension as well, and to reduce that we use soap, which allows the water to wet things. You can't use soap with solder because it wouldn't stand the heat, so you need something else. Most fluxes for heavy jobs are made from nasty chemicals like hydrochloric acid or phosphoric acid. These are too corrosive to be used with electronics, so what is normally used is some sort of rosin flux. Although you can get this in a pot, by far the best thing is to use Multicore solder, where the flux is built into the solder wire as five very thin strands. That way, the right amount of flux is always delivered with whatever amount of solder you use.

COMPLYING WITH ENVIRONMENTAL REGULATIONS

There is a further complication nowadays with the advent of the Reduction of Hazardous Substances (RoHS) Act, which bans the use of certain metals and plasticizers in certain classes of electrical equipment in the European Union, the most prominent of which is lead. Lead causes RoHS compliance problems at levels above 0.1 percent of a substance. In fact, some people think RoHS is entirely about being lead-free, but it's not. You can get lead-free solders, but they are expensive because they have a large amount of silver in them and they are difficult to work with; they also tend to produce a product with a shorter lifetime. They require a hotter iron and so are potentially more harmful to the components.

Lead-free solders also don't "wet" as well, which means it's harder to get the right molten state you need in order to achieve the flow around the joint. Tin whiskers often grow out of the joints, causing shorts years later. Home-built electronics are not required to be lead free in the United States or Europe, and there is no measurable health effect in using solder that contains lead. RoHS was mainly brought in to stop lead from accumulating in landfill sites from mass consumer electronics and potentially polluting the water supply. Although there is no evidence that this can happen, it is banned under the "precautionary principle." In the European Union (EU) or the UK, you are under no legal or health requirements to use lead free solder. However, if you start making stuff to sell in the EU or UK, you're legally required to make sure it's RoHS-compliant. This is like brewing beer at home: You can brew as much as you like, but you can't sell any. However, in 2018, a new set of regulations known as REACH (short for Registration, Evaluation, Authorisation, and Restriction of Chemicals) were enacted. REACH covers a wide range of substances and says that the supply of solder containing lead at a concentration above the relevant limit is restricted to professional use only. In other words you can use solder with lead, but in theory, as an individual, you shouldn't be able to buy it. When the UK left the EU, all existing regulations were still part of UK law, but this is constantly being updated (though nothing relevant to a home user has changed). Go to www.gov.uk/search/all?keywords=Rohs&order=relevance& page=1 for the latest information.

It's always sensible to wash your hands after soldering and avoid putting solder in your mouth. The same goes for the soldering iron when it is on. Mike was once responsible for this sort of compliance in one job he had, so he had to know the legislation and standards on these matters.

We recommend using a good quality 60/40 tin/lead solder alloy, with a diameter of 0.7mm and a built-in rosin-based flux core. Anything else is making life difficult for yourself. We've found that solder with self-cleaning fluxes or non-fuming fluxes are harder to work with, as well as being more expensive. Couple the right kind of solder with a good soldering iron, preferably a temperature-controlled one with a fine tip.

TIP

It is often said that you can use any old tool to learn on, and then get a good tool when you get good at using it. This is rubbish. As a beginner, you are fighting how to do the job, so you don't want to be fighting your tools as well. A good iron includes a stand and a place for a sponge for removing flux that has accumulated during the soldering process. If it were not removed, then it would form a glassy layer that would prevent the iron from making good thermal contact with the solder of joint. Use a proper soldering iron sponge — a natural one that won't melt on contact with the iron. Do not use a plastic foam sponge because your iron will go straight through it.

Making a soldered joint

The first thing you should do when making a soldered joint is to make a mechanical joint. For example, if you're joining two wires, bend each end into a hook and squeeze them together lightly with pliers.

Wipe the tip of the iron on a damp sponge, and melt just a spot of solder on the tip. This "wets" the tip and allows good thermal contact to take place between the tip and the work. Then apply the iron, solder, and wires all together. The secret is then to look at the joint and the solder closely, to see how it sits. Remove the solder, but keep the iron on the joint until you see the solder flow around the joint and seep into the cracks. Only then is the joint hot enough for you to withdraw your iron. It is a quick process and needs a bit of practice.

Many beginners make the mistake of putting too much solder on a joint. Try to use as little solder as possible. A joint is rarely bad because of too little solder, but it's often bad because of too much. When you are done, you see a small amount of black flux residue around the iron tip. Wipe that away on a damp sponge before returning the iron to its stand. Do not move the joint as the solder sets. It should take about 3 seconds but could be longer if a large area of metal has to be heated up. When set, the solder should be shiny. If it goes frosty, that's an indication of a bad joint due to not heating the joint to a high enough temperature, so try again and keep the iron on slightly longer.

REMEMBER

A good-quality iron is ready immediately for the next joint. A poor iron needs a few seconds to come up to temperature again after making a joint.

TIP

Use some sort of fume extractor when soldering. A simple fan works to guide the curl of smoke from the iron away from your face. Air currents from the warmth of your body tend to attract the flux. Try not to breathe it in. This is more important as you spend a long time (hours at a time) with a soldering iron in your hand. The fumes are from the flux in the solder; they are not lead fumes.

Although Chapter 16 contains projects that can be made without the use of a soldering iron by using solder-less bread boards, we recommend that you never take that approach for a permanent project. When a project is meant to last more than a day or two, you should always solder it up.

Looking at Ready-Made Add-On Boards

There are basically two types of ready-made boards: those designed for making it easy to get access to the GPIO pins and those with components that have already been soldered up. We will look at some of the latter type here and leave the former types to the start of Chapter 16.

Since the introduction of the Raspberry Pi in 2012, many companies have produced ready-made boards with all sorts of components already built on. They normally come with sample code to show you how to use them, and many users just stick to that. However, this can be a wasted opportunity because you can always do more with a board than is shown in these examples. (Chapter 17 shows you some examples of this.) Many of these boards contain sensors that allow your Pi programs to measure things. New boards are constantly being developed and produced.

Boards come in three styles:

>> Separate boards that connect to the GPIO pins via a ribbon cable or your own wires.

>> Boards that plug into all the GPIO pins and cover most of the area of the Raspberry Pi board. These are sometimes called *shields* or *plates*.

>> Boards similar to the shield/plate variety just mentioned but that contain in addition an identification and sometimes software so that the Pi can read what they are on start-up and install some software and prepare the GPIO pins automatically. These are called HATs — short for Hardware Attached on Top. Read more about HATs at www.raspberrypi.org/blog/introducing-raspberry-pi-hats/.

The Sense HAT

The Sense HAT was specifically designed for the Astro-Pi mission and two rug-gedized versions were flown on the International Space Station from December 2015 running code written by school children. It has an 8-x-8 RGB LED matrix, a five-button joystick, and sensors to measure acceleration, magnetism, tempera-ture, pressure, and humidity — as well as a gyroscope (see Figure 15-10). The Sense HAT also has an extensive Python library associated with it, which allows for easy access to this board. You can find comprehensive coverage of how to use the Sense HAT on the Raspberry Pi Foundation's website at `www.raspberrypi.org/learning/getting-started-with-the-sense-hat/`.

FIGURE 15-10:
The Sense HAT.

The Trill sensors

Trill sensors (see Figure 15-11) are a series of differently shaped multitouch sen-sors designed primarily for user interaction for controlling musical instruments, but they can be used in any situation where you need to use a proportional touch control. They have a complete microprocessor built into each sensor to handle the complex task of detecting touch proximity and are far and away the best such technology on the market. The shapes are a bar, a ring, a square, and a hexagon, along with a craft option where you can make your own shaped sensor from any conducting material, including metal foils or even fruit. Normally, these sensors are programmed in C, but Mike has written a Python library for them: `https://magpi.raspberrypi.org/issues/102`.

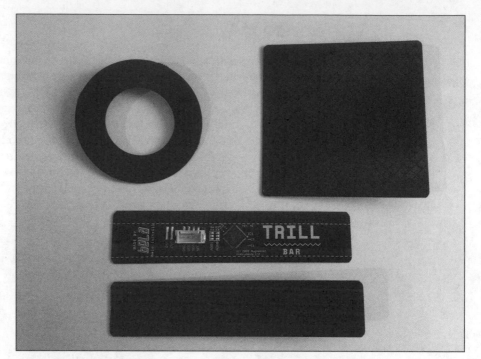

FIGURE 15-11:
Trill sensors
have two bars,
a square,
and a ring.

The LED SHIM

The LED SHIM consists of 28 RGB LEDs in a row (see Figure 15-12). It's a halfway house between a HAT and a breakout board and is designed to slip over the Raspberry Pi's GPIO pins and use a friction contact, so that no soldering is required. This is great if you aren't ready to take up a soldering iron yet. You can plug another board over the top of the LED SHIM as well. Of course, you can also solder directly to the board and run wires to it because it only actually uses two GPIO pins, plus a power and ground.

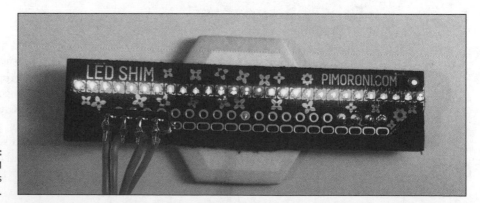

FIGURE 15-12:
The LED SHIM
shown with wires
soldered on.

Other boards

The class of boards you can get that do require soldering are often grouped with the name *breakout boards.* These boards have a tiny chip that is too small for normal mortals to solder, surrounded by a few external components to make them work. These include such diverse sensors as air quality, pressure, temperature, and humidity, as well as GPS location and distance sensing. A good but not comprehensive selection can be found at `https://shop.pimoroni.com/collections/electronics`.

There are many other boards available from small start-up manufacturers as well as web-based projects for you to build. You can find a good starting point for information on many of these at `https://elinux.org/RPi_Expansion_Boards`.

IN THIS CHAPTER

» Learning what GPIO pins you can use

» Seeing how you can control GPIO pins in Scratch and Python

» Making a GPIO pin flash an LED and read a push button

» Building a working electronic dice display

» Building a working model of a pedestrian crossing

Chapter **16**

Taking Control of Your Pi's Circuitry

Chapter 15 tells you all about what general purpose input/output (GPIO) pins are, but in this chapter we want to describe how to access them physically and how to control them with software. We use both Scratch and Python to do this, but ultimately Python is the much more capable language for input/output control.

Accessing Raspberry Pi's GPIO Pins

The GPIO pins are the key to enabling the Raspberry Pi to take control of any external circuit. They can be used as an output to switch on an LED or as an input to sense the state of an external push button. These connections into the computer, along with the fixed voltage power pins, are on a dual-row, 40-pin header plug. There are many ways to physically access these pins, and often the least expensive way isn't always the easiest or the most convenient way. But what you need to know first is which signals are on which pins. A top-down view of the pins and their signals is shown in Figure 16-1.

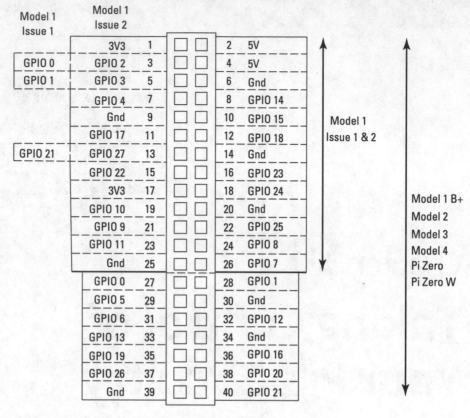

FIGURE 16-1:
GPIO header pins and their function.

	3V3	1	2	5V
GPIO 0	GPIO 2	3	4	5V
GPIO 1	GPIO 3	5	6	Gnd
	GPIO 4	7	8	GPIO 14
	Gnd	9	10	GPIO 15
	GPIO 17	11	12	GPIO 18
GPIO 21	GPIO 27	13	14	Gnd
	GPIO 22	15	16	GPIO 23
	3V3	17	18	GPIO 24
	GPIO 10	19	20	Gnd
	GPIO 9	21	22	GPIO 25
	GPIO 11	23	24	GPIO 8
	Gnd	25	26	GPIO 7
	GPIO 0	27	28	GPIO 1
	GPIO 5	29	30	Gnd
	GPIO 6	31	32	GPIO 12
	GPIO 13	33	34	Gnd
	GPIO 19	35	36	GPIO 16
	GPIO 26	37	38	GPIO 20
	Gnd	39	40	GPIO 21

Model 1 Issue 1

Model 1 Issue 2

Model 1 Issue 1 & 2

Model 1 B+
Model 2
Model 3
Model 4
Pi Zero
Pi Zero W

GPIO 0 = ID_SD   For I2C bus 0 - no pull up resistors fitted
GPIO 1 = ID_SC   I2C address 50 (0x32) reserved for HAT EEPROM

Sean McManus

There are two ways to label the pins. Figure 16-1 shows what is known as BCM mode, which corresponds to the numbers used in the Broadcom data sheet that defines the hardware. Another system was used in the early days — the BOARD system — but it is falling out of favor now. BOARD referred to pins mainly by a mixture of special function names and arbitrarily assigned "free" GPIO pin names. The idea was that, as the *pinout* (what connections are on what pins) on the header changed, the names could remain constant.

As it was, only the first issue of the Pi had a different pinout to all the rest, and those were only for three pins. These are shown in Figure 16-1, on the left side. Model 1 before the B+ had only 26 pins, whereas all subsequent models have 40 pins. You cannot buy the earlier models nowadays, so you have no need to worry; the extra information is only given in case you come across an older Pi.

Now, it can be quite daunting to have to identify a single pin on a 40-pin header to connect to it, so you can get or make a template to place over the pins that labels them with their names. Figure 16-2 shows one of these templates in action.

FIGURE 16-2:
GPIO template identifying the pins.

This particular template is made of a thin, printed circuit board material that can sit over the pins as you connect to them. We highly recommend this type if you're going for the option of individual wires to the GPIO pins. You can also find full-scale drawings online, so you can print one out on paper and just push the pins through. Or stick the paper on thick cardboard, and drill holes through for the pins.

When making your own template, do not use any conducting material, like aluminum.

REMEMBER

Soldering the GPIO pins onto Pi Zero or Pi ZeroW

The Pi Zero has exactly the same GPIO pins as all the other models of the Raspberry Pi, but the challenge is that the 40-way pin header doesn't come prefitted. To use the GPIO pins, therefore, you need to buy and fit the header pins. This presents an opportunity for you because it allows you to have a header socket here instead of pins, if you want — or, in fact, any other style of connector. You could even solder wires directly into these pins, if you want to build the Pi Zero permanently into a system.

What you should *not* do is attempt to use these holes by simply pushing wires into them. This is an unreliable way of getting a connection that inevitably leads to damaging your Pi.

Normally, this process would involve soldering, which some people might find off-putting, so the people at Pimoroni have come up with a solution involving a compression eyelet header and a hammer. See `https://shop.pimoroni.com/products/gpio-hammer-header`.

However, a header isn't too difficult to solder up yourself with a fine-tipped hot iron. Though we cover soldering in general in Chapter 15, here are some extra tips about soldering on a header strip. Push the header sockets into the holes and turn the board upside down so that it rests on the pins. Use a large lump of adhesive putty, not to be confused with epoxy putty, to make everything level and stable. Place an iron so that its tip touches both the pin and the printed circuit board hole, and then apply a little solder. You should see the solder flow round the hole. Keep the iron in place and after a half a second or so, you'll see the solder sucked into the hole by the capillary action of the hole and molten solder. When you see that happen, remove the iron. Don't move the joint until it cools. If the solder isn't pulled in, you either have too much solder, your soldering iron isn't hot enough.

Now, before you solder another pin, ensure that the whole header strip is lying flat on the board. If it isn't, melt the joint again, and push the header strip flat. (*Remember:* You can't correct a crooked connector after more than one pin has been soldered.) Then proceed and solder the other joints.

If you have trouble with later joints, it's probably because your soldering iron has cooled a bit after making a joint. If so, allow a few seconds between joints for it to warm up again. Wipe the iron on a damp sponge after each joint, to remove any flux.

Though this all sounds complex when written down this way, in practice it's much easier and quicker. Figure 16-3 shows a close-up of a header in the process of being soldered.

Getting at all the pins with one connector

Though a template is useful for identifying individual pins, you might want to transfer all the pins to a solderless breadboard, which consists of rows of sockets you can plug wires and other components into. This is a popular way of making a temporary circuit that is quick to change. One way of doing this is with a device called a *cobbler,* which consists of a ribbon cable and a printed circuit board. Figure 16-4 shows one of the various types you can buy — a T-Cobbler Plus, from Adafruit (`www.adafruit.com`). Many others are available, too — just search the web.

FIGURE 16-3:
Soldering a
header pin
to a Pi Zero.

FIGURE 16-4:
A typical cobbler
connector, for
bringing out all
the pins to a
breadboard.

With the Pi 400, the GPIO socket is at the rear of the keyboard and mounted at a right angle compaired to the traditional board. This arrangement will still work with a cobbler-type ribbon cable, but perhaps a better arrangement is the *Flat HAT hacker* (`https://shop.pimoroni.com/products/flat-hat-hacker`), which allows you the same horizontal mounting as you get on a bare PCB Pi. Figure 16-5 shows it attached to the Raspberry Pi 400. Note that you'll need to buy a couple of stand-off pillars so that the board is stable and not putting any strain on the GPIO pins. These use M2.5-size screws.

Connecting things together

Once you know which pins are what, you have to connect them to other components to make a circuit. The best way to do this is to use jumper wires with pre-crimped connectors. Flexible wires often have their connectors crimped on, so the joint between them is made by squashing together the wire and connector. This method is more reliable than a soldered joint because no point of stress is created at the point in the wire where the solder ends. The point of stress is where the flexing wire breaks from metal fatigue. To be able to crimp successfully, the connector has a specially designed end to allow the wire and connector to join securely when squashed together with a special tool.

The proper name for the pin header on a Raspberry Pi is a *Dupont connector*; the part that slips on the pins is known as a *female* connector, and the part that plugs into a breadboard is a *male* connector. So, for connecting the header on the Pi into a breadboard, you need a female-to-male jumper wire. These often come as rainbow ribbon wires, which separate easily by simply pulling them apart. You can crimp the connectors yourself with the help of a crimping tool, but we recommend that you get them already crimped. For interconnections between points on a breadboard, you can use some male-to-male jumper wires, but most people use solid-strand wire with the insulation cut back a little. Do not put too fat a wire into a breadboard — you don't want to stretch the connectors and make it too loose to fit components later. To this end, we recommend using wire with a 0.5mm diameter — otherwise known as 24 SWG (Standard Wire Gauge, used in the UK) or 24 AWG (American Wire Gauge, used in the U.S. and Canada). It just so happens that, at this size, the two standards use the same numbers for the same size wires. Or, in the international IEC 60228 standard, which specifies a wire's cross-sectional area, it's 1.5mm².

Though electricity doesn't care about the color of the wire it's flowing through, by convention red or orange is used for a positive wire, and black, green, or blue is used for a negative or ground wire. Some Far East suppliers use white for negitave.

Your First Circuit

The first thing anyone should do with a Pi system is to make an LED flash — because it's the hardware equivalent of the "hello world" program. To light an LED, you need two components: the LED itself and a resistor to keep the current down to a safe level for the Pi's GPIO pins. The circuit diagram is shown in Figure 16-6, along with a diagram of the physical layout.

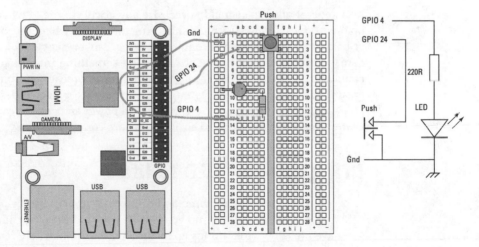

FIGURE 16-6:
An LED and switch, wired to the GPIO pins.

Note: We have also added a push button input switch to this diagram, which we will use later for controlling the behavior of the LED.

Notice two points here: the orientation of the LED and the orientation of the switch.

LEDs are sensitive to the polarity of the current you put through them. The positive connects to the anode (the top flat bit of the triangle in the symbol), and the cathode (the bar in the symbol) goes to the negative. That way, the LED lights up;

wire it the other way round, and it won't. So, faced with an LED, how do you know which wire is which? Well, on a new LED, the anode normally has the longest lead, but what happens if the leads have been snipped? Well, examine closely the rim of the LED and you will see a portion of the round housing that is not curved but flat next to one lead. That lead is the cathode, or the negative end of the component. You can see the flat on the layout diagram.

With the push button input switch, you must get it the right way round, but for a different reason. Basically, there are only two contacts on a switch, and it doesn't matter which way the current flows through it. However, many popular, small push buttons, known as *tack* (short for *tactile*) *switches,* have four leads for mechanical stability, and each pair is connected together. In the layout and on the switch, you can see that the leads are on opposite sides of the rectangular base, with the other two sides not having a lead protracting from them. It's the two leads on the protracting side that make contact when the button is pressed. So you need to orient the switch correctly — otherwise, it will act as if it were permanently pressed. If you ever get confused, resort to wiring the switch up to opposite corners, and that will ensure you wire it correctly.

TIP

As a general rule, you should never wire any circuit to the Pi when it's powered-up. Also, you should never plug something into the Pi when it's powered-up. This is because a circuit with incomplete wiring can present conditions that could damage the Pi or its components. Plugging something into an already-powered connector is known as *hot-plugging,* and special precautions must be taken to ensure the order in which connections are made. A device designed for hot-plugging, like a USB connector, will have the power and ground connectors longer than the signal connectors, to ensure these are connected first.

Bringing your LED to life

Once you have your circuit wired and connected, it's time to bring it to life with software. Go ahead and boot up your Pi. (If you find that your Pi won't boot up, you have probably wired things up incorrectly, so immediately remove the power and check the circuit.) We show you how to use Scratch and Python 3 to make your LED blink.

Using Scratch 3.0

Scratch is a computer language we cover in more detail in Chapter 9 and 10, but the basic idea is that it uses graphic function blocks that the user drags and joins together to construct a program. It is very popular with children under 10 because it involves little in the way of typing.

Scratch 3 doesn't come with the blocks to control the GPIO pins installed, so before you can use them you have to load the Raspberry Pi Simple Electronics extensions. You do this by clicking the blue button in the lower-left corner of the screen with the plus (+) sign and the icon of a short and a long block. Then scroll down in the window that pops up and click the Raspberry Pi Simple Electronics box. This will return you to the Scratch Desktop window and you'll have four new blocks in green to use:

>> **When Button Is:** This is a conditional statement that controls what to do depending on the state of a button or, in other words, a GPIO pin. The condition in which the button has to be in order to run the blocks following it depends on the button number (selected by the first drop-down menu) and the state of this button, pressed or released (selected by the second drop-down menu).

>> **Button Is:** A logic statement that gives a True or False depending on the state of the button. This is a lot like the When Button Is block, but it's designed to be incorporated into your own If conditional blocks.

>> **Turn LED:** An action command that sets an LED or, in other words, a specified GPIO pin set as an output (defined by the first drop-down menu) to a state of on or logic HIGH or off or logic low (as defined by the second drop-down menu).

>> **Toggle LED:** Very much the same as the Turn LED block, but the state the LED is set to is the opposite of the state it currently is.

To build the Scratch code to blink an LED, shown in Figure 16-7, follow these steps:

1. **Open Scratch, go to Events, and drag out the When Green Flag Clicked block from the control panel.**

   This gives a place for the code to start from.

2. **Add the Forever Loop from Controls and attach it to the When Green Flag Clicked block.**

   To do this, drag the block up to the underside of the Green Flag block until it snaps onto it. The two blocks will move as one.

3. **Scroll down to the Raspberry Pi Simple Electronics tab, drag in a Turn LED block, and then use the drop-down menu to select the LED number 4, and using the drop-down menu, select the State to On.**

   We want to make pin 4 change between high and low or on and off. Because of the way the LED is wired, when the pin is high (in other words, the voltage on that pin is +3V3), the current will flow through the LED and resistor to make the LED light up. When the pin is low (the voltage is 0 volts), the LED is off.

4. **Add a Wait 1 Seconds block from the Control section, and change the time from 1 to 0.3 seconds by clicking the 1 and typing** 0.3.

Add it to the last block by dragging it underneath until it snaps into position.

5. **Duplicate these last two blocks, or drag some more in, and change them like the first two, except set the second Turn LED block to off; attach these two blocks to the previous two.**

6. **Drag all these boxes into the Forever loop.**

The Scratch program should look like Figure 16-7. To try it out, click the green flag and watch the LED blink.

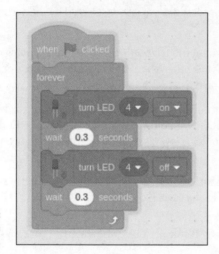

FIGURE 16-7:
Scratch program
to blink the LED.

REMEMBER

You can change the numbers in the wait blocks to change the blink speed. Note that the two wait block delays do not have to be the same; you can have a quick 0.3-second flash every second, if you want.

Control the flashing speed with an input

Connecting a GPIO pin to use as an input is simple: Just wire a push switch between the input GPIO pin and ground. Then you can use the When Button block or the Button Is block. Follow these steps:

1. **Start like the previous example and drag out a When Green Flag Clicked and a Forever loop.**

This time, we're going to add a delay before changing the LED, and that delay will depend on whether the push button is pressed.

2. So, drag out an If . . . Else block and drag a Wait Seconds into both the if and else arms of this block.

3. Next to the If, add a Button Is block and set the pin to 24 and the State to Pressed.

4. Also change the time in the Wait block of the else condition to 0.3 by clicking and typing on the 1.

5. Follow this with a Turn LED block and set the pin to 4 and the State to Off.

6. Duplicate these blocks and change the Turn LED to Off.

After you have done this, you should see the program shown in Figure 16-8.

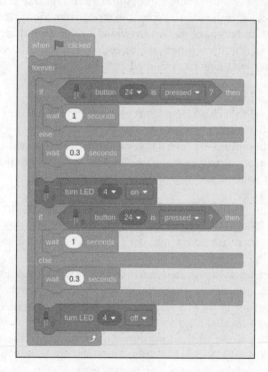

Click the green flag, and you should see the LED blink rapidly. When you press and hold the button, the blinking slows down. Now, here's a test: Change the bottom if so that it tests for being released not pressed. Before you try it, see if you can predict what will happen. Then make the change and see if you were right. Finally, see if you can you use the toggle LED block to eliminate nearly half of those blocks. In electronics, *toggle* means "change the state of," so becuase this is binary and there can be only two states, it's easy to see what it will do.

There is an other set of extensions called Raspberry Pi GPIO extensions, which basically give you the same operations as the Raspberry Pi Simple Electronics extensions, but with logic levels High and Low instead of Pressed or Released, and Set GPIO to Output instead of Turn LED. In addition, there is a Set GPIO to Input Pulled to give you a choice of pulling up, down. or not pulling at all, as opposed to the automatic assumption of pulling up. There isn't much difference here because, in practice, the Raspberry Pi Simple Electronics extensions can do all you would normally want.

Using Python

The Python language was introduced in Chapter 11 and, unlike Scratch's graphic-based program blocks, it uses entirely text-based instructions. Its great power is that the basic Python language can be extended to do more things by the use of *libraries.* These are prewritten functions, or code modules, that can be written in Python or any other language to extend what Python can do.

When using Python to access the GPIO pins, you have a number of different librar-ies you could choose that can give you access to them. They can provide not only normal input/output access but also access to some of the special functions or capabilities of certain pins. We have deliberately written the next two examples in in a style to match, as closely as possible, the two Scratch examples we have just presented, so that you can see how similar they are.

REMEMBER

In the early days of the Raspberry Pi, the only access to the GPIO pins was if you were running in Supervisor, or Root, mode. Fortunately, this has now been changed so that you can run in normal User mode with most libraries.

Although there are many integrated development environments (IDEs) in which to develop and run programs in Python, Thonny Python is popular with beginners and favored by the Raspberry Pi Foundation, so we'll use it here. Load it from the Programming menu by clicking the Raspberry icon at the top left of the screen, moving the mouse pointer to the Programming option, and then moving across and down to the Thonny Python IDE item and clicking on it. The Thonny editor window opens. To create a file to put your code into, click the new file icon in the top-left corner, and type in Listing 16-1. Remember that, in Python, the case of a word matters, as do the spaces at the start of a line, so be sure to get them right; otherwise, you'll get errors when trying to run your code.

LISTING 16-1:	**LED Blink**

```
#!/usr/bin/python3
import time
import RPi.GPIO as io # using RPi.GPIO
```

```
io.setmode(io.BCM)
io.setwarnings(False)
io.setup(4,io.OUT) # make pin into an output

print("LED blinker - By Mike Cook")
print("Click the Stop/Restart back end icon quit")
while True:
    io.output(4,1)
    time.sleep(0.3)
    io.output(4,0)
    time.sleep(0.3)
```

Save the file and run it. Your LED blinks just like the Scratch example.

Take a look at Listing 16-1 line by line. The program starts by importing the support packages you need. In this case, it's the time package we use to achieve a delay, or Wait, function, and the RPi.GPIO package to access the GPIO pins. Note here that a lot of examples you see in other places use as GPIO in the import statement — but we prefer the simpler and shorter as io because it reduces the amount of typing and we don't have to keep switching to uppercase. Then you have to tell the GPIO library what sort of numbering system you want to use. We're using the now-standard BCM system. Then we turn off the warnings — these are annoying and don't give much useful information. The next line tells the GPIO pins that you want to use Pin 4 as an output. The setup method takes two values: the pin number and a number that tells the library to make that pin an output. This is conveniently hidden by the library, by using a predefined constant that's defined in the library — that's why it's prefixed with io. Next, the loop forever of Scratch is carried out in Python with the use of the while True statement, with all statements in this loop indented. The actual number of spaces used for indents can be anything, but it must remain constant for any section (4 is a standard value often used).

This loop then commands the GPIO Pin 4 to be a 1, turning the LED on. Then there's a delay (or sleep) for 300mS, followed by turning the LED off by making Pin 4 go low. Finally, there's another delay, so you can see the LED in the off state. A common mistake is to leave out this last delay; as a result, the LED looks to be on all the time. So, controlling a GPIO output is very simple.

Now, on to the second example of a switch-controlled blink speed. This is shown in Listing 16-2, so open a new file and type it in, or choose File⇨ Save As and modify the original Listing 16-1 to match Listing 16-2.

LISTING 16-2: **An LED Blink Rate Controlled by a Push Button**

```python
#!/usr/bin/python3
import time
import RPi.GPIO as io

io.setmode(io.BCM)
io.setwarnings(False)
io.setup(4,io.OUT) # make pin into an output
io.setup(24,io.IN, pull_up_down=io.PUD_UP) # make pin into an input

print("LED blinker - By Mike Cook")
print("Click the Stop/Restart back end icon quit")
while True:
    if io.input(24) == 0:
        time.sleep(1.0)
    else :
        time.sleep(0.3)
    io.output(4,1) # LED on
    if io.input(24) == 0:
        time.sleep(1.0)
    else :
        time.sleep(0.3)
    io.output(4,0) # LED off
```

Listing 16-2 has the same commands for the output pin, but the input pin setup is new. This command has three parameters: the number of the pin to use, a number in the form of a constant defined by the library, specifying that this pin should function as an input, and, finally, an optional command telling the computer to activate the internal pull-up resistor (although, to our ears, PUD_UP sounds more like a mother calling children to tell them their pudding is being served). The input from the push button is read by the input(24) method, which returns a value of 0 or 1, depending on whether the pin is connected to the ground. This returned value is then compared to 1, and if it's equal, a 300mS delay is made; otherwise, it produces a 1-second delay. The LED is then turned on, and the conditional delay code is repeated before the LED is turned off.

Using GPIO ZERO

When it comes to accessing the GPIO pins in Python, the GPIO Zero library is sometimes used by beginners, especially if they want just simple operations (although it is capable of more sophisticated operatons). Don't confuse GPIO Zero with the Pi Zero — the two are not related. The GPIO Zero library takes the class method approach to control, as opposed to the function method approach of RPi.

GPIO and other, similar libraries. Pins become Python objects, which must be set up before use. Despite this complication, this system is easy to use for simple work. For example, for our LED blink example, we can use the code in Listing 16-3.

LISTING 16-3: **Python LED Blink**

```
#!/usr/bin/python3
import time, os
import gpiozero as io # using LED zero

led = io.LED(4) # make pin 4 into an output

print("LED blinker using gpiozero - By Mike Cook")
print("Click the Stop/Restart back end icon quit")
while True:
    led.on()
    time.sleep(0.30)
    led.off()
    time.sleep(0.30)
```

Listing 16-3 at first looks very similar to the code in Listing 16-1 but only because we have written it to be like this. The GPIO Zero library uses only the BCM method of pin numbering, so there's no need to tell the library what system to use. Just like in Scratch, an output pin is known as an *LED*, irrespective of whether that pin controls an LED, a motor, or a chip. So, the line that makes the pin an output is io.LED(4). This code segment returns a reference to the class object. Classes and objects are topics we cover in Chapter 17, where we describe how to write one, but basically, it's a way to get the same piece of code to handle different specific objects — in this particular case, different GPIO pins. In order to know what it has to do, each thing/pin must be declared separately, and when it is, the program gets a code number to use to identify which specific thing it has to handle. That code number is placed into a variable called led in the Listing 16-3 example, but could be called anything. (This variable is known as the *instance reference*.) The class has methods associated with it — things it can do, in other words — and these are called up by writing the instance reference variable name, followed by a period or dot, followed by the method name. So, turning an LED on or off is a simple process — you just write led.on() or led.off(). However, in a way, this is a bit limiting. For example, you cannot send a number in a variable that's used to turn the LED on or off — it has to be specifically spelt out as a method name. This is a limitation of how GPIO Zero is written, not a limitation of using classes. There are ways round this, but at this point, the simple system gets rather complex.

In fact, Listing 16-3 is written in a way that it looks a lot like the earlier listings. The same thing could simply be written as shown in Listing 16-4.

LISTING 16-4: **A Simple LED Blink**

```
import gpiozero
led = gpiozero.LED(4)
led.blink()
```

And that's all you need. We aren't sure what you can learn from this, apart from the wiring skills needed to attach the hardware, but it's helpful for beginners. That last statement can have some parameters in it to control the blink rate — so, for example, to exactly match our other examples that last line could be:

```
led.blink(on_time = 0.3, off_time = 0.3)
```

This library has even more tricks up its sleeve. If you want to fade that LED up and down instead of just blinking, that last line could say

```
led.blink(on_time = 2.0, off_time = 2.0, fade_in_time = 1.0, fade_out_time = 1.0 )
```

This is great if you only want to do that, and for a very young beginner, that's exactly what you want to do.

The blinking speed controlled by a push button can be written as shown in Listing 16-5.

LISTING 16-5: **A GPIO zero example of an LED Blink Controlled by Push Button**

```
#!/usr/bin/python3
import time, os
import gpiozero as io # using LED zero

led = io.LED(4)          # make pin 4 into an output
push = io.Button(24)       # make pin 24 into an input

print("LED blinker switch using gpiozero - By Mike Cook")
print("Click the Stop/Restart back end icon quit")
while True:
    if push.is_pressed:
        time.sleep(1.0)
    else :
```

```
        time.sleep(0.3)
    led.on()
    if push.is_pressed:
        time.sleep(1.0)
    else :
        time.sleep(0.3)
    led.off()
```

The input works in a similar way to the output. In this case, you make an instance of the Button class and put its reference in the variable push. Then you use the is_pressed method of this class to determine the time delay in the blinking. By default, the input pull-up resistor is enabled when you create the class reference.

There are other options contained in this input class. It can specify whether a press is to be auto-repeated or specify a *debounce time* — the time after a change to ignore further changes. (We take another look at debouncing later in this chapter.) As well as the is.pressed method, other class methods include is_held, wait_for_press, wait_for_release, is_held, when_held, when_pressed, and when_released. Rather than return any information about the input, the when_pressed, and when_released methods cause a function to run when the button is pressed or released. This other function runs in a separate thread — in effect, another separate program — with this thread and the main code being swapped in and out alternately until the function is complete. This gives a beginner access to complex concepts that they need to understand only when something goes wrong. So, our controlled blink-rate program could be written as shown in Listing 16-6.

LISTING 16-6: **GPIO Zero-Specific LED Blink Controlled by Push Button**

```python
#!/usr/bin/python3
import gpiozero as io # using LED zero
from signal import pause
led = io.LED(4) # make pin 4 into an output
push = io.Button(24) # make pin 24 into an input

def blinkFast():
    led.blink(on_time = 0.3, off_time = 0.3)

def blinkSlow():
    led.blink(on_time = 1.0, off_time = 1.0)

push.when_pressed = blinkFast
push.when_released = blinkSlow

pause()
```

The pause function, in effect, ends the program, and the only thing that happens is when the callback functions, as they are called, are invoked or triggered on the push or release of the button.

REMEMBER

A lot of the programming work involved in using GPIO Zero involves leafing through the documentation to see what simple functions the authors have implemented for you. GPIO Zero enables you to get results quickly, and without needing to understand much of what is going on — which is helpful if a function does what you need. However, we can't help but feel that although some things can be done simply, the skills learned are not exactly transferable. As you develop more complex electronic projects, you might find GPIO Zero to be too limited for your needs or require more advanced concepts than you know about.

Starting Out with a Dice Display

After you have the basic toolkit of dealing with the GPIO pins under your belt, it's time for a bit of fun with some projects. First off, let's look at making a computer-controlled dice.

A dice display

Now, before you get pedantic about our closing statement in the previous section and state that there's only one dice, so it should be called a *die*, let us point out that in modern standard English, *dice* is both the singular and the plural. To throw the dice could mean one, or more than one, dice. If you disagree an online search will verify this definition.

A dice display is a good project to start with because it not only makes a useful and interesting replacement when you want to use one in playing a game but also serves to introduce some important programming concepts. Basically, it's simply seven LEDs and a push button, so it isn't much different from the circuit we describe earlier in this chapter. However, the arrangements of the LEDs is vital to the final effect, and it's important to get the correct LEDs in the correct spatial position. Figure 16-9 shows the basic schematic.

The actual GPIO pins you use for the LEDs doesn't matter because the relationship between the LEDs and the pins is defined in software — if you get it wrong, you can correct it in the pattern lookup table. However, for a trouble-free experience, follow our GPIO usage. Note that the schematic is quite clean and straightforward. It follows the rules of positive signals at the top and ground signals at the bottom, with a minimum number of wires (in this case, none) crossing. Also, the signal flow is from left to right. This is a good example of a schematic. When it comes to the physical layout, Figure 16-10 shows one way of wiring this up on a breadboard.

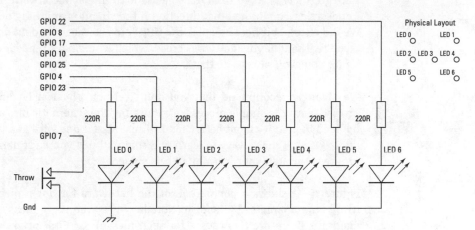

FIGURE 16-9:
LED dice schematic.

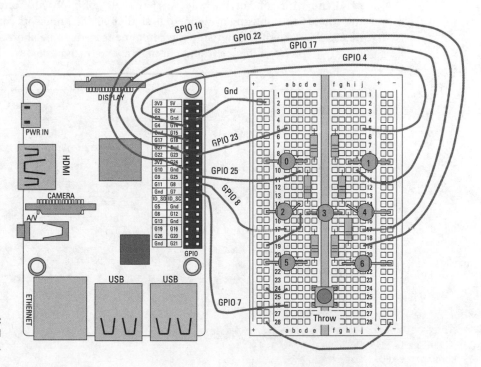

FIGURE 16-10:
LED dice physical layout.

You will instantly see that this layout is altogether more cluttered, which makes it difficult to see what is happening, although it does show you where to place all the parts and wires. Note that LED 2 and LED 4 have their legs bent slightly to allow them to physically line up with LED 3 and get a good match to the die pattern. Though it's easy to go from the schematic to the layout in your head, it's almost impossible to go in the other direction if the circuit is anything other than trivial. We did cheat a bit here and pick the GPIO pins so that we could draw the physical layout without having any wires cross, which would have made it even more cluttered. You can't always do that.

TIP

We strongly recommend that you don't rely on physical layout diagrams, but instead learn the simple process of converting the schematic into a layout. There's no need to plan it all out beforehand — just take it one wire at a time and build it up. It's a skill that's easy to learn and that will pay you back handsomely as you progress through your learning of this subject.

Figure 16-11 shows a photograph of the hardware for our project. We ended up using 3mm diameter LEDs for it. Note how it's even more messy than the layout diagram. There are two causes for such messiness: First of all, you can't easily bend the wires out of the way so that they don't obscure anything. Second, the whole point of using breadboard is so that you can reuse its components. This means you generally don't snip the component leads to be short and close to the board — you leave them sticking out high above the board.

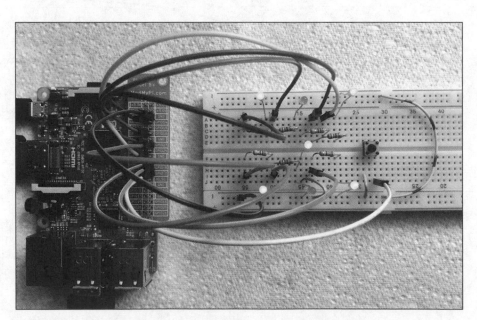

FIGURE 16-11:
Photo of the LED dice project.

The project

When doing any project, no matter how small, it always pays to take it one step at a time and test as you go. So, the first thing to do is to test your hardware before you get into anything fancy. You can simply extend the LED blink program to blink all the LEDs or, if you aren't up to that, simply take the LED blink program and change the GPIO number to one number you're using here. Run that and see the correct LED flashes, and then stop the program and change to another number. Repeat until all the LEDs (as well as the push button) have been tested.

The project splits up into two main parts: Generate the random number between 1 and 6 and then display it.

The numbers

There are two ways to generate the random numbers: Use the computer's random-number generator or use some random event. The problem with the random-number generator is that it's not really a random-number generator. It's a *pseudo* random-number generator — the sequence of numbers is fixed and will repeat every time you run the program, unless you seed it by picking a random starting point in this sequence. Because the sequence is very long, this approach will appear to be a random number. (The Python random-number generator uses as the seed the length of time the system has been powered on in the absence of a seed instruction.)

However, a better way is to use something that is by its nature truly random: In this case, we use the length of time the user has pressed the button, indicating a new throw. You might not think this is random, because the user could just make a long press or a short press and that would influence the number that's produced. However, this doesn't take into account that the timing of a press can be very accurate and that all six numbers are cycled through many thousands of times a second. Controlling a button press with microsecond accuracy is simply not possible for us slow humans, so it's a good source of random numbers. Therefore, in our project, the number is generated by counting from 1 to 6 while a push button is held down.

When a push button is first pushed, the contacts come together and then bounce apart and then come together in a rapid sequence of contact make-and-break, rather like a ping-pong ball falling onto a table. This happens on all mechanical switches, but some designs are worse than others. A lot of the time in programming, this isn't important because other stuff happens after a contact is first made. Other times, it could be a problem, but the solution can be quite simple — just delay for a few milliseconds after the first contact before looking at the input again. This strategy — known as *debouncing* an input — is what we show you how to do in this project.

The display

Once you have generated a number, it must be displayed in the pattern of a conventional dice. This involves turning on a specially selected pattern of LEDs. This pattern is, of course, different for each number. Now, you could use a series of if statements to test for each possible number, and when you find a match, write a series of statements to turn on and off the appropriate LEDs. An example of this for one number using the GPIO Zero library would look like the code in Listing 16-7.

LISTING 16-7: **Fragment of Code to Set a Pattern**

```
# the variable dice_number contains the number to display
if dice_number == 3: # make the three pattern
    led0.off()
    led1.on()
    led2.off()
    led3.on()
    led4.off()
    led5.on()
    led6.off()
if dice_number == 4: #if the number is four then
    led0.on()
    #......... and so on
```

You can see that this listing is easy to read, but it contains the most awful, excessively ornate style, that is repetitive and is much longer than necessary. (This is typical of what a beginner might produce.)

REMEMBER

This bad style is not confined to the GPIO Zero library — you can write bad code like this with any library.

You might wonder why it matters, as long as the code works. Well, if you can understand the concepts of doing it properly, you can apply them in situations where code like this will not work or is far too long. Even so, before you can begin to write any code, you need to know what LEDs to turn on and off to display each number. Figure 16-12 shows the patterns we have decided to use to represent the dice pattern. Note that the main choice is which diagonal to use for representing 3 and 2. We think it's best using different diagonals for the two numbers, but we have no preference for which diagonal to use.

LED 0 ○   LED 1 ○

LED 2 ○   LED 3 ○   LED 4 ○

LED 5 ○   LED 6 ○

	Display Number	LED 6	LED 5	LED 4	LED 3	LED 2	LED 1	LED 0	Binary Pattern	Hex Pattern	Decimal Pattern
	1	off	off	off	on	off	off	off	0b0001000	0x08	8
	2	on	off	off	off	off	off	on	0b1000001	0x41	65
	3	off	on	off	on	off	on	off	0b0101010	0x2A	42
	4	on	on	off	off	off	on	on	0b1100011	0x63	99
	5	on	on	off	on	off	on	on	0b1101011	0x6B	107
	6	on	on	on	off	on	on	on	0b1110111	0x77	119

FIGURE 16-12: LED dice display patterns.

Given the physical arrangement of LEDs shown, each number needs to be displayed by having certain LEDs on and off. This pattern can be summarized as a sequence of 0s and 1s and is shown in the Binary Pattern column in Figure 16-12. *Binary* is a way of writing numbers that exactly matches how those numbers are stored in the computer as a sequence of 1s and 0s. A single digit is called a *bit*. Everywhere the LED is on, there's a 1 in the binary pattern, and there's a 0 when it's off. This pattern can define the LED pattern, functioning as a concise store of the pattern you need. In a program, it's way easier to express this pattern as a hexadecimal number, because with a little practice, you can convert from a pattern to a number in your head — and it's then easy to read the code. *Hexadecimal* is a number system that groups a byte, or 8 bits, into two 4-bit groups. Each of the two groups is then expressed as a single character — from 0 to 9 and then the letters A to F. This gives 16 different characters that represent the 16 different patterns of 1s and 0s you can get with four bits. This is shown in the Hex Pattern column of Figure 16-12, as far as the computer is concerned, if you start a number with 0x, it treats what follows as a hexadecimal number, and if you start it with 0b, it treates it as a binary number.

REMEMBER

In Figure 16-12, we show — for the sake of completeness — the equivalent number in *decimal* notation, which is the counting system we humans use. No one in their right mind would dream of using it in a program — although it would work because it follows the correct pattern, it's almost impossible to know what the pattern is by just looking at the number. In other words, it makes for bad code.

Keep in mind that the actual numeric value of this pattern number has no meaning at all — the meaning is in the binary bit pattern. Variables don't always have to contain a number; what a variable stands for is determined by the context in which you use it. For example, you can add one to a bit pattern, but the result, while still being a valid number, is meaninglessas as bit pattern. Often, beginners ask how to convert a number into binary, but there's no need because any number stored in the computer is already in binary format. Only when you need to do a task like printing it out do you need to convert a number from binary into something else, and the `print` statement does that for you. All you need now is some way to read that "bit pattern number" and turn it into the action of switching the LEDs on and off. In effect, you have to unpack that information and turn it into a pattern — and do it efficiently without turgid code.

So let's break this down and see if you can do this for only one bit of the pattern —
let's say the least significant, or right-hand, bit. If it is 1, turn on the LED, and if
it's 0, turn it off. You can handle that task with this line:

```
io.output(LEDnumber,pattern & 1)
```

The variable LEDnumber has, as you might suspect, the LED number you want to
control, and the variable called pattern has the bit pattern you're trying to produce. The & symbol means *the arithmetic AND operation.* The AND operator takes
two numbers and considers the contents of each bit position individually. If both
numbers contain a 1 in the same position, the result is a 1. Otherwise, the result
for that position is 0. So, when you're ANDing the number 1 with the pattern
number, the result is to *zero,* or remove, all bits except the least significant one,
which is left alone. In other words, it isolates a single bit in the pattern variable.
So if the least significant bit in the pattern variable was a 0, that operation returns
a 0. If it was a 1, that statement returns a 1 — which is just the number you need
to put in the output command to turn that LED on or off. (We say that the number
1 here is a *mask* because it masks out bits you don't want.)

Now, what about the LEDnumber that changes every time we want to look at a different bit in the pattern? Instead of a single number here, we actually need a list.
The list is made up of the GPIO numbers that control each LED. So the first element in the list is the GPIO number that controls LED 0, which, according to the
schematic, is GPIO 23. Finally, you need a way to move the bit pattern in the pattern variable one place to the right so that the same instruction will do the same
thing for the next bit. You can do this with a shift-to-the-right operator (>>), so
you can have a little loop that generates the pattern:

```
LEDnumber = [23,4,25,10,17,8,22]
for i in range(0,len(LEDnumber)):
    io.output(LEDnumber[i],pattern & 1)
    pattern = pattern >> 1
```

And that's all there is to it — no bad code at all. It's short, concise, and easy to
write.

Finally, we have added the small, fun feature of a dice roll display. This shows
several random dice patterns in quick succession, to give the impression that the
dice is being rolled.

We're all ready to show you how to write the full code. This is shown in
Listing 16-8.

LISTING 16-8: **LED Dice**

```python
#!/usr/bin/python3
# Electronic dice. By Mike Cook
import time, random
import RPi.GPIO as io

LEDnumber = [23,4,25,10,17,8,22]
dicePattern = [0,0x08,0x41,0x2A,0x63,0x6B,0x77]
pushButton = 7

def main():
    print("Electronic Dice Click the Stop/Restart back end icon quit")
    initGPIO()
    number = 1
    while 1:
        displayRoll()
        displayNumber(number)
        number = generateNumber()

def displayRoll(): # pattern to show when rolling
    for i in range(0,20):
        displayNumber(random.randint(1,6))
        time.sleep(0.1)

def displayNumber(number):
    pattern = dicePattern[number]
    for i in range(0,len(LEDnumber)):
        io.output(LEDnumber[i],pattern & 1)
        pattern = pattern >> 1

def generateNumber(): # wait for a push
    throw = random.randint(1,6)
    while io.input(pushButton) == 1:
        pass
    time.sleep(0.030) # debounce delay
    while io.input(pushButton) == 0:
        throw += 1
        if throw >6: # wrap round the number
            throw = 1
    return throw

def initGPIO():
    io.setmode(io.BCM)
    io.setwarnings(False)
    for pin in range (0,len(LEDnumber)):
        io.setup(LEDnumber[pin],io.OUT) # make pin into an output
    io.setup(pushButton,io.IN,pull_up_down=io.PUD_UP) # make pin into an input
```

(continued)

LISTING 16-8: *(continued)*

```
# Main program logic:
if __name__ == "__main__":
    main()
```

When this code runs, it loads the libraries, sets the global variables, defines all the functions, and then, finally, starts running the `main` function. This is where the action is coordinated, and it's a good idea to make it as short as you can. The `main` function prints out a message, initializes the GPIO pins and number to display, and then starts an infinite loop, displaying the roll pattern and the dice number and then generating the next number when the button is pressed again.

REMEMBER

Beginners often try to cram everything into the `main` function, but using some well-named functions makes it much easier to see what's going on. These functions could, of course, be called anything, but for readability, make them as descriptive as you can. This benefits not only others reading your code but also, more importantly, *you* in six months' time, when you have forgotten what is going on.

One thing that might puzzle you is that the dice pattern list starts off with a 0. This was not shown in Figure 16-12, so why is it there? Well, all lists are numbered starting at 0, and the first pattern you're interested in is the pattern for a throw of 1. So, in order to align the position of the pattern information with the position in the list, you must start out with a blank entry — so that, for example, the information for the number 4 is in the fourth position on the list. The rest of the code ensures that 0 will never be required to be displayed.

Looking more closely at Listing 16-8, the `displayRoll` function is simply a loop that calls the `displayNumber` function 20 times, giving it a random number between 1 and 6. It shows each number for one-tenth of a second. The `displayNumber` function should be familiar to you from earlier mentions in this chapter. The one function that may need close examination is `generateNumber`. It starts out generating a random number called `throw`, and then it waits in a `while` loop until the push button is pressed. When this is detected, a 30ms delay ensures that any contact bounce is finished before another `while` loop looks to see if the button has been released. Though the program is in this loop, it increments the `throw` variable, and when the maximum value of 6 is exceeded, the `throw` variable is reset to one. The last line returns this `throw` value to the calling line. The line in the `main` function

```
number = generateNumber()
```

ensures that this function's returned value is placed in the variable called `number`.

Taking it further

We encourage you to play around with all the code you find in this book, by making your own changes and extensions. One area that's fun to play with is the roll animation. Instead of showing random numbers, you can make any sort of display you like in the same way as you defined the patterns for the dice display. As a start, we have written an alternative function for the display roll, as shown in Listing 16-9. You should use this code to replace the function of the same name in Listing 16-8. You should save any changes you make using a different filename if you want to keep them.

LISTING 16-9: **An Alternate displayRoll Function**

```
def displayRoll(): # pattern to show when rolling
    rollPattern = [0x01,0x03,0x07,0x0F,0x1F,0x3F,0x7F,0,0]
    for roll in range(0, len(rollPattern)):
        pattern = rollPattern[roll]
        for i in range(0, len(LEDnumber)):
            io.output(LEDnumber[i], pattern & 1)
            pattern = pattern >> 1
        time.sleep(0.2)
```

This has a new list of patterns to be used in the roll — called, appropriately enough, rollPattern. You can write your own numbers in here and use as many of them as you want. The code ensures that all pattern numbers you write in here are displayed. Can you change the numbers in the rollPattern list so that a single lit LED goes round the dice display? You could also change the speed of the dice roll display, making it longer between display changes as it comes to an end. If you want to use binary numbers to define the pattern, you can use

```
rollPattern = [0b1,0b11,0b111,0b1111,0b11111,0b111111,0b1111111,0,0]
```

Note that there's no need to specify the leading 0s, personally, we find binary much harder to read in a line of code than hex because it's easy to lose your place amongst all those 1s. If you like this project, we encourage you to put all the electronics in a box and connect it up to the Pi with a multiway header and ribbon cable, or use the much cheaper Pico processor in the box.

Pedestrian Crossing

The dice project covered in the preceding section is all about displaying patterns. The Pedestrian Crossing project is about time and sequences. It simulates a UK pedestrian crossing — basically, a traffic-light–controlled crossing initiated by the pedestrian's pressing a button. These crossings are also common in the United States, but a little of the detail may be different from state to state, so let's first go through the sequence we're trying to simulate.

The crossing has a three-light traffic control — red, amber, and green — and the pedestrian has a two-light system — Cross (or Walk), which is green, and Don't Cross (or Don't Walk), which is red. The normal state of the crossing is that the green light is on for the traffic and all the other lights are off. There's a button at the crossing for the pedestrian to request a stop, and a traffic sensor is buried in the road. This sensor controls how quickly the crossing sequence will start after the pedestrian presses the button. If no traffic has been detected recently, the crossing sequence starts immediately; otherwise, there's a delay. This ensures that a busy road isn't stopped too many times by pedestrians or jokers, whereas a quiet road will not make pedestrians wait long before they can cross. Once the Cross Request button is pressed, the Don't Cross light immediately comes on.

When the crossing sequence begins, the green traffic light turns to amber and then to red. Then the green crossing light comes on and, at many crossings, a sounder "bleeps" to help blind people cross. After a period of time, it may be safe to continue crossing but not be safe to start a crossing, so the green cross light and the amber traffic light flashes, and the sounder no longer bleeps. Then the Don't Cross light comes on, with the traffic light remaining amber, and, finally, the traffic light goes to green and all pedestrian lights go out.

This sounds straightforward on the face of it, but the traffic sensor needs to be monitored through this sequence, to ensure that you don't miss any traffic that may drive up and stop during the crossing sequence and allow the next crossing to occur too quickly. This means you can't just use a simple `time.sleep` delay to control the sequence; you need what is known as a *state machine* — this is a way of juggling two or more processes while making it look like they're happening at the same time.

REMEMBER

A state machine is an important technique — and one that you'll need over and over as you take on more projects.

The idea of a state machine is that most things do not need the processor's full attention. The LED blink program, for example, spends virtually all of the time in the sleep function, just waiting for time to pass. Instead of burning that time, the idea of a state machine is to spend it looking after other tasks. In order to do this, you have to have some idea of when your tasks need attention, so you incorporate a variable that specifies the time each task needs to be looked at next. This is made possible by the `time.time()` method of the time library, which returns a *floating-point* number (one with decimal places) of the time (in seconds) that the Pi has been switched on. The thing is that you don't actually care what this number is, because you use it as a relative measure to find out when to do the task.

Let's look at a simple example first — blinking two LEDs independently.

First of all, go ahead and wire up two LEDs and resistors, connecting them to GPIO 23 and 24, as shown in Figure 16-13.

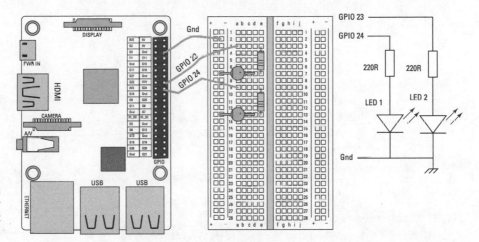

FIGURE 16-13: LED wiring for Independent blinking.

Now we'll describe how to blink these two LEDs at totally independent rates. The state of each task is given by a `state` variable, with the two values 0 or 1. The `state` variable shows if the light is on or off so that you know what you need to do for the next stage of the blinking task. The time between each change of state of the LEDs is controlled by a variable that gives the time the change must occur. This time is calculated by adding the current time to the length of time you want to elapse before the next LED state change. This can be the same for all states or different for each state. The example in Listing 16-10 has the same time for the on/off times of LED1, but a separate on/off time for LED2.

LISTING 16-10: **Blinking Two LEDs at Different Rates**

```python
#!/usr/bin/python3
# Two blinks By Mike Cook
import time
import RPi.GPIO as io

led1pin = 23
led2pin = 24
led1BlinkRate = 0.5 # the speed of blinking
led2onTime = 0.5 ; led2offTime = 0.51
led1State = 0
led2State = 0
blink1Time = time.time()
blink2Time = time.time()

def main():
    print("Two blinks Click the Stop/Restart back end icon quit")
    initGPIO()
    while 1:
        if time.time() > blink1Time:
            blink1()
        if time.time() > blink2Time:
            blink2()

def blink1():
    global blink1Time, led1State
    if led1State == 0:
        io.output(led1pin,1)
        led1State = 1
    else:
        io.output(led1pin,0)
        led1State = 0
    blink1Time = time.time() + led1BlinkRate

def blink2():
    global blink2Time, led2State
    if led2State == 0:
        io.output(led2pin,1)
        led2State = 1
        blink2Time = time.time() + led2onTime

    else:
        io.output(led2pin,0)
        led2State = 0
        blink2Time = time.time() + led2offTime
```

```
def initGPIO():
    io.setmode(io.BCM)
    io.setwarnings(False)
    io.setup(led1pin,io.OUT) # make pins into outputs
    io.setup(led2pin,io.OUT)

# Main program logic:
if __name__ == "__main__":
    main()
```

The program spends most of its time in the `while` loop in the `main` function, checking to see if it's time to call either of the two functions, `blink1` or `blink2`. (*Note:* You can, of course, have as many functions here as you want.) When the functions are run, they do what they need to do given that task's current state, which, in our simple case, involves turning the LED on or off. Then you advance the `state` counter and, finally, set the time when this function/task needs to be done again. If the task is being advanced at a constant rate (like the blinking LED1), then this is done at the end of the function. If each state needs to last a specific length of time, that is set when the `state` count is advanced.

When you run the program, you see both LEDs blinking in unison; but, as time goes by, the two become increasingly out of sync until they are seen to be blinking alternately, and, in another minute, they will drift back in sync. This is because the full cycle of LED2 is 0.01 seconds longer than LED1.

Play with these times to see how it changes things. Can you change the code and hardware to add two more LEDs into the mix?

The Pedestrian Crossing hardware

After you have all the background information you need to make the Pedestrian Crossing project, let's start, as always, with the hardware. Figure 16-14 shows how it should be wired up.

The diagram is quite similar to what we have shown you in this chapter, except that it's arranged differently. First, there are two push buttons for the traffic sensor and cross request. Next are five LEDs connected to resistors and GPIO pins. Note here, though, that you need different colours of LED — two red, two green, and a yellow. Finally, there's the sounder. We're using a piezo electric sounder, but a word of warning is in order here: There are two types of piezo electric sounders, and there seems to be no universally accepted word to differentiate between the two. The sort you want is one that generates sound when you apply a voltage to it. These could be called *self-drive* sounders, but sometimes they are not. Other

times, they can be referred to as *buzzers,* but don't confuse them with electrome-
chanical buzzers, which take a lot of current — too much current for the Pi.
A suitable one would work from 3V and take less than 10mA. The type you *don't*
want is the one where you have to supply an electrical pulse train to it before it
makes a noise. This type is sometimes called a speaker-type sounder. A simple
test is to wire the sounder to the 3V3 output and ground of the Pi and then make
and break the contact. If it only clicks, it's a speaker type. If it makes a sound, it's
the type you want to use. (To be honest, in development, we replaced the sounder
with an LED and a resistor, to reduce the annoying bleeps.)

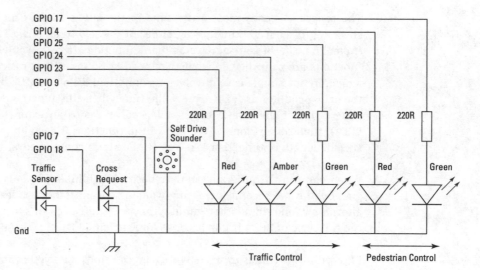

FIGURE 16-14:
Pedestrian
Crossing
schematic.

The layout of this circuit on a breadboard is shown in Figure 16-15. We have
arranged the traffic-control LEDs like traffic lights, and the cross indicators above
each other like you would find on a real-life crossing.

The Pedestrian Crossing software

Now let's apply the multi task/state machine principle to our crossing project. We
need one task to look at the traffic sensor — if we see traffic activity, we want to
note the time it took place. Then we need one task to look at the cross request
button — if that button is pressed and the crossing sequence isn't under way, we
need to start that sequence. Finally, if the crossing sequence has reached the
"cross phase," we need a task to turn the sounder on and off, to make the bleeping
noise. (Note that we have reduced the times in the crossing sequence to be easy to
look at and see what is going on. In real life, some of the times would be much
longer.) The software for this project is shown in Listing 16-11.

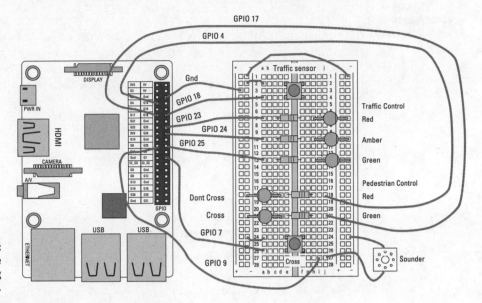

FIGURE 16-15:
Layout of the
Traffic Crossing
circuit.

GPIO 17

GPIO 4

Gnd

GPIO 18

GPIO 23

GPIO 24

GPIO 25

Traffic sensor

Traffic Control

Red

Amber

Green

Pedestrian Control

Red

Green

Dont Cross

Cross

GPIO 7

GPIO 9

Cross

Sounder

LISTING 16-11: ## Pedestrian Crossing Project

```python
#!/usr/bin/python3
# Pedestrian crossing By Mike Cook
import time, random
import RPi.GPIO as io

LEDcontrol = [23,24,25,4,17,9]
# GPIO for Traffic Red, amber, Green & noCross, cross sounder
crossRequest = 7
trafficSensor = 18
nextSequenceTime = time.time()

def main():
    global lastTraffic, state, bleep
    print("Pedestrian crossing simulator Click the Stop/Restart back end icon quit")
    initGPIO()
    io.output(green,1) # Green light on to start
    state = 0
    bleepTime = time.time()
    bleep = False
    lastTraffic = time.time()
    while 1:
        checkTraffic()
        if checkRequest() and state == 0 :
            io.output(noCross, 1) # turn on the no cross light
            if time.time() - lastTraffic > 10.0: #cross immediately
```

(continued)

LISTING 16-11: *(continued)*

```
                            state = 1
                    else:
                        time.sleep(10.0) # let traffic flow for a bit
                        state = 1
                crossSequenceFunction()
                if bleep and time.time() > bleepTime:
                    bleepTime = time.time() +0.3
                    io.output(sounder, not(io.input(sounder)))

def checkTraffic():
    global lastTraffic
    if io.input(trafficSensor) == 0:
        lastTraffic = time.time()

def checkRequest():
    request = False
    if io.input(crossRequest) == 0:
        request = True
    return request

def crossSequenceFunction():
    global nextSequenceTime, countFlash, state, bleep
    if state == 0:
        nextSequenceTime = time.time() + 2.0
        return
    if time.time() > nextSequenceTime :
        if state == 1: # show amber
            #print("doing state", state)
            io.output(green, 0)
            io.output(amber, 1)
            state = 2
            nextSequenceTime = time.time() + 2.0 #show amber time
        elif state == 2: # show red
            #print("doing state", state)
            io.output(amber,0)
            io.output(red,1)
            state = 3
            nextSequenceTime = time.time() + 2.0 # show red time
        elif state == 3: # show cross light
            #print("doing state", state)
            io.output(noCross, 0)
            io.output(cross, 1)
            bleep = True
            state = 4
            nextSequenceTime = time.time() + 5.0 #crossing time
        elif state == 4: # change to amber clear crossing
            #print("doing state", state)
            io.output(amber, 1)
```

```
                    io.output(red, 0)
                    bleep = False
                    io.output(sounder, 0) # turn off sounder
                    state = 5
                    nextSequenceTime = time.time() + 1.0
            elif state == 5: # flash amber and cross
                    #print("doing state", state)
                    io.output(amber, not(io.input(amber)))
                    io.output(cross, not(io.input(cross)))
                    nextSequenceTime = time.time() + 0.2 #flashing speed
                    countFlash +=1
                    if countFlash > 20 : #clear crossing time 20 * flash speed
                        countFlash = 0
                        state = 6
            elif state == 6: # hold amber
                    io.output(cross, 0)
                    io.output(noCross, 1)
                    io.output(amber, 1)
                    state = 7
                    nextSequenceTime = time.time() + 2.0 # hold amber time
            elif state == 7: # put on red light
                    #print("doing state", state)
                    io.output(amber, 0)
                    io.output(green, 1)
                    io.output(noCross, 0)
                    state = 0
                    nextSequenceTime = time.time() + 1.0


def initGPIO():
    global red, amber, green, noCross, cross, countFlash, sounder
    io.setmode(io.BCM)
    io.setwarnings(False)
    for pin in range (0,len(LEDcontrol)):
        io.setup(LEDcontrol[pin],io.OUT) # make pin into an output
        io.output(LEDcontrol[pin],0) # set to zero
    io.setup(crossRequest,io.IN, pull_up_down=io.PUD_UP) # make input
    io.setup(trafficSensor,io.IN, pull_up_down=io.PUD_UP)
    red = LEDcontrol[0]
    amber = LEDcontrol[1]
    green = LEDcontrol[2]
    noCross = LEDcontrol[3]
    cross = LEDcontrol[4]
    sounder = LEDcontrol[5]
    countFlash = 0

# Main program logic:
if __name__ == "__main__":
    main()
```

Stepping through this program, first note that the GPIO pins are initialized and that each pin in the list is given its own name. This makes the program easy to read, although we could have used a "magic number" — a number that appears for no immediately apparent reason — as an index in the GPIO list. In the infinite `while` loop in the `main` function, traffic activity is first measured and then the Cross Request button is looked at. The function `checkRequest` returns a value of `True` if the button is being pressed. That means the function can be called directly inside an `if` statement — there's no need for any intermediate variables. Also, the `if` statement looks to see if a cross sequence is in operation by looking at the `state` variable. (It's 0 if nothing is running.) If a cross request has been made, a delay occurs if there has been some traffic activity in the last ten seconds, and then the `state` variable is changed to a value of 1.

Next in the `main` loop, the `crossSequenceFunction` is called, which only returns if the `state` variable is 0; otherwise, the function looks to see if it's time to change the state, and returns if it isn't. Finally, in the `main` loop, the `bleep` variable is looked at because it controls when the sounder should make a noise. This variable is set at State 3 of the `crossSequenceFunction` and cleared at State 4. The `crossSequenceFunction`, for most steps, simply turns on or off the various lights according to the required sequence. The exception to this is in State 5, where both the cross light and the amber light need to flash. We do this by toggling the two lights. (*Toggling* means turning off a light that is on, and turning on a light that is off.) Rather than use a variable to tell you if you set the light on or off last, you can use a little trick: Read the state of a GPIO output pin. This then gives you the value you last set it to, and you know you want to use the inverse of this state. Therefore, the line

```
io.output(amber, not(io.input(amber)))
```

sets the amber light to the opposite state to its current state. Before you can get out of State 5 and advance it to State 6, you must have ten flashes of the lights. Finally, in State 7, you set the traffic lights to green for go and return the state machine variable back to State 0, which indicates that the cross sequence is no longer in progress.

Taking it further

There are a number of `print` statements we have commented out, to show the state of the Cross sequence; you might like to uncomment them to see what is going on, although the LEDs should tell you. You can go and measure the times on an actual crossing and replace the times used in the program with realistic times. You can extend the program so that there's a more intelligent traffic-control system, by counting the cars over a set interval — say, the last 20 seconds — and graduating the Wait Before Crossing sequence delay is lengthened accordingly.

Figure 16-16 shows three LEDs mounted as a miniature traffic light. Using this sort of thing adds a bit more realism to the project. Simply search the web for "5mm LED traffic light," although the one in the picture actually has 8mm LEDs — but that's search engines when they try to sell you things.

FIGURE 16-16:
A better traffic light display.

Chapter **17**

Lots of Multicolored LEDs

hapter 16 shows you how to control LEDs in terms of both patterns and sequences. Now we're going to get colorful and look at multicolored LEDs. We use different colors of LED in Chapter 16's Pedestrian Crossing project, where each LED has its own color. Another type of LED contains three different colored LEDs in the same package. This type is known as an RGB LED because the three colors are *red*, *green*, and *blue*.

RGB LEDs come in three basic forms: common anode, common cathode, and separate connections. This last type, the least common, only comes in a small, 6-pin, surface-mount package. (This type is used in the Pimoroni Sense HAT, mentioned in Chapter 15.) The internal configuration of the three types is shown in Figure 17-1.

The dotted box around each LED is the way a schematic shows that they're all in one package. Note that the common anode and common cathode types have four wires going into them instead of the normal two wires for a single-color LED. Each LED can be treated by the software as a separate LED. To use common anode LEDs, you have to wire them up so that the GPIO pins act as current sinks, that is making the connection to ground. The wiring for the two common "something" LED types is shown in Figure 17-2.

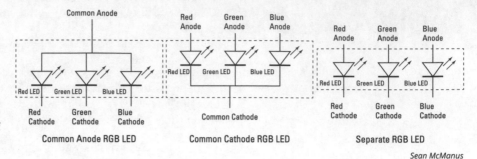

FIGURE 17-1:
Three types of basic RGB LEDs.

Sean McManus

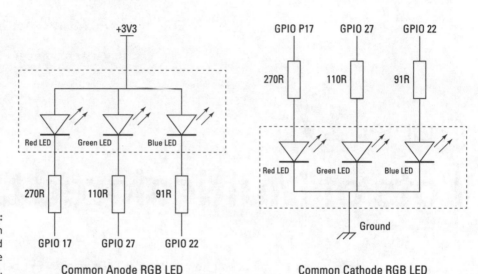

FIGURE 17-2:
Wiring common anode and common cathode LEDs.

With a common cathode, the GPIO pins should be set to a logic high to turn the LED on, and to a zero to turn it off. (For more on logic highs and zeros, see Chapter 16.) However, with a common anode LED, this situation is reversed, with a logic low turning on the LED by sinking current and making a complete circuit that way. When confronted by Figure 17-2, a beginner typically has this question: "Do I need to have three resistors? Why can't I just have one in the common line?" — to which the answer is, "Yes, you need all three resistors."

Here's why. If you only have one resistor, then the voltage level that the resistor will drop depends on the current running through it. That means the brightness will change depending on which LEDs are turned on. Even worse, if the red LED is on, there's not enough voltage across the other two LEDs to turn them on. This is because each color of LED drops a different voltage when it is on, and red drops the lowest voltage. It will appear that the other colors are not working — not good.

Making Colors

With the wiring schematics out of the way, it's time to turn to the fun stuff: mixing colors. Yes, you can turn on just one LED at a time, and that will give you one of the three colors — red, blue, or green. However, turn two on and the colors will mix in a way known as *additive mixing*. The primary colors are red, green, and blue. Mixing them together gives you other colors, the so-called *secondary* colors. So a red-and-green light together make a yellow one. A green-and-blue light make a cyan color, and a red-and-blue light make magenta. Turn on all three LEDs together and you make white — or, more precisely, they make a white tint.

In theory, all three LEDs on together will make white, but in practice this depends on the exact brightness of the three separate lights. They have to be identical to make a pure white; otherwise, the white looks tinted, which is not altogether a bad thing. It's not an easy task to do this, because each color has a different forward voltage drop — you need different resistors to ensure the same current through each LED. Not only that, each color of LED converts current to brightness with a different efficiency, which complicates things tremendously.

There are new types of LED that include a white LED alongside the normal red, green, and blue LEDs. These can be used to produce a better range of pastel colors and more subtly tinted white lights.

You might be used to mixing colored paint, but keep in mind you get different results doing this than mixing light. Paint mixing is known as *subtractive mixing* because each paint color you put into the mix takes out (or subtracts) some other color. This is how your color printer works. For subtractive mixing, the primary colors are cyan, magenta, and yellow; red, green, and blue being the secondary colors.

Using diffusers

The light from three LEDs in the same package will still be seen as three separate points of light, unless there is some sort of diffuser, which allows the light to mix evenly. For individual RGB LEDs, this is sometimes provided by the package or body of the LED itself. Viewing distance alone can provide enough diffusion to mix the colors, but often a diffuser of some sort will help. Diffusers also reduce the brightness per unit area of the LED, making it much easier to take a good color picture.

You can use anything that is translucent as a diffuser. Our favorite is a very thin styrene sheet — about 0.5mm thickness is fine and is easy to work with, because it can be cut with scissors. (A good alternative is a simple sheet of paper.) If you have several LEDs and want to see the light from each distinctly, then you have to have each one surrounded by a light baffle — sometimes known loosely as an *egg box*, or *waffle box*. Without the baffle, the light from each LED mixes with those adjacent to it and gives a soft focus effect that is not at all unpleasant. The degree of diffusion you get is proportional to not only the diffusing material but also the distance of that material from the LED. In most cases, a few millimeters is fine.

TIP

You can turn the clear-plastic housing LED into a diffuser by rubbing it gently with very fine sandpaper or wire wool. Even better is to use a foam-backed sanding block, because it gets round the curves much better than paper. These LED housings are made of resin, so solvents like acetone do not affect the surface.

Making more colors

The trick to making more colors than the simple primary and secondary colors is to have different brightness of each of the three colors. In that way, many more subtle colors can be made. So how can you control the brightness of an LED? Well, the answer might not be immediately apparent, but what you need to do is to turn the LED on and off very rapidly. If you do this fast enough — that is, faster than about 30 times a second — then the eye/brain sees this as a light that is constantly on and not flickering. Furthermore, you perceive the brightness of the LED according to the relative length of the On and Off times. That is, if the LED is on and off for equal times, the LED appears to be only half as bright. This rapid switching technique is known as PWM — short for *Pulse-Width Modulation* — and is the way you control the LED's brightness. The waveforms are shown in Figure 17-3.

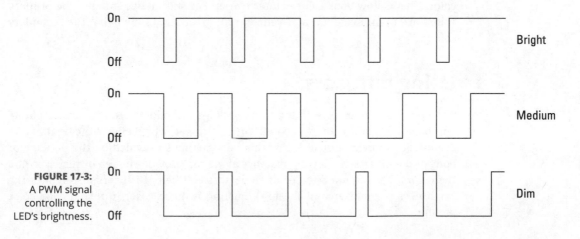

FIGURE 17-3: A PWM signal controlling the LED's brightness.

You can see that the three PWM signals go on and off at the same speed; however, the one that spends more time being on is brighter than the one that spends only half the time being on. Finally, the last waveform has a little time on but a long time off and produces a dim LED. The ratio of the On time to the Off time is known as the *duty cycle* of the waveform. Note that the frequency of this PWM signal does not matter once it is above the rate where you see it flicker.

The RPi.GPIO library has the ability to make the GPIO pins output a PWM signal, and the library can set both the frequency and duty cycle. If you wire up an RGB LED according to Figure 17-3, you can test out the colors an RGB LED can produce with the code in Listing 17-1.

LISTING 17-1: | **RGB Color Test**

```python
#!/usr/bin/python3
import time
import RPi.GPIO as io

io.setmode(io.BCM)
io.setup(17,io.OUT) # make pins into an output
io.setup(27,io.OUT)
io.setup(22,io.OUT)

ledR = io.PWM(17,60) # Set up outputs as PWM @ 60Hz
ledG = io.PWM(27,60)
ledB = io.PWM(22,60)

ledR.start(0) # start off the PWM
ledG.start(0)
ledB.start(0)

print("RGB color cycle an LED using RPi.GPIO - By Mike Cook")
print("Press Ctrl+C to quit ")
try:
    while(1):
        print("Start cycle")
        time.sleep(2.0)
        for stepR in range(0,100,5):
            for stepG in range(0,100,5):
                for stepB in range(0,100,5):
                    ledR.ChangeDutyCycle(stepR)
                    ledG.ChangeDutyCycle(stepG)
                    ledB.ChangeDutyCycle(stepB)
                    time.sleep(0.1) # Whole cycle 8000 times this
                    ledR.ChangeDutyCycle(0)
```

(continued)

LISTING 17-1: *(continued)*

```
                    ledG.ChangeDutyCycle(0)
                    ledB.ChangeDutyCycle(0)

except KeyboardInterrupt:
    pass

ledR.stop(0) #stop the PWM
ledG.stop(0)
ledB.stop(0)
io.cleanup() # Restore default GPIO state
```

When you walk through this listing, you see that the first thing the code does is set three GPIO pins to be outputs and then set them up to produce a PWM signal. The value 60 in these opening lines of code is the frequency 60 Hz, which is the frequency the PWM signal will go at. The duty cycle goes from 0, which is off all the time, to 100, which is on all the time. The main part of the program consists of three nested for loops, which ensure that all combinations of red, green, and blue are produced in duty cycle steps of five. It takes 800 seconds — just over 13 minutes — to do a complete cycle where 8,000 colors are produced. It might not look like that many colors when you test your LED output running this code, but they are all there. It's just that many of them from this demonstration might look the same. This has to do with the way people perceive colors — they're much more sensitive to the difference between two colors than to the colors themselves.

The Way Forward

The problem with individual LEDs used to be that they took up a lot of the Raspberry Pi's resources, both in terms of GPIO pins — three per LED — and the software needed to generate the PWM cycles. This problem has been solved in the last few years with the advent of LEDs that contain their own PWM generators that you can control from the computer. You only need to tell these sorts of LEDs once what color to produce and they will carry on producing it until you tell them to stop. Even better, these LEDs can be *chained:* Once the LED has its instructions from the computer, it passes any further instructions to the next LED down the line.

TIP

The two major types of LED with these capabilities are the WS2812b and the APA102C. The Adafruit company brands these as NeoPixels and DotStar, respectively. (They are known generically as *addressable* LEDs.) Note that the SK9822 LED is identical to the APA102C.

Admittedly, these LEDs aren't much to look at. They're most commonly packaged in 5mm-square surface-mount packages, as shown in Figure 17-4. The tiny black squares are the actual chips containing the PWM generator and the memory to hold the RGB values. The blank-looking areas are the parts where the light is generated.

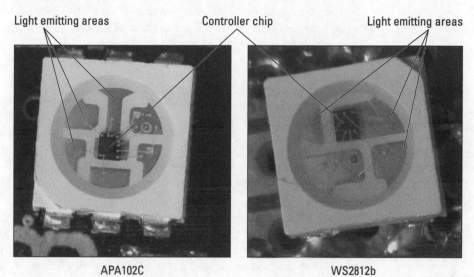

FIGURE 17-4:
DotStar
(APA102C) and
NeoPixel
(WS2812b) LEDs.

You can get NeoPixels in conventional leaded LED packages — not the surface mount type, in other words — as well as the DotStar's 5mm-square packages. Both are available as individual parts as well as long strips of LEDs. But the major difference between the two lies in how they take their commands from the computer. The NeoPixel requires one signal wire, and the DotStar requires two. Given that fact, you might be forgiven for thinking that the NeoPixels are easier to use. The problem is twofold:

>> This one signal wire needs to be controlled to give a precisely timed pulse.

>> The Raspberry Pi, with its Linux operating system, is not good at precise timing.

There are ways around this with the help of various libraries, but each workaround comes at a price. The original library from Adafruit only works with the original Model 1 Raspberry Pi as well as the Pi Zero. The new method Adafruit uses to drive its Neopixels uses its Circuit Python framework, an attempt to unify all its many drivers. This works well on all models, but it requires the Pi to be in supervisor mode before it will work.

The DotStar LEDs requires two signals and uses the sequence of transitions that these signals make to drive the data transfer process. This is ideal for a system like the Raspberry Pi, where, because of Linux, there might be a longer-than-expected delay before the next line of your code is executed. The APA102C LEDs are wired together in a chain, as shown in Figure 17-5, with the signals regenerated by each LED. Therefore, the signals never "get tired," or, as we say in electronics, *degrade*. The DotStar LEDs also generate a faster PWM signal than the Neopixel ones, making them better for persistence of vision projects where the LEDs are moved rapidly or provide illumination for videos.

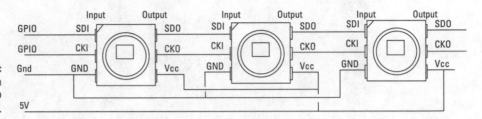

FIGURE 17-5:
Wiring of an APA102C LED chain.

Each LED has a Data In and a Data Out line as well as a Clock In and Clock Out. So the inputs of the first chip are all that is connected to the GPIO pins of the Raspberry Pi. Both types of LED (Neopixel and Dot Star) work on the principle that once an LED has received and stored its own data, any further data it receives is passed on to the next LED. So, after the first set of data is sent to the first LED, there has to be some sort of reset condition in place in order to recognize situations where new data is meant for the first LED and is not just data for LEDs further down the chain. This reset is done by pausing the data stream for greater than 50uS (micro seconds) on the WS2812b LEDs and by sending a stream of start and end pulses in the case of the APA102C LEDs. (The wiring of the WS2812b is very similar to the wiring of the APA102C except that there's only one data wire in and out.)

Figure 17-6 shows the *timing diagram* for each type of LED — a picture of how a logic signal will change in order to indicate logic one, logic zero or the end of a data set.

For the WS2812b, in order for the single wire to write a Logic One to the LED, the signal wire makes a jump to a Logic One and stops there for 3.5uS before jumping down to a Logic Zero for 0.9uS. For a Logic Zero, the signal jumps to a one for 0.9uS and then drops to low for 3.5uS. This repeats until all the data has been transferred to the LEDs. The end of the data transmission is signaled by a zero being held for at least 50uS. These times have to be +/– 0.15uS to ensure correct operation.

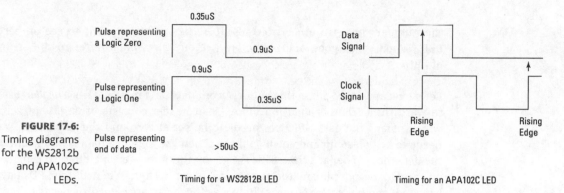

The APA102C, on the other hand, has two signals, called *clock* and *data,* which can go at any speed. But the important point is that when the clock signal *rises* — transitions from a zero to a one — whatever logic level is on the data input, at that moment, is treated as the logic input to the LED. Doing this requires nothing special about the timing; you just have to put the GPIO pins at the correct logic levels in the right order.

Programmers have a special name they use to describe implementing a protocol in this way — it's known as *bit-banging.* (The technical term for this protocol is SPI, or *Serial Peripheral Interface.*) There's some hardware on the processor chip to do this, but it requires a specific set of pins running in one of the alternative modes. The great thing about bit-banging is that you can use any pins. Let's see how we can bit-bang this protocol on the GPIO bus.

Bit-banging the APA102C protocol

The LEDs used in our example are connected together in a string, where one LED takes its input from the output of the previous one. The first LED in the string takes its data from the Pi. Each LED along the string can be thought of as having a number or an address. When we send out data, we have to send it to all the LEDs, but we need to change the data only for those LEDs that we want to change. If the data for an individual LED is the same as last time, it will not change. That way, we can change individual LEDs by only changing data for those LEDs we want to change and leaving the data for the others, the same as last time.

The APA102C — in addition to the normal red, green, and blue PWM values, which control the brightness — has *another* brightness-controlling number, which controls the current down the LED. This is the equivalent of changing the resistor value. Valid brightness-control numbers here range from 0 to 31. Each LED needs data consisting of four numbers: brightness plus red, green, and blue values. This data needs to be stored in the computer in what is known as a *buffer.* In Python, we can implement this buffer as a list. We can then manipulate this list — change the numbers in it, in other words — to reflect what we want to see on each LED. When we're good and ready, we fire out the whole buffer to the string of LEDs.

Let's look at what to put in this buffer, A complete data message has a *header*, used to warn the LEDs that data is coming. That header consists of 32 bits of zero, which means the data GPIO pin needs to be set at zero and the clock GPIO pin needs to be pulsed up and down 32 times. Then the message data is pulled out of the list, one number at a time. This is a 32-bit number — each bit in turn is placed on the data output pin and the clock signal is set to High. When all the LEDs have been fed, another series of zeros, known as the *footer*, are sent. This needs to consist of a number of pulses — at least 32 plus half the number of LEDs, to be precise. The code in Listing 17-2 is a fragment of code that would do this.

LISTING 17-2: **Bit-Banging the Data to the LED**

```
io.output(da,io.LOW) # set data pin low
for i in range(0,32): # send header
  io.output(ck, io.LOW) # pulse the clock pin
  io.output(ck, io.HIGH)
for i in range(0,numLeds): # send data
    d = ledArray[i] # get a single LED's worth of data
    for j in range(0,32):
      io.output(ck, io.LOW)
      if d & 0x80000000 :
        io.output(da, io.HIGH)
      else:
        io.output(da, io.LOW)
      d = d << 1
        io.output(ck, io.HIGH)
io.output(da, io.LOW)
for i in range(0,33+(numLeds/2)): # send footer
  io.output(ck, io.LOW)
  io.output(ck, io.HIGH)
```

Note that this is a fragment only and not a complete program. Note also that this code uses the shift operator (<<), which moves the *data word*, or pattern, one place to the left so that the next byte to send is in the most significant bit of the word. This is the same technique we show you how to use in forming the dice pattern in Chapter 16.

Creating a class

The bit-banging code is so useful that it could be included in all programs where you want to use these LEDs. However, in order to make it convenient to use, we can make this code into a class so that you don't have to keep including these lines in your own programs. This is just like the RPi.GPIO library, which is written as a class and installed in Python. You can write your own classes and have them included in the language or, as we do here, just have the class file in the same folder as the program that uses it and then call it up at the start of the program. As well as outputting the data to the GPIO pins, we can bundle other useful stuff in the class, like ways to set the brightness, set an LED's color, and set up the data buffer. These functions of a class are known as *methods* and can be simply invoked. Listing 17-3 shows the implementation of a bit-banging class for the APA102 LED. Save it in a file called apa102bang.py in the same folder as the other code from the rest of this chapter.

LISTING 17-3: **Bit-Banging the APA102 Class**

```python
#!/usr/bin/env python3
# Class for driving APA102 LEDs
#By Mike Cook

import RPi.GPIO as io
io.setwarnings(False)

class Apa102bang(): # Define our class

    def __init__(self,numberLeds,data,clock,bright):
        self.setBrightness(bright)
        self.da = data
        self.ck = clock
        self.numLeds = numberLeds
        io.setmode(io.BCM)
        io.setup(self.ck,io.OUT)
        io.setup(self.da,io.OUT)
        io.output(self.ck, io.HIGH)
        io.output(self.da, io.HIGH)
        self.ledArray = [self.br<<24 for i in range(0,self.numLeds)]

    def setBrightness(self,brightness):
        if brightness > 31:
            brightness = 31
        if brightness < 0:
            brightness = 0
        self.br = brightness | 0xE0
```

(continued)

LISTING 17-3: *(continued)*

```python
    def setLed(self,pos,col):
        if pos < self.numLeds and pos >= 0:
            self.ledArray[pos] = (self.br<<24)|(col[2]<<16)|(col[1] <<8)|col[0]

    def setAll(self,col):
        for i in range(0,self.numLeds):
            self.ledArray[i] = (self.br<<24)|(col[2]<<16)|(col[1] <<8)|col[0]

    def show(self):
        io.output(self.da,io.LOW)
        for i in range(0,32): # send header
            io.output(self.ck, io.LOW)
            io.output(self.ck, io.HIGH)
        for i in range(0,self.numLeds):
            d = self.ledArray[i] # send data
            for j in range(0,32):
                io.output(self.ck, io.LOW)
                if d & 0x80000000 :
                    io.output(self.da, io.HIGH)
                else:
                    io.output(self.da, io.LOW)
                d = d << 1
                io.output(self.ck, io.HIGH)
        io.output(self.da, io.LOW)
        for i in range(0,33+(self.numLeds>>1)): # send footer
            io.output(self.ck, io.LOW)
            io.output(self.ck, io.HIGH)
```

Notice how the class name is nearly the same as the filename? The only difference is that the filename is in all lowercase letters. Keeping this distinction is important because it allows classes to be recognized correctly. Each method is defined just like a function with a def statement, and the whole class is initialized with a def __init__ function. The other thing you will notice is that a lot of the variables start with self. This is to tell the compiler that this variable is one that belongs to the class itself and that these can actually have different values, if the class has more than one instance.

You can usually tell what each method will do just by looking at the function names. For example, when you use the setBrightness method, all that happens is a variable is set that you can then use to set the brightness for future LED settings. (Note that it will not affect the current brightness of all the LEDs.) You could write a method that does set the current brightness, if you want — that would involve adding the new brightness level to all LEDs in the list and calling the show method.

To see this in action, we use the Pimoroni Blinkt! LED strip, a low-cost strip of eight APA102C LEDs that fits over all 40 GPIO header pins (even though it only uses two GPIO pins in addition to power [5V] and ground). Listing 17-4 shows a program designed to use the class. (Be sure to save this to the same folder as the class file.)

LISTING 17-4: **Using the APA102 Class**

```
#!/usr/bin/env python3
#demo1 using Apa102bang class driver

import time
from apa102bang import Apa102bang

length = 8 ; brightness = 4
dataPin = 23 ; clockPin = 24 # Blinkt! wiring
leds = Apa102bang(length,dataPin,clockPin,brightness)

def main():
    print("APA102 demo - Click the Stop/Restart back end icon quit")
    while True:
        for i in range(0,8):
            leds.setLed(i,(120,0,0)) # red
            leds.show()
            time.sleep(0.06)
        time.sleep(0.6)
        for i in range(0,4):
            leds.setLed(i,(0,120,0)) # green
            leds.show()
            time.sleep(0.06)
        time.sleep(0.6)
        for i in range(4,8):
            leds.setLed(i,(0,0,120)) # blue
            leds.show()
        time.sleep(0.06)
        time.sleep(0.6)
for i in range(0,8):
            leds.setLed(i,(0,0,0)) # black
            leds.show()
            time.sleep(0.06)
        time.sleep(0.6)

# Main program logic:
if __name__ == "__main__":
    main()
```

SOURCING SOUNDS

A good source of sounds used to be those sounds used by the Scratch language, which were kept in the directory /usr/share/scratch/Media/Sounds. Unfortunately, this is no longer the case with the newer releases of the language. However, with a small bit of work, you can get access to them again. Follow these steps:

1. **Use the File Manager to create a folder at** /home/pi **called** SoundResources.

    This is where you'll put the sounds.

2. **Open the Chrome web browser, and in the address bar, type** https://download-ib01.fedoraproject.org/pub/fedora/linux/releases/33/Everything/source/tree/Packages/s/scratch-1.4.0.7-21.fc33.src.rpm.

3. **After the file is finished loading, using the File Manager go to the** Downloads **folder, right-click the newly downloaded file, and choose the Extract Here option.**

4. **Right-click the** scratch-1.4.0.7.src.tar.gz **file and again choose the Extract Here option.**

5. **Double-click the** scratch-1.4.0.7.src **folder, and then double-click the** Media **folder.**

6. **Right-click the** Sounds **folder and choose Copy.**

7. **Navigate to the** SoundResources **folder you just created, right-click, and choose Paste.**

    You've now copied all the sounds from Scratch into your folder.

8. **Select all the newly created files in your** Downloads **folder, right-click, and choose Move to Trash.**

9. **If you don't like having things in your Trash, right-click the Trash and choose Empty Trash.**

The code starts off by defining the number of LEDs you have, the brightness you want to run them at, and the pins used for the data and clock lines. These lines are determined by the way the Blinkt! is made, but these could be any two GPIO pins, if you want to experiment with more LED strips connected to the Pi. The leds variable is set to be the class reference — use this reference followed by a dot and the method's name to call up any method. The program simply sets each LED in turn to the three primary colors and back again to black. Note how in the setLED method the colors are given as three values in a tuple. It is the show method that actually displays the contents of the data buffer on the LEDs. If you don't call this method, the LEDs do not change.

Though this is all very interesting, the results of running this program are not very spectacular — so let's make a game with this small LED strip.

Rainbow Invaders

Rainbow Invaders plays a bit like the old Space Invaders game, only in one dimension. The alien invaders drop bombs of various colors onto your base in an attempt to destroy you, but you can neutralize these bombs by sending up a beam of the same color. You have three buttons to choose from — red, green, and blue — and pushing a button launches a beam of that color. Sounds easy, right? Not so fast. The invaders can also drop bombs of the secondary colors as well as white, so you have to press the right combination of your three buttons to generate the right color beam. If three bombs get through, then your base is destroyed. As your base gets more hits, its color changes. The higher up a bomb is when your beam hits it, the higher you score. Add a few sound effects and this is the game.

To get this game off the ground, the first thing you have to do is to build the hardware. In this case, this means adding three push buttons to the Blinkt! LED strip. Unfortunately, this is a bit trickier that it could be. This is because the LED strip takes up all the GPIO pins — even the ones it doesn't use. There are two ways round this problem: You could solder wires on the back of the Blinkt! and bring these out to your push buttons, or you could use an extender board, like the Black HAT Hack 3R or the Mini Black HAT Hack 3R. Figure 17-7 shows where to solder the wires to the back of the Blinkt!, and Figure 17-8 shows the Black HAT Hack 3R extender board.

Your base in the game is the righthand LED, but with the Black HAT board, you can position this LED so that your base is at the bottom of a vertical line.

We wrote the software using the Pygame framework that comes preloaded with the Pi's operating system, because it's a great framework for handling sound effects. You should store your sound samples in a directory you create and then name sounds, in the same directory as this game file. Go to your SoundResources directory, then into the Sounds one. Copy the sounds ComputerBeeps2, Laser1, and Screech one at a time from the Electronic directory; and then put them in the sounds directory you just created. Do the same for the Pop sound from the Effects folder of your sound resources. The code for the game is in Listing 17-5.

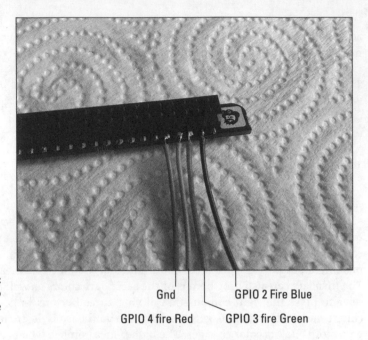

FIGURE 17-7:
Soldering wires to the back of the Blinkt! strip.

Gnd

GPIO 4 fire Red

GPIO 2 Fire Blue

GPIO 3 fire Green

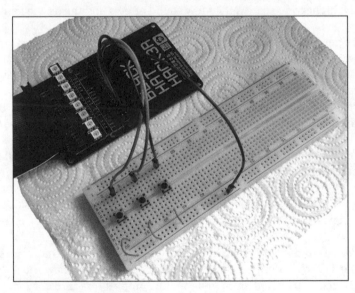

FIGURE 17-8:
Black HAT Hack3R, giving two sets of GPIO pins.

LISTING 17-5:	**Rainbow Invaders Game**

```
#!/usr/bin/env python3
# Rainbow Invaders by Mike Cook

import RPi.GPIO as io
import pygame
```

```
import time, random
from apa102bang import Apa102bang

def main():
    global beam,gameOn,bombHeight,bombCol
    print("Rainbow Invaders Press Ctrl+C to quit")
    init()
    gameOn = True ; yourScore = 0
    bombHit = False ; bombHeight = 8
    hits = 3 ; sky.setLed(0,base[hits])
    sky.show() ; time.sleep(gameSpeed)
    nextBombMove = time.time()+gameSpeed
    bombCol = (0,0,0) ; setBombCol()
    while 1:
        while gameOn:
            if time.time()>nextBombMove:
                bombHeight -=1
                showBomb(bombHeight)
                nextBombMove = time.time()+gameSpeed
            yourScore = fireBeam(bombHeight,yourScore)
            if bombHeight == 0 :
                sky.setAll((0,0,0))
                sky.setLed(0,(255,0,0))
                sky.show() # explosion
                gameSound[3].play() # base hit
                time.sleep(2.5) # time to play
                hits -= 1
                if hits == 0:
                    gameOn = False
                else:
                    print("Base hit:-",hits,"from destruction")
                    sky.setLed(0,base[hits])
                    setBombCol()
                    bombHeight = 8
        sky.show()
        print("Base destroyed:- your score",yourScore)
        time.sleep(gameSpeed * 10)
        print("Push any button for a new game")
        while io.input(fireR) and io.input(fireG) and io.input(fireB):
            pass
        print("New game starting")
        hits = 3
        gameOn = True
        bombHeight = 8
        sky.setLed(0,base[hits])
        sky.show()
        setBombCol()
        while (not io.input(fireR)) | (not io.input(fireG)) | (not io.input(fireB)):
            pass
```

(continued)

LISTING 17-5: *(continued)*

```
def getBeam():
    m = 128
    b = ((not io.input(fireR))*m,(not io.input(fireG))*m,(not io.input(fireB))*m)
    return b

def showBeam(col,limit):
    for i in range(1,limit):
        sky.setLed(i,col)
        sky.show()
        time.sleep(0.05)

def showBomb(pos):
    global bombCol
    showBeam((0,0,0),8)
    sky.setLed(pos,bombCol)
    sky.show()

def fireBeam(far,score):
    global beam,bombCol,bombHeight
    if getBeam() != beam: # laser fire
        beam = getBeam()
        if beam != (0,0,0):
            gameSound[1].play()
        showBeam(beam,far)
        if beam == bombCol : #hit OK
            time.sleep(0.2)
            gameSound[2].play() #hit sound
            score += far
            time.sleep(1.5) # allow hit to finish
            bombHeight = 8
            setBombCol()
    return score

def setBombCol():
    global bombCol
    gameSound[0].play() # bomb incoming
    while 1 :
        r = random.randint(0,1) * 128
        g = random.randint(0,1) * 128
        b = random.randint(0,1) * 128
        if not(r==0 and g==0 and b==0) and bombCol != (r,g,b) : break
    bombCol = (r,g,b)

def init():
    global fireR,fireG,fireB,beam,sky,gameSpeed,base,gameSound
    random.seed()
    io.setmode(io.BCM)
```

```
fireR = 4 ; fireG = 3 ; fireB = 2
io.setup([fireR,fireG,fireB],io.IN, pull_up_down=io.PUD_UP)
beam = getBeam()
length = 8 ; brightness = 4
sky = Apa102bang(length,23,24,brightness)
gameSpeed = 0.3 # the step speed of game in seconds
pygame.mixer.quit()
pygame.mixer.init(frequency=22050, size=-16, channels=2, buffer=512)
base = [(0,0,0),(32,0,32),(0,64,0),(0,128,128)] # base colors
soundEffects = ["ComputerBeeps2","Laser1","Pop","Screech"]
gameSound = [pygame.mixer.Sound("sounds/"+ soundEffects[sound] + ".wav") ↩
    for sound in range(0,4)]

# Main program logic:
if __name__ == "__main__":
  try:
    main()
  except KeyboardInterrupt:
    pass
# turn off all LEDs
sky.setAll((0,0,0))
sky.show()
```

The main function, consisting of two large loops, is where most of the action happens. The inner loop `while gameOn` runs a round of the game, where bombs are sent down until there have been three hits to the base — in other words, when a bomb reaches a height of zero. The first `if` statement checks to see whether enough time has passed to move the bomb another step closer to the base. The call to the `fireBeam` function checks to see whether buttons are being pressed to generate a beam, and if they are, the beam is fired off and a laser sound is triggered. If the beam color matches the bomb color, then the hit sound is triggered and the current bomb height is added to your score. The round is then restarted by setting the bomb height back to 8 and a new bomb color is generated. Then the program returns to the main loop.

If a bomb is found to have reached the base, then the base is turned red and the explosion sound is triggered. The `hits` variable is decremented, and if it has reached zero, the round is over and the `gameOn` variable is set to `false`, thus ending the current loop. Then the final lines of the `while 1` loop displays the score and waits until you have pressed a button before starting the game over again.

The array of LEDs here is called `sky`, which is descriptive enough, and the overall speed of the game is controlled by the `gameSpeed` variable in the `init` function. So far, we've been advising you to stop Thonny by clicking the stop icon, which is the best way to stop. However, there are other ways to quit a running program. When you quit the game by pressing Ctrl+C. This a Keyboard Interrupt occurs, this allows us to do some tidying up after we've finished. In this case, all LEDs are turned off to clean up the display. However, this method of stopping sometimes leaves Thonny hanging (unresponsive); when this happens, click the stop icon as well.

TIP

You can tinker with the game speed and the number of hits your base can stand, but you can also make the game a lot harder. If you add a fourth push button, you can make this a half-brightness button. That means to match the bomb color, you have to press the right combination of red, green, and blue and also the brightness. You also need to feed this variation in brightness into the generation of the bomb color. It isn't hard, but it does require a bit of thinking.

Keepy Uppy

Another game you can play on exactly the same hardware is Keepy Uppy, the well-known football pastime and skill demonstration where you have to keep a football in the air all the time, just by kicking it up. The ball is a moving LED, but unlike the game in the previous section, it's a fixed color and you have to press a single button during the time when the ball is on your "foot" — the bottom LED, in other words. If you time it right, the ball moves up again until it reaches the top and then descends. A further challenge in playing is that if the player's actual button press is in the second half of the time allotted (that is, *just* in time), then the game speed increases, making it harder to kick the ball as soon as it can be kicked. Kick it in the first half of the allotted time and the game speed reverts to the original speed. This makes it more interesting and difficult to play. The code for the game is in Listing 17-6.

LISTING 17-6: **Keepy Uppy Game**

```
#!/usr/bin/env python3
# KeepyUppy by Mike Cook

import RPi.GPIO as io
import time, random
from apa102bang import Apa102bang
stripLength = 8

def main():
  global beam,gameOn,ballHeight,ballCol
  print("Keepy Uppy")
  init()
  gameOn = True
```

```python
ballDirection = -1; ballHeight = stripLength
foot = 0 ; kicks = 0
currentGameSpeed = gameSpeed
leds.setLed(0,footCol[foot]) # foot color
leds.show()
time.sleep(gameSpeed)
nextBallMove = time.time()+gameSpeed
lastKickMade = not(io.input(kick))
while 1:
  while gameOn:
    kickMade = not(io.input(kick))
    if kickMade and not(lastKickMade) :
      if ballHeight == 0: # kick
        ballDirection = ballDirection * -1
        #speed up if you are late pressing
        if nextBallMove - time.time() < currentGameSpeed/2 :
          currentGameSpeed = currentGameSpeed / 2
        else:
          currentGameSpeed = gameSpeed
        nextBallMove = time.time()-1.0
        foot = 1
        kicks += 1 # add to your score
      else:
        foot = 2
    if time.time()>nextBallMove:
      leds.setLed(ballHeight,(0,0,0))
      ballHeight +=ballDirection
      leds.setLed(0,footCol[foot])
      leds.setLed(ballHeight,ballCol)
      foot = 0
      nextBallMove = time.time()+currentGameSpeed
    if ballHeight >= stripLength-1 :
        ballDirection = -1
    if ballHeight < 0: #missed the ball
        gameOn = False
    lastKickMade = kickMade
    leds.show()
  print("Ball missed your score",kicks)
  while not(io.input(kick)):
      pass
  time.sleep(gameSpeed* 5)
  print("Press kick for a new game")
  while not(io.input(kick)):
      pass
  kicks = 0
  gameOn = True
  ballHeight = stripLength
```

(continued)

LISTING 17-6: *(continued)*

```
        ballDirection = -1
        leds.show()
        currentGameSpeed = gameSpeed
        while io.input(kick):
            pass

def init():
    global kick,ball,leds,gameSpeed,footCol,ballCol
    random.seed()
    io.setmode(io.BCM)
    kick = 4
    io.setup(kick,io.IN, pull_up_down=io.PUD_UP)
    brightness = 4
    dataPin = 23 ; clockPin = 24 # Blinkt! wiring
    leds = Apa102bang(stripLength,dataPin,clockPin,brightness)
    gameSpeed = 0.3 # step speed of game
    ballCol = (128,128,0)
    footCol = [(32,32,32),(128,0,0),(0,0,128)] # foot colors

# Main program logic:
if __name__ == "__main__":
  try:
    main()
  except KeyboardInterrupt:
    pass
# turn off all LEDs
leds.setAll((0,0,0))
leds.show()
```

Here, again like the previous program, the main structure of the code can be found in the two `while` loops in the `main` function. One issue that needs to be addressed is that the push button could just be held down so that it would register as being pressed all the time — and so always kick the ball at the right time. To prevent this from happening, you have to implement a state-change detector on the button. This involves looking at the current state of the push button and comparing it with the previously read value. Only if the previous value indicated Unpushed and the current value equals Pushed does the program recognize that the button has just become pushed. We say this is an *edge detection* because, rather than look at the simple state of the push button, we can detect when the state changes. (If this is drawn on a diagram, this transition looks like an edge.)

For each successful kick, you get one point added to your score, just like in the real game. The `nextBallMove` variable is the time the ball moves next. By subtracting the current time from this, you can tell how quickly the kick has been detected in the time period allotted to make a successful kick. This is used to speed up or restore the original speed of the ball movement.

TIP

If you want to take things further, one thing you can do is make the strip longer. The game was designed for an 8-LED strip, but simply by changing the strip-Length variable at the start of the code and adding a longer strip, you can make the ball go higher. You can get much longer strips than the simple Blinkt! 8-LED strips, as you shall soon see. You can also try using some sounds with the game.

LEDs Galore

The great thing about LEDs like the APA102C is that you can control lots of LEDs with just the two connections, "almost" without limit. So, before you consider projects that involve hundreds of LEDs, let's look at some of those limits, and some ways round them.

Current limits

Current is the first and biggest limitation when trying to drive lots of LEDs. Sure, one LED only draws a small amount of current, but any small number when multiplied by a big number starts to be important. Suppose that you're running your Raspberry Pi off a 2A power supply —roughly twice the peak current that the Pi needs. You should never take the maximum rated current out of a power supply. Instead, always keep the maximum current at about 80 percent of rated current. With a 2A power supply, that ends up being 1.6A. It is therefore reasonable to say that you have 0.6A, or 600mA (milliamps), of current capacity in a typical setup. If each LED takes 60mA, that is 20mA for each color. With that setup, you can drive a maximum of 10 LEDs with a white color at full power.

Now, APA102C LEDs are very bright and would be overpowering at close range. Luckily, there are two ways to control the brightness, and if it's controlled, you use less current and so can drive more LEDs. The first way is to scale down the PWM values, so instead of using a value of 255 for a red, you'd simply use 128 and that will use half the current.

WARNING

Easy enough to say, but programming mistakes can easily happen and put unintended values into the PWM control. It's also the case that using PWM doesn't limit the peak current the LED takes — only the *average* current. Nevertheless, the PWM control is capable of splitting the 20mA current per LED into 78uA (micro-amp) steps.

A better way of controlling the brightness of the APA102C is by using the brightness control. This applies to each LED individually along with the PWM values. So far, we have just showed you how to set the brightness at a fixed level and then ignored it — this is a good strategy because it means you have a great deal of protection from programming errors. The brightness control actually limits the current through the LEDs and is controllable in 31 steps. Full current is given by a brightness value of 31, and other values of brightness give the number of 31ths of full current. So, if a single LED takes a maximum of 60mA, then with a brightness of 8, it will only take a maximum 15.48mA. So, the spare 600mA can drive 38 LEDs. Even so, a brightness of 8 is very bright when looking directly at an LED. A brightness of 2 pushes up that number to 155 LEDs and still looks good.

Anything over 155 LEDs and you will have to resort to an external power supply. It's quite easy to use: You just supply the 5V and ground to the LEDs and make sure the ground of the power supply is connected to the ground of the Raspberry Pi.

Signals and memory

The two other limitations you have to deal with concern signals and memory. Fortunately for you, there's no shortage of memory on the Raspberry Pi, unlike other embedded controller boards, like the Arduino. There is cause for concern, however, when it comes to signals. The Blinkt! board used earlier in this chapter connected the GPIO signals directly into the chain of LEDs. The problem is that the LEDs, because they're powered with 5V, need a 5V logic-level signal to drive them. It just so happens that the LED's specifications state that the minimum level sufficient to be seen as a Logic One is 0.7 times the supply voltage. For a 5V supply voltage, this means a logic high of 3.5V, which is just over the 3.3 volts you get from a GPIO pin. It turns out that it does seem to work anyway, but it isn't guaranteed to work in all conditions and temperatures. That means a driver circuit is often included to boost the clock and data signals to 5V. This becomes more important the further away the first LED is from the Pi — and the more interference there is in the local environment from things like motors or fluorescent light fittings. Figure 17-9 shows a signal driver you can make yourself with a 74LS14 or a 74AHCT14 chip.

Basically, this chip acts as an inverting buffer, but we don't need to invert the signals, so we just pass each signal through two buffers to keep it the right way up. As we now have a 5V signal, there's a chance that, if the supply voltage on the LEDs drops, then a series resistor will keep the signals from damaging the first LED. There are ready-built drivers around, with most based on the 74HTC125 chip. Note the large capacitor across the external power supply. The value is not too critical, and you can make it much bigger than shown here.

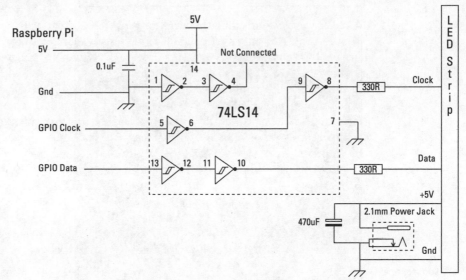

FIGURE 17-9:
APA102C driver
circuit.

Display update

Finally, you have to deal with the update question, or how fast the patterns shown by the LEDs can be changed. Using the bit-banging technique we described in the "Bit-banging the APA102C protocol" section, earlier in this chapter, it can take between 0.45 and 0.23mS to send out the data to a single LED. This time difference occurs because Linux steals time from any running program, and so the time for a refresh depends on what the operating system is doing at that moment. If this interruption occurs in that crucial time when the program is bit-banging the data out to the LEDs, then the bit banging data refresh takes longer. This means that for 144 LEDs, you can update the pattern they show about 15 times per second, at the slowest. Once updated, they require no further intervention from the Raspberry Pi until you need to update them again.

Getting more LEDs

There are various ways you can get more LEDs to play with, and perhaps the most exciting is the series of shapes you can get from the RasPiO Inspiring programmable LED boards. Shapes range from straight strips, circles, semicircles to triangles and squares. There's also a driver board for producing the correct voltage signals and, as a bonus, a socket on this driver board allows you to plug in an analogue-to-digital converter chip. That means you can measure analogue voltages for items like control knobs and sensors. Figure 17-10 shows the triangle and straight strip.

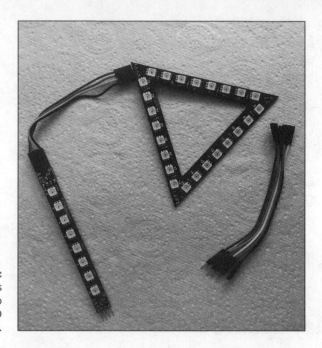

FIGURE 17-10:
Some shapes
from the RasPio
Inspiring LED
strip range.

The great thing about these shapes is that they're easy to set up. Each strip has a plug and socket on it, for input and output, thus enabling you to easily chain together as many strips as you want. The triangle is our favorite — use two or three to make a 3D LED pyramid. In addition to working great with your Raspberry Pi, they can be used on a number of other controllers.

LED strips

Another popular way of getting lots of these LEDs already wired together is to get them on a flexible, printed circuit strip. These come in various lengths, from half a meter to 4 meters. Be sure to get the APA102 type and not the cheaper WS2812b type, to ensure that they work with all the code in this book. You can get 30, 60, or 144 LEDs per meter — as you'd expect, the cost is related to the number of LEDs. These circuit strips also come in a variety of options — black or white backing PCB, bare, coated in silicon, or in a waterproof silicon tube. You can easily cut these strips into lengths, if need be, with a sharp blade.

TIP

In addition to the RGB LED format, the APA102 comes in a white-only LED variant. These are good for domestic lighting and can be dimmed very easily.

LED matrix

One more form you can get these LEDs in is an LED matrix. Adafruit sells a number of different configurations — the biggest (and most expensive) model is a

240mm ridged disk. If that's not what you need, you can get an 8 x 32 grid, 16 x 16 grid, or 8 x 8 grid, all on a flexible backing board. However, our favorite is a high-density 8 x 8 grid, on a ridged board. Its main feature is that, instead of using the 5 x 5mm standard LED, it uses a tiny 1.8 x 1.8mm package. That means the whole grid can be fitted into a display an inch square. These tiny LEDs only take 40mA when fully on, so they take less current than the normal-size LEDs. However, this does add up to a maximum of 2.5A. At that amount of current, the board will get quite warm as well. Nevertheless, they can be very bright, so keep the brightness down. You need to do a bit of soldering to connect the wires and the supplied capacitor to the back of the board, but these are all marked. We also wired the 5V and ground output to the 5V and ground input, to get a better power distribution on the board. This is shown in Figure 17-11.

FIGURE 17-11:
Adafruit's high-density matrix display.

Such a small matrix has many applications, one of which is to make a smart decorative brooch. We thought we would like to have a go at this, and so we came up with some colorful moving-display patterns.

We're all spoiled these days when it comes to graphics, because all their *primitive* functions — things like drawing lines and squares — are built into most languages. However, when faced with a display like this matrix, it is necessary to write your own.

When we address the LEDs in a strip, they're numbered from zero to the length of the strip, and physically placing these LEDs so it looks like a matrix is just the same. However, when thinking about a matrix, it's much more convenient to consider the *x* and *y* address of an LED; so one of the fundamental routines when driving a matrix is to convert an *x* / *y* coordinate pair into an LED number. The actual conversion involved depends on how the LED chain is bent into a matrix. The Adafruit matrix uses what is called a *row bottom-up* raster arrangement, which is a zigzag arrangement where the next LED of a line is followed by the first LED on the next line up. The LEDs start off at zero in the lower-left of the matrix (known as the *origin*) and increase along the *x*-axis as the LED number increases. When the end of the *x* LEDs is reached, the next LED number is directly above the original origin. This repeats along the second row, and the third, and so on until the end of the strip. (The other fundamental way of wiring a matrix is known as a *serpentine* raster, where the strip zigzags up the display like a snake. Here the last LED in a line is followed by the last LED of the next line, with this next line going in the opposite direction to the first one. You can see these two different schemes in Figure 17-12.)

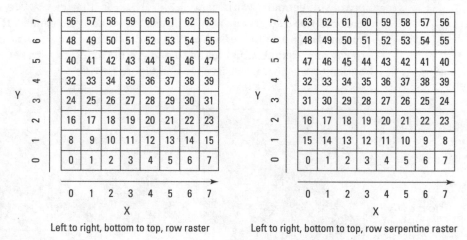

FIGURE 17-12:
Making a matrix with raster wiring and serpentine raster wiring.

Converting to LED numbers for the simple raster is quite easy and is given by

```
LedNumber = X + (Xmax * Y)
```

given that X and Y are the coordinates you want to get the LED number for and that Xmax is the number of LEDs in a row. Of course, this assumes that both coordinates are within the confines of the matrix. In practice, it is necessary to check the coordinates before calculating the LED number.

The serpentine raster is physically easier to wire up, but a little more complicated to work out an LED number for. This is because the preceding conversion only works on odd-numbered rows. For even-numbered rows, you need to use this:

```
LEDNumber = (Xmax - X - 1) + (Y * Xmax)
```

That means any conversion routine must first test to see if the y-coordinate is indicating an odd or even row before choosing the formula to use. This is simply done by looking at the least *significant* bit — the rightmost bit in the number — of the y-coordinate.

Other graphics primitives can help you draw things on a matrix. Perhaps the most fundamental is *line drawing*, where, given a start and an end coordinate pair, you can draw a line of lit LEDs between them. Of course, on such a small matrix, only horizontal, vertical, and 45-degree diagonal lines will look good; all other lines will be stepped, or jagged. The trick here is to work out the difference between the start and end points; the axis with the biggest difference is given an increment value of 1, with the smaller axis change given an increment value of the smaller length over the larger length. Then the x-y coordinate to plot starts off at the initial point of the line, and subsequent points are found by repeatedly adding the increments to the respective coordinates. One of the increments will be 1, and the other a fractional value. It is only when the size of this fractional coordinate exceeds a whole number that the coordinate is changed.

We implemented two other graphics primitives: the square and the filled square. In fact, the code for only an *outline* of a square is longer than a filled square. We could have used the line primitive to implement these two functions, but drawing horizontal and vertical lines is much quicker with a simple loop because no division operation is needed.

We made three different pattern displays that change automatically after a fixed amount of time. The program is shown in Listing 17-7.

LISTING 17-7: **Matrix Broach**

```python
#!/usr/bin/env python3
# Adafruit 8X8 matrix broach

import time, random
from apa102bang import Apa102bang

dataPin = 23 ; clockPin = 24
lenX = 8 ; lenY = 8
length = lenX * lenY ; brightness = 1
half = int(length / 2)
patternDuration = 8.0
leds = Apa102bang(length, dataPin, clockPin, brightness)

def main():
    print("APA102 8 by 8 matrix broach - Press Ctrl+C to quit")
    nextChange = time.time() + patternDuration
    pattern = 1
    while True:
        if pattern == 1:
            pattern1()
```

(continued)

LISTING 17-7: *(continued)*

```
          if pattern == 2:
              pattern2()
          if pattern == 3:
              pattern3()
          if time.time() > nextChange:
              nextChange = time.time() + patternDuration
              pattern +=1
              if pattern > 3:
                  pattern = 1

def pattern1():
      for side in range(1,lenX):
          for i in range(0,lenX):
              square(i,i,side,randCol())
              leds.show()
              time.sleep(0.1)
          time.sleep(0.6)
          leds.setAll((0,0,0))
      leds.setAll((0,0,0))
      time.sleep(0.6)

def pattern2():
    start = 0
    for side in range(lenX-1,0,-2):
        square(start,start,side,randCol())
        leds.show()
        time.sleep(0.08)
        start += 1
      #time.sleep(0.6)
      #leds.setAll((0,0,0))

def pattern3():
    col = randCol()
    for t in range(1,8):
        line(0,0,t,0,col)
        hold()
    for t in range (1,8):
        line(0,0,8,t,col)
        hold()
    for t in range (8,0,-1):
        line(0,0,t,8,col)
        hold()
```

```
            for t in range(7,0,-1):
                line(0,0,0,t,col)
                hold()

        def hold():
            leds.show()
            time.sleep(0.08)
            leds.setAll((0,0,0))

        def matrix(x,y,col):
            pixel = x + y*8
            if pixel < length and x < lenX:
                leds.setLed(pixel,col)

        def squareFill(x,y,side,col):
            for xp in range(x,x+side):
                for yp in range(y,y+side):
                    matrix(xp,yp,col)

        def square(x,y,side,col):
            for xp in range(x,x+side):
                matrix(xp,y,col)
            for xp in range(x,x+side+1):
                matrix(xp,y+side,col)
            for yp in range(y,y+side):
                matrix(x,yp,col)
            for yp in range(y,y+side):
                matrix(x+side,yp,col)

        def line(xs,ys,xe,ye,col):
            xl = xe - xs
            yl = ye - ys
            if xl > yl:
                xinc = 1.0
                yinc = yl/xl
            else:
                yinc = 1.0
                xinc = xl/yl
            x= float(xs) ; y = float(ys)
            while x != xe or y != ye:
                matrix(int(x),int(y),col)
                x += xinc
                y += yinc
```

(continued)

LISTING 17-7:   *(continued)*

```
def randCol():
   r = 0; g =0; b=0
   while r+g+b == 0:
      r = random.randint(0,2) * 64
      g = random.randint(0,2) * 64
      b = random.randint(0,2) * 64
   return (r,g,b)

# Main program logic:
if __name__ == "__main__":
  try:
    main()
  except KeyboardInterrupt:
    pass
# turn off all LEDs
leds.setAll((0,0,0))
leds.show()
```

The program uses a variable called pattern to determine what to display. When it has been displayed for the time given by the patternDuration variable, the pattern value is changed. This patternDuration variable is only looked at when the pattern function returns, so long patterns like pattern1 will always complete at least one cycle of the pattern.

The three patterns use the different pattern primitives to display a sequence of patterns in random colors given by the randCol function. This function generates a mix of two levels of color for each of the three components but excludes black. The first pattern generates a sequence of squares whose lower corners are on the diagonal of the display. These squares start off small and then increase each time the diagonal is filled, until it uses an 8 x 8 square.

The next pattern displays a series of nested squares in changing colors. And the final display shows a "straight" line sweep, from the bottom corner round the display counterclockwise.

TIP

Given the fact that you've created a broach, it can be made into a wearable device by running it off a battery-powered Raspberry Pi Pico controller. You can customize the timing of the display by adjusting the sleep times. Note that the solid square function is not used; you can add that to one of the patterns to make them more complex. However, our main hope is that you will write your own patterns and have even more of them added to the sequence.

The Raspberry Pi Pico controller uses microPython, so it won't run with the listings shown here. If it's teamed up with a Pimoroni Pico Unicorn backpack, it also has a different aspect ratio consisting of 16 by 7 LEDs. In addition, it has four push buttons, which allows us to use them to change the patterns. You'll find some extra code to do this on this book's website.

Figure 17-13 shows the schematic for powering a Pico controller using two AA batteries and a switch. We used a two-cell case with switch that we cut off the end of a string of battery-powered Christmas tree lights that we bought from a thrift shop for about $1.50. (This requires soldering the positive to pin 39 and the negative to pin 38.) We also added a blob of hot glue on the wires after the soldered connection to act as strain relief to prevent the wires from snapping as a result of repeated flexing. This circuit would work with three cells, but don't use four. Figure 17-14 shows a photograph of our wiring.

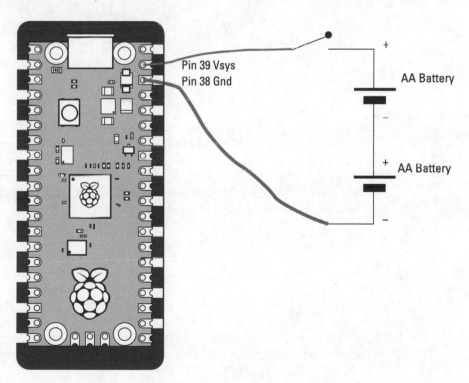

FIGURE 17-13: Wiring up a battery-powered Pi Pico controller.

FIGURE 17-14:
Our battery-powered Pi Pico controller.

Chapter **18**

Old McDonald's Farm and Other RFID Adventures

The radio frequency identification (RFID) card is becoming ubiquitous. It's being used these days for not only building-access control but also travel cards, prepayments at cafeterias, and even antiforgery protection on event tickets. RFID cards can also be used in amusement park wristbands, ski lift passes, and even blood donor cards. When your pet is *chipped*, a small glass–encapsulated RFID tag is injected just under the skin at the neck of the animal.

Given all the things that can be done using RFID, you might think RFID cards are complex devices, and you would be right. However, if you pack away the complexity in a very smart chip and hide the interaction with that chip in a good class driver, actually using them in a project is remarkably easy. There are lots of fun things you can do with RFID cards, and we're going to show you a few of them in this chapter.

CHAPTER 18 **Old McDonald's Farm and Other RFID Adventures** 391

How RFID Work

There are basically three different types of RFID systems available on the market, mainly distinguished by which frequency range they use. All systems consist of two parts: a tag or card and a reader. The reader extracts binary bits from a tag or card using radio waves, so no wires are needed between the reader and the tag or card. These tags normally known as *passive* tags — tags that apparently require no source of power — although a small number are *active* tags that require fitting with a small watch battery. These active tags are used when you need a much longer read range.

The reader sends out a radio signal, and the tag picks it up and uses the power in that radio signal to activate a microchip. That microchip then sends back a sequence of pulses, which the reader interprets as a number. The way the tag sends the data back is different on different types of tag, but the main way is transmitting the data back to the reader on a different frequency.

Here's a list of available tags, with their frequency ranges:

» **125–135 KHz tags:** These tags are the ones used in chipping pets and also in a lot of access control systems. They have a limited storage capacity and are normally restricted to holding a serial number of only 64 or 128 bits. The tags and cards used most widely today conform to the EM4100 / EM4200 standard. Most of the tags are read-only, but a few (the Hitag tags, for example) allow you to store data using a special peripheral to change the data on the chip.

» **13.56 MHz tags:** These are the so-called *smart* cards, capable of storing not only a serial number but also some data that can be read or written. There is a measure of security built into the cards so that the data is accessible only to those who can provide the encryption key. There are many different types of cards, but by far the most common is the one known as a MIFARE classic card.

» **UHF (860–960MHz):** Unlike the other two RFID types, UHF readers are capable of reading more than one tag at the same time. They are also quite long-range, typically 10 to 30 feet. They're designed for bulk inventory taking. Each item on a palette can be recorded and counted at the same time. The readers have very high-power transmitters in them, so much so that they're a health hazard, requiring strict time limits for workers operating them in order to avoid long-term radiation exposure. There are other, higher-frequency systems in this class as well.

We use the MIFARE classic card in the projects in this chapter — to give it its Sunday name, the ISO/IEC 14443 A/MIFARE mode protocol. The readers are cheap, and the tags come in various shapes ranging from key fobs to cards. It's the cards that are most convenient for these projects, but it's not essential to use these. Figure 18-1 shows the block diagram of what an RFID system looks like electronically.

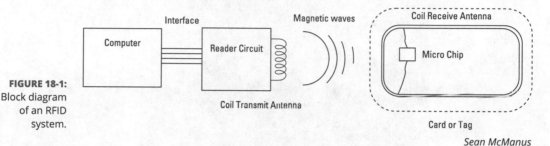

FIGURE 18-1: Block diagram of an RFID system.

TECHNICAL STUFF

You might have come across the acronym *NFC*, which stands for near field communication. NFC is increasingly used in smartphones to pay for things by waving your phone close to a reader. The phone is essentially a programmable RFID tag, but using a protocol that's different from normal tags. Your phone effectively becoming a virtual RFID tag. Because it isn't easy to program your phone to be a virtual tag, we don't cover phones in this book, but be aware they are the same sort of thing as conventional tags.

The antennas used are formed from coils of very thin wire or even metal foil. Readers for these cards come in all price ranges and interfaces. We use one of the cheapest — the RFID-RC522 — which can be had for less than $13 for three on popular electronics and auction sites. They're based on the MFRC522 chip from NXP Semiconductors (formerly Phillips), and though this chip is capable of being connected to a computer in a number of different ways, the way these low-cost boards are designed, they're restricted to an SPI interface only. In Chapter 17, we explain how we could "bit-bang" the SPI protocol, but in this chapter we use the Raspberry Pi's built-in hardware SPI interface, which can use only specific pins on the GPIO connector.

REMEMBER

When you get these RFID readers, they come with a choice of two types of header pins. You need to solder on the right-angled pins. Then you can either make up a lead or mount the reader vertically on a breadboard, as shown in Figure 18-2.

We feel it's much better to mount it vertically because it's away from the metal forming the internal clips of the breadboard, and metal affects the resonant frequency of the antenna coil — and thus the tag read range. Figure 18-3 shows both the schematic and layout diagram of how to wire up the reader to the Raspberry Pi.

FIGURE 18-2:
Mounting a
reader on a
breadboard.

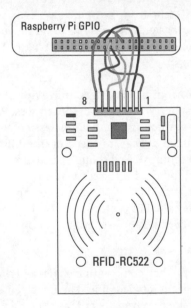

Raspberry Pi GPIO

8   1

Raspberry Pi
GPIO connector
3V3 (17)   8 3V3

Physical pin number

$\overline{CS}$
GPIO 8 (24) ───── 1 SDA
GPIO 10 (19) ── Din ── 3 MOSI
GPIO 9 (21) ── Dout ── 4 MISO   RFID-522
GPIO 11 (23) ── CLK ── 2 SCK
GPIO 25 (22) ───── 7 RST

                                        6 GND

Gnd (20) ───────────────

FIGURE 18-3:
Schematic
and layout
for attaching
to a Pi.

RFID-RC522

TIP

For a really neat job, mount the reader in a wooden or plastic box, preferably using nylon screws because close metal reduces the reading range.

After you attach the reader to the Pi, it's time to get the software you need to read it. First off, you have to install the `python-dev` system by typing **sudo apt-get install python-dev** into a terminal window of a Raspberry Pi connected to the Internet.

There is a fair chance that you have the latest version already installed, but it's best to check. Next, you need to install the `SPI-Py` library, which allows you to use the hardware SPI as a C extension for Python. First, go to `https://github.com/lthiery/SPI-Py`. Then click the green Code button and select Download ZIP from the popup menu.

When the file has finished downloading, open File Manager and click the `Downloads` folder. Right-click the `SPI-Py-master.zip` file, and select Extract Here. You'll now see a new directory called `SPI-Py-master`.

Open up a command window and type in:

```
cd Downloads/SPI-Py-master
```

and finally type in

```
sudo python3 setup.py install
```

This installs the code that allows you to use the SPI hardware from Python 3. Finally, go to the Desktop Raspberry icon menu and choose Preferences, then select the Raspberry Pi Configuration application. When the application opens, click on the Interfaces tab and make sure that the SPI interface is enabled. If it isn't, click to enable it and reboot your Pi.

A MIFARE card's structure

A MIFARE card consists of 64 blocks of data, with each block 16 bytes long. Some blocks are for user data, and others hold authentication keys, UID numbers, and manufacturers' ID numbers. (A unique identification [UID] number, is 4 bytes long. In fact, despite its name, a UID might not actually be unique, but chances of finding a duplicate are many thousands of times less likely than your winning a big lottery. So, for all practical considerations, it can be considered unique.)

Here's a short list of the first 12 sectors of a card:

```
Sector 0  [203, 58, 164, 213, 128, 8, 4, 0, 98, 99, 100, 101, 102, 103, 104, 105]
Sector 1  [0, 0, 0, 0, 0, 0, 0, 0, 0, 0, 0, 0, 0, 0, 0, 0]
Sector 2  [0, 0, 0, 0, 0, 0, 0, 0, 0, 0, 0, 0, 0, 0, 0, 0]
Sector 3  [0, 0, 0, 0, 0, 0, 255, 7, 128, 105, 255, 255, 255, 255, 255, 255]
Sector 4  [0, 0, 0, 0, 0, 0, 0, 0, 0, 0, 0, 0, 0, 0, 0, 0]
Sector 5  [0, 0, 0, 0, 0, 0, 0, 0, 0, 0, 0, 0, 0, 0, 0, 0]
Sector 6  [0, 0, 0, 0, 0, 0, 0, 0, 0, 0, 0, 0, 0, 0, 0, 0]
Sector 7  [0, 0, 0, 0, 0, 0, 255, 7, 128, 105, 255, 255, 255, 255, 255, 255]
Sector 8  [0, 0, 0, 0, 0, 0, 0, 0, 0, 0, 0, 0, 0, 0, 0, 0]
Sector 9  [0, 0, 0, 0, 0, 0, 0, 0, 0, 0, 0, 0, 0, 0, 0, 0]
Sector 10 [0, 0, 0, 0, 0, 0, 0, 0, 0, 0, 0, 0, 0, 0, 0, 0]
Sector 11 [0, 0, 0, 0, 0, 0, 255, 7, 128, 105, 255, 255, 255, 255, 255, 255]
```

The first sector's first four bytes contain the UID that, along with the other bytes in that sector, cannot be changed. Attempts to write to it will fail. The sectors containing all zeros are the sectors you can write data to. Each group of four sectors is preceded by a sector that contains two keys that allow read and write access to the following sectors. You don't write directly into these, but writing to the sectors can change these values. They're initially set up, as shown here, with the default key values. When you plan a project, be aware that you can only write data to those sectors that are not key sectors.

Key sectors not to write to are

0, 3, 7, 11, 15, 19, 23, 27, 31, 35, 39, 43, 47, 51, 55, 59, 63

All you need to do to get started using RFID is to include the Rc522rfid class file in the folder you're using for your RFID projects. This can be downloaded from the book's website. (See the Introduction for more on how to access the site.) We have not printed it here because it's long, and you won't get much from reading it. It also automatically blocks you from trying to write to key sectors.

Talking to the reader chip over the SPI interface is somewhat complex, and not really necessary to understand in order to use an RFID reader in a system. It involves mainly putting numbers into internal chip registers or memory locations in the correct sequence, as laid out in the chip's data sheet. However, we have added five methods at the end of the class in order for you to be able to easily tap in and communicate with the reader. The methods you need to know about are described in this list:

>> **RC522_WaitForCardRemoved**: Ensures that the code pauses until a card has been removed from the reader — reading of a card can be quick. This is

complicated by the fact that, with a card held permanently on a reader, every other call to the detect a card function returns incorrect information, saying that there is no card on the reader. This is a side effect of how the chip works, and so to work around it, this method waits until two successive attempts at reading a card both return no card present. (Note that this is known as *blocking* code in that, once it's called, nothing else can happen until the card is removed.)

» **RC522_ReadCard:** Reads and returns the UID number of the card at the reader. If there's no card or there's a reading error, this method returns a value of –1.

» **RC522_GetCard:** Waits until there is no card at the reader and then waits again until a card is presented and read, and then it returns the card's UID number. This is blocking code in that nothing else can happen until a valid card is read.

» **RC522_GetSector:** Reads a 16-byte sector block of data from a card and returns it. You supply the sector number, the authorization key, and the sector number you want to read. This is blocking code in that nothing else can happen until a valid card is read.

» **RC522_WriteSector:** Writes a 16-byte block of data to a sector number. You supply the sector number, the authorization key, and the sector number you want to write to. This is blocking code in that nothing else can happen until a valid card is presented. It will not, however, allow you to write directly to key sectors.

So, let's see how this works in practice. Listing 18-1 shows some code to simply read out the UID of a card.

LISTING 18-1: **Reading the Card's UID**

```
#!/usr/bin/python3
# Token reader by Mike Cook

from rc522rfid import Rc522rfid

rfidReader = Rc522rfid()

print ("Card token reader")
print ("Click stop when done.")
while 1: # repeat forever
    cardNumber = rfidReader.RC522_GetCard()
    print(hex(cardNumber))
```

The UID, as a 16-bit number, is printed out in hexadecimal format. (Printing it in decimal format doesn't make sense, because there will be a mixture of positive and negative numbers, depending on the state of the most significant bit.) So you can now look at the card's number, which is not so exciting, but what if you could get that card number to make something happen — like play a sound file?

A simple RFID jukebox

We can simply use the RFID cards to trigger playing music from the default Music folder, or from anywhere else, if you like. Copy any MP3 files you may have lying around into the Music folder on the Raspberry Pi. Next, get as many RFID cards as you have files, and, using the code in Listing 18-1, write down the card number of each one on a sheet of paper and put the cards next to their numbers. Then load up the program in Listing 18-2. Note you will have to change the entries in the files list, to the names of the MP3 files you actually have here.

LISTING 18-2: **Simple RFID Jukebox**

```python
#!/usr/bin/python3
# Simple RFID Jukebox by Mike Cook

from rc522rfid import Rc522rfid
import pygame

pygame.mixer.quit()
pygame.mixer.init(frequency=22050, size=-16, channels=2, buffer=512)
rfidReader = Rc522rfid()
tokens = [0xf6690ebb, 0xcb3aa4d5, 0x9c35fddd, 0x8cd32dde, 0xc682ade, 0x1cb413de]
files = [ "File1", "File2", "File3", "File4", "File5", "File6" ]

def main():
    print ("Simple RFID Jukebox reader")
    print ("Click stop when done.")
    while 1: # repeat forever
        cardNumber = rfidReader.RC522_GetCard()
        pygame.mixer.music.stop()
        for i in range(0,len(tokens)):
if cardNumber == tokens[i]:
        try :
            pygame.mixer.music.load("/home/pi/Music/"+str(files[i])+".mp3")
            pygame.mixer.music.play()
            print("Playing:- "+str(files[i])+".mp3")
        except :
            print("No file:-/home/pi/Music/" + str(files[i]) + ".mp3")
```

```
# Main program logic:
if __name__ == '__main__':
    try:
        main()
    except KeyboardInterrupt:
        pass
```

You have to change the tokens list to the token numbers of the cards you have, and the files list to match the sound files you have. The program automatically appends the .mp3 prefix. So, when you present a card, the file plays through the Pygame sound system. What is going on here is that the connection between a file and an RFID card's number is fixed in the code by the position of the filename and the card's number in a list. That means the first number in the list will trigger the first file in the list.

This is all well and good, but it means that every time you want to change things, you have to edit the code. It would work much better if the card itself carried the filename.

A better RFID jukebox

You can make the card carry the filename, but you need a program to handle this task. In other words, you need some way of *enrolling* the cards. In the example in the previous section, this enrolling was done by reading the card's number and putting it in a list. It also involved copying down the filenames in the target folder and adding them to a list. In a much better system, all that could be done by an enrolling program.

The idea here is that there are two programs: one to play files in response to a card and the other to save the filename into the data segments of the card. The enroll program we came up with reads all the files in your Music folder and then offers them for selection one by one, giving you the choice of whether to enroll them. If you choose to enroll a card, the filename, including any extension, is stored in segments 4 and 5 of the card. (This means the filenames must be no longer than 32 bytes.) The enrolling program is shown in Listing 18-3.

LISTING 18-3: **RFID Jukebox Enrolling Program**

```
#!/usr/bin/python3
#RFID Jukebox Enroll -
#     puts music files names on sectors 4a & symbol 5
#     by Mike Cook
```

(continued)

| LISTING 18-3: | *(continued)* |

```
from rc522rfid import Rc522rfid
import os

rfidReader = Rc522rfid()

def main():
    print ("RFID Jukebox card enroller")

print ("Click the Stop/Restart back end icon quit when done.")
getFiles()
for file in range(0,len(soundList)):
    action = ""
    print( soundList[file] )
    print("Enroll (e) or just Return to Skip")
    action =input()
    if action == "e" :
        enroll(soundList[file])

def enroll(name):
    print("Enrolling", name, "place a card on the reader")
    data1 = []
    data2 = []
    for i in range(0,16): # fill with all zeros
        data1.append(0)
        data2.append(0)
    for i in range(0,len(name)):
        if i < 16:
            data1[i] = ord(name[i])
        else:
            data2[i-16] = ord(name[i])
# This is the (default) key for authentication
key = [0xFF,0xFF,0xFF,0xFF,0xFF,0xFF]
rfidReader.RC522_WriteSector(key,4,data1)
rfidReader.RC522_WriteSector(key,5,data2)
print("finished enrolling\n")

def getFiles():
    global soundList
    path = os.path.abspath("/home/pi/Music/")
    #get a list of files
    soundList = [fn for fn in next(os.walk(path))[2]]
    list.sort(soundList) # put in alphabetical order
    print (len(soundList),"files found")

# Main program logic:
if __name__ == "__main__":
    main()
```

The `getFiles` function targets a specific folder in the `path` variable and then searches that folder, making a list of all the files it finds using the `os.walk` method. The list is then sorted into alphabetical order to get an ordered list to use for enroll selection. Then the `main` function displays these filenames one at a time, and you can press e (and then Enter) to enroll the card or just press the Enter key to skip to the next file. The `enroll` function takes the filename and splits it into two 16-byte blocks before writing them to the card.

Having enrolled a card like this, it's best if you mark the card in some way to show which file you have enrolled. You can do this in many ways, from a simple hand-written stick-on label to a complex design with the artist's name — and maybe even artwork from the album. Print this out on a color printer and stick it to the card with spray glue. Finish off the card by covering it with the sort of transparent plastic film that is used to cover and protect books, much used in libraries.

After you have your cards, you need a new program to play them back. This is shown in Listing 18-4.

LISTING 18-4: **RFID Jukebox Enrolled Card Playback**

```python
#!/usr/bin/python3
# RFID Jukebox2 with track names on the card by Mike Cook

from rc522rfid import Rc522rfid
import pygame

pygame.mixer.quit()
pygame.mixer.init(frequency=22050, size=-16, channels=2, buffer=512)
rfidReader = Rc522rfid()

def main():
    print ("RFID Jukebox using enrolled cards")
    print ("Press Ctrl+C to quit when done.")
    while 1: # repeat forever
        toPlay = readFileName()
        pygame.mixer.music.stop()
        try :
            pygame.mixer.music.load("/home/pi/Music/"+toPlay)
            pygame.mixer.music.play()
            print("Playing:- "+toPlay)
        except :
            print("No file:-/home/pi/Music/"+toPlay)
        rfidReader.RC522_WaitForCardRemoved()
```

(continued)

LISTING 18-4: *(continued)*

```
def readFileName():
    # This is the (default) key for authentication
    key = [0xFF,0xFF,0xFF,0xFF,0xFF,0xFF]
    block1 = rfidReader.RC522_GetSector(key,4)
    block2 = rfidReader.RC522_GetSector(key,5)
    if block1 != -1 and block2 != -1:
        fileToPlay = getString(block1,block2)
        return fileToPlay
    else:
        return("Error in reading card")

def getString(s1,s2): # make string from two sectors
    name = ""
    i = 0
    notDone = True
    while notDone and i < 32:
        if i<16:
            c = s1[i]
        else:
            c = s2[i-16]
        i+=1
        if c !=0:
            name = name + chr(c)
        else:
            notDone = False
    return name

# Main program logic:
if __name__ == "__main__":
    try:
        main()
    except KeyboardInterrupt:
        pygame.mixer.quit()
```

The main function waits until a card is presented and then stores the filename
on that card, into the variable toPlay. It then stops any sound already being
played and attempts to play the new file. If the file doesn't exist, a simple message is displayed and the program waits until the card has been removed. The
readFileName function uses the default key to get sectors 4 and 5 from the card
and the getString function assembles these into a filename variable.

Taking it further

You can change the folder that the files are in quite easily. However, a project you might consider is a story time bear: a teddy bear or another cuddly toy with the RFID reader embedded into it — in its paw or chest, for example. Then your child has a set of bedtime story cards and can choose which one is read to them. For added realism, you can also embed a speaker into the toy.

Dressing Up a Paper Doll

Paper dolls have been around for a long time, but they're still popular with children of all ages. With the help of RFID technology, you can make an up-to-date version of this toy that outperforms the original. The idea is that you have a basic doll and a number of different outfits and accessories. These are normally pasted onto thin cardboard and cut out. Then, by use of tabs on the clothes, you can dress your doll in a mix-and-match style. With RFID, you can have a card specify each item of clothing and construct your dressed doll on the screen with a degree of seamless integration you just can't get with the paper version.

There are many clothing sheets to cut out online that are free for personal use, as well as patterns you can buy. You can also draw them yourself — if you're better at drawing dolls than we are. Given the fact that we're not too hot at this drawing business, we teamed up with Imani Osmon, a young artist, to provide the artwork we use as an example here. (We still think it may be just as exciting finding or drawing your own.)

You start off with the basic sheet, and you'll need a drawing package in order to prepare your artwork. Your package should be able to scale images and also add a transparent background. Being able to touch up and change colors could also be useful, though it's not essential. This prep work can be done on a laptop using established packages like Photoshop, or on the Pi itself using Inkscape or Gimp. (For more on using GIMP, see Chapter 7.)

If you decide to go with GIMP, here's what you need to do:

1. **Start off with the full sheet, and scale it so that the doll will fit into a Pygame window on your monitor.**

The sheet should be about 600 pixels high.

2. **Start chopping out the clothes.**

   Use GIMP's Select rectangle to pick out each piece, and then choose Edit⇨Copy from the main menu. Don't worry if adjacent bits of graphics are also included at this stage. Then choose File⇨Create⇨From Clipboard to get that piece on its own. Then work on the image with a square paintbrush, to remove any tabs and other bits of image until it sits in a white background. Next, use the Fuzzy Selection tool to select this white background. Choose Layer Transparency⇨Add Alpha Channel from the main menu, and then choose Edit⇨Clear. Finally, choose File⇨Export As to save this as a .png file, using the default checked options. Finally, close the image window and click the Discard Changes option.

3. **Repeat this process until the doll and all the accessories have been saved in separate** .png **files in a folder called** doll, **which you have created inside your working** RFID **folder.**

4. **Rename the** doll **file to start with an ampersand.**

   This is important because the ampersand ensures that this file appears first in an alphabetically ordered list. We use this so that the program knows it's the doll file.

Now it's time to enroll the RFID cards. We use two blocks here — one to hold the filename and the other to hold the position for drawing the file on screen. Now type in and run the code in Listing 18-5.

TIP

If the graphics files are put on through a network or downloaded, then some computers can place invisible files into folders that the program here might mistake for a graphics file and try to load. Apple's macOS puts a file called .DS_Store and a folder called .AppleDouble in every folder it sees. On an enroll program, this file will produce an error message saying unsupported image format. If this happens, use the File Manager to navigate to the folder, then go to the View menu and turn on the Show Hidden option. You can then delete this normally hidden file — the hidden folder does no harm.

TIP

When you run this program, you might get a "libpng warning" — this is nothing to worry about and can be safely ignored. It comes about from the image editing software sometimes adding extra chunks (called *metadata*) to the images.

LISTING 18-5: **Dress Up Doll Enroll**

```
#!/usr/bin/python3
# Dress Up Enroll by Mike Cook


from rc522rfid import Rc522rfid
import pygame
```

```
import os
pygame.init() # initialise graphics interface

os.environ['SDL_VIDEO_WINDOW_POS'] = 'center'
pygame.display.set_caption("Dress Up")
pygame.event.set_allowed(None)
pygame.event.set_allowed([pygame.KEYDOWN,pygame.QUIT])
pygame.key.set_repeat(500, 80)


screenWidth = 240
screenHeight = 550
screen = pygame.display.set_mode ([screenWidth,screenHeight], 0, 32)
needsRedraw = True
enrol = False


rfidReader = Rc522rfid()

def main():
    global xPlot,yPlot,done,enrol,needsRedraw
    print ("Dress up doll - Enrolling Cards")
    print ("Use the cursor keys to get the garment into position")
    print ("Space - next one, e - Enroll card")
    init()
    xPlot = int(screenWidth / 4)
    yPlot = int(screenHeight / 4)
    while 1:
        checkForEvent()
        for i in range(1,len(clothes)):
            done = False
            needsRedraw = True
            while not done:
                checkForEvent()
                if needsRedraw:
                    drawScreen(i)
                if enrol :
                    enrol = False
                    print("enrolling")
                    enrolCard(cList[i])

def drawScreen(i):
    global needsRedraw
    pygame.draw.rect(screen, (210,210,210),(0,0,screenWidth, screenHeight), 0)
    screen.blit(clothes[0],(0,0))
    screen.blit(clothes[i],(xPlot,yPlot))
    pygame.display.update()
    needsRedraw = False
```

(continued)

LISTING 18-5: *(continued)*

```python
def init():
    global clothes,cList
    path = os.path.realpath(__file__)
    path = os.path.dirname(path) + "/doll/"
    #get a list of files
    cList = [fn for fn in next(os.walk(path))[2]]
    list.sort(cList) # put in alphabetical order
    print("files found",cList)
    clothes = [ pygame.image.load("doll/"+cList[i]).convert_alpha()
                for i in range(0,len(cList))]

def enrolCard(name):
    print("Enrolling", name, "place a card on the reader")
    data1 = [] ; data2 = []
    for i in range(0,16): # fill with a sectors worth of zeros
        data1.append(0) ; data2.append(0)
    for i in range(0,len(name)):
        data1[i] = ord(name[i])
    #save position of this item
    data2[0] = xPlot >> 8
    data2[1] = xPlot &amp; 0xFF
    data2[2] = yPlot >> 8
    data2[3] = yPlot &amp; 0xFF

    # This is the (default) key for authentication
    key = [0xFF,0xFF,0xFF,0xFF,0xFF,0xFF]
    rfidReader.RC522_WriteSector(key,8,data1)
    rfidReader.RC522_WriteSector(key,9,data2)
    print("finished enrolling\n")

def terminate(): # close down the program
    print ("Closing down")
    pygame.quit() # close pygame
    os._exit(1)

def checkForEvent(): # see if we need to quit
    global xPlot,yPlot,needsRedraw,done,enrol
    event = pygame.event.poll()
    if event.type == pygame.QUIT :
        terminate()
    if event.type == pygame.KEYDOWN :
        if event.key == pygame.K_ESCAPE :
            terminate()
    if event.key == pygame.K_SPACE :
        done = True
        enroll = False
```

```
        if event.key == pygame.K_e :
            done = True
            enrol = True
        if event.key == pygame.K_UP :
            yPlot -= 1
            needsRedraw = True
        if event.key == pygame.K_DOWN :
            yPlot += 1
            needsRedraw = True
        if event.key == pygame.K_RIGHT :
            xPlot += 1
            needsRedraw = True
        if event.key == pygame.K_LEFT :
            xPlot -= 1
            needsRedraw = True

# Main program logic:
if __name__ == '__main__':
    main()
```

You need to make only one change: Set the dimensions of the window to the dimensions of your doll image. Double-click on the doll file to bring up a viewing window that specifies the image size in pixels in the top window bar. Make sure those numbers are used to set the values of the screenWidth and screenHeight variables in the code.

When you run the code, you will see your doll and the first piece of clothing. Use the cursor keys to maneuver this piece into the exact spot where you want it to appear on the doll. Then place a card on the reader and press the E key, to store the filename in Sector 8 and the position in Sector 9. Remove the card and press the spacebar for the next piece of clothing.

TIP

Keep pressing the spacebar to cycle through the clothes until you find the one you want to enroll. This way, if you find a piece that isn't quite right, you can go and enroll just that piece again. Also, if you add to the files in the doll folder, you can just enroll the new piece on a new card.

Finally, print out each piece of clothing onto paper and use spray glue to mount it onto each card. Cover the card with a transparent plastic film and trim off any surplus with a sharp blade, holding the card face down on a cutting mat. Some of our cards are shown in Figure 18-4.

FIGURE 18-4:
Doll clothes
RFID cards.

Reproduced by permission of Imani Osmon

Runway time

Now it's time to actually dress up the doll. The program for doing this is in
Listing 18-6.

LISTING 18-6: **Dress Up Doll**

```python
#!/usr/bin/python3
# Dress Up by Mike Cook

from rc522rfid import Rc522rfid
import pygame
import os

pygame.init() # initialise graphics interface

os.environ['SDL_VIDEO_WINDOW_POS'] = 'center'
pygame.display.set_caption("Dress Up")
pygame.event.set_allowed(None)
pygame.event.set_allowed([pygame.KEYDOWN,pygame.QUIT])
pygame.key.set_repeat(500, 80)

screenWidth = 240
screenHeight = 550
screen = pygame.display.set_mode([screenWidth, screenHeight], 0,32)
xPlot=[0] ; yPlot=[0] ; garment=[0] # list of what to display
rfidReader = Rc522rfid()
```

```python
def main():
    global xPlot,yPlot,done,enroll
    print ("Dress up doll")
    print ("Space bar to start over")
    init()
    drawScreen()
    while 1:
        checkForEvent()
        readCard()
        if needsRedraw:
            drawScreen()
            rfidReader.RC522_WaitForCardRemoved()


def readCard():
    global needsRedraw
    status = -1
    while status == -1:
        checkForEvent() # allow pygame a look in while waiting
        status = rfidReader.RC522_ReadCard()
    # This is the (default) key for authentication
    key = [0xFF,0xFF,0xFF,0xFF,0xFF,0xFF]
    block1 = rfidReader.RC522_GetSector(key,8)
    block2 = rfidReader.RC522_GetSector(key,9)
    if block1 != -1 and block2 != -1: # if data OK
        x = block2[0] << 8 | block2[1]
        y = block2[2] << 8 | block2[3]
        lookAtNew(block1,x,y)
        needsRedraw = True
        return -1
    else:
        return("Error in reading card")


def lookAtNew(s1,x,y): # look at the new element
    global garment, xPlot, yPlot
    name = ""
    i = 0
    notDone = True
    while notDone and i < 16:
        c = s1[i]
        i+=1
        if c !=0:
            name = name + chr(c)
        else:
            notDone = False
    #Look up name in list of files
    for i in range(1,len(cList)):
        if cList[i] == name: # find the name on the card
            # remove if already in the list
            if i in garment:
```

(continued)

LISTING 18-6: *(continued)*

```python
                place = garment.index(i)
                del garment[place]
                del xPlot[place]
                del yPlot[place]
            else: # otherwise add it
                garment.append(i)
                xPlot.append(x)
                yPlot.append(y)

def drawScreen():
    global needsRedraw
    pygame.draw.rect( screen, (210,210,210), (0,0, screenWidth, screenHeight), 0)
    for g in range(0,len(garment)):
        screen.blit(clothes[garment[g]], (xPlot[g],yPlot[g]))
    pygame.display.update()
    needsRedraw = False

def init():
    global clothes,cList
    path = os.path.realpath(__file__) # path of this program
    path = os.path.dirname(path) + "/doll/" # path of images
    #get a list of files
    cList = [fn for fn in next(os.walk(path))[2]]
    list.sort(cList) # put in alphabetical order
    clothes =
    [ pygame.image.load( "doll/"+cList[i] ).convert_alpha()
    for i in range(0,len(cList))]

def terminate(): # close down the program
    print ("Closing down")
    pygame.quit() # close pygame
    os._exit(1)

def checkForEvent(): # see if we need to quit
    global xPlot,yPlot,garment
    event = pygame.event.poll()
    if event.type == pygame.QUIT :
        terminate()
    if event.type == pygame.KEYDOWN :
      if event.key == pygame.K_ESCAPE :
        terminate()
      if event.key == pygame.K_SPACE :
        xPlot = [0] ; yPlot = [0]
        garment = [0] # clear lists
        drawScreen()

# Main program logic:
if __name__ == "__main__":
    main()
```

Again, make sure that the values of the `screenWidth` and `screenHeight` variables are set to the same values as in the enroll code. Presenting an enrolled card places that piece of clothing on the doll; presenting it again removes it. Pressing the spacebar starts again from scratch. Note that the clothes are drawn in the order that the clothes are presented. This might look odd if one item is meant to partially cover another. The doll should be dressed as in real life, with garments that are worn under other items first. It's fun to build up your own collection of dolls and accessories. Figure 18-5 shows some of the dressed-up dolls we created.

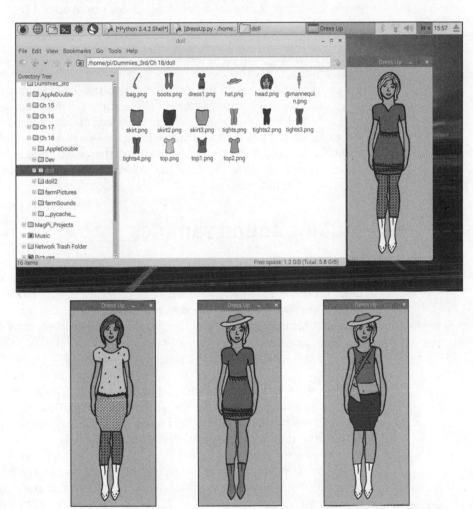

FIGURE 18-5:
Dolls we created.

Reproduced by permission of Imani Osmon

Old McDonald's Farm

So far in this chapter, we have shown you how to associate information on an RFID tag to a sound with the jukebox, and how to associate an image with a dress-up paper doll. Now we combine the two techniques to create the ultimate in children's songs: "Old McDonald Had a Farm." The unique algorithm used to generate the chorus of this song ensures that the length of the song grows exponentially with every animal added. Also the point of this song is to include not only well-known farmyard animals but also unusual ones. (Truth be told, Mike's wife said that is only a boy thing.)

As with the other projects in this chapter, we present just a start that can easily be extended, almost without limit, simply by placing more files in folders. Just like the Dress Up Doll project, there are two programs: an enroll program and a play program. The idea is that fragments of the song are recorded, and the program puts together these fragments into the ever-lengthening song. What's more, the computer never forgets what is on the farm, and each verse is sung with the same enthusiasm of the first. (You can't always say that about singing "Old McDonald's Farm" in real life.) The child, or the parent themselves, should record at least some of the sound samples.

Making sound samples

The sound samples can be recorded on a laptop and moved over to the Pi by using a USB memory stick. However, with a bit of work, it's possible to do the whole thing on the Pi itself. The Raspberry Pi can record sound, but it has no built-in microphone interface, so you have to provide one in the form of a USB sound card or USB microphone. You can purchase one for less than $10 (about £8) and you can find suitable ones listed at https://elinux.org/RPi_Verified Peripherals#Class_compliant_USB_sound_cards.

It's best to look at the installation details for the specific card you have. After it's plugged in and connected to a microphone, go to the Desktop menu and choose the Preferences submenu, and then select Audio Device Settings. From the drop-down menu that appears, choose your USB sound card.

Next, you need some software to record the sound. By far, the best free software to do this is Audacity. To install it on your Pi, go to a terminal window and enter the following:

```
sudo apt-get update
sudo apt-get install audacity
```

After Audacity installs, it's available from the Desktop menu, under the Sound & Video entry. The screen is shown in Figure 18-6.

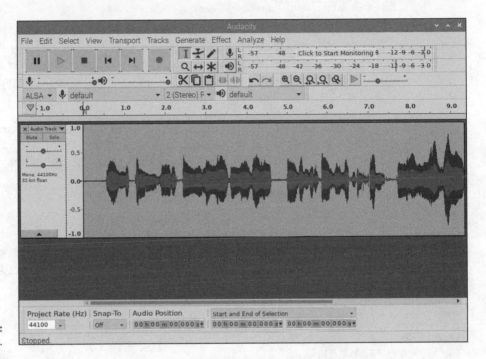

FIGURE 18-6:
Audacity.

TIP

You can find versions of Audacity for the Macintosh and Windows operating systems as well.

Audacity can not only record sound but also edit it — and even add effects. The controls are just like a tape recorder, with a round, red Record button and a square Stop button. If you've never used it, have a play with it first: Just click the Record button and say something (you'll see the waveform plotted), and then press the Stop button, press Rewind, and press the triangular Play button.

REMEMBER

Whenever you make a recording, you need to "top and tail it" — that is to say, remove the silences at the start and end. In Audacity, use the Magnifying Glass icon to enlarge the waveform, and then click and drag over the opening silence to highlight it. Then you remove it by simply pressing the Delete key on the keyboard. Do the same for the end silence. With that out of the way, you then need to make sure that a sample begins and ends on the zero line, running through the middle of the displayed waveform. If it does not, you will hear clicks when you play it back. To ensure that it does begin and end on the zero line, highlight a small section at the start of the sample, and select Fade In from the Effects menu. Do the same for the end of the sample, but this time select Fade Out. Finally, select the whole sample and select Normalize from the Effects menu. With the top two options checked, select O. This last step ensures that all samples have the same overall sound level.

To save your sample, start by choosing File ⇨ Export Audio from the main menu, and then choose Ogg Vorbis File from the drop-down menu in the lower right. Type in your chosen filename with a `.ogg` extension and select Save. The `.ogg` format is Python's native sound format and is the best supported. Close the Audacity window and do not bother with the Save Changes option.

So, after seeing how to make a sample, you need to know what samples to record. Make a folder called `farmSounds` in your working `RFID` folder, and inside that, make another one called `fixed`. You will store the skeleton of the song there.

Table 18-1 shows the sound samples you need to make.

TABLE 18-1 ## Old McDonald's Farm Sound Samples

Sample Name	Sung Words
start	"Old McDonald had a farm ee eye ee eye oh"
and	"and on this farm he had some"
ee_eye_oh	"ee eye ee eye oh"
witha	"with a"
here	"here"
anda	"and a"
there	"there"
herea	"here a"
therea	"there a"
everywherea	"everywhere a"
well	"well"
end	"Old McDonald had a farm ee eye-e-e, ee eye-e-e, o-o-o-h" (and feel free to extend those last few notes)

Back in the `farmSounds` folder, you need two samples for each animal you use. For example, if you have a chicken, you need a sample of you singing the word *chickens*, called `chicken`, and then another sample of the noise that animal makes, having the name `chickenSound`. So in general that is two files: "name" and "nameSound."

It's important that you use this exact naming scheme or the program will not work.

Making the graphics

You have lots of choices when it comes to how you want your farm to look. You can buy a packet of farm animal stickers and scan them in or photograph them and use the stickers for the RFID cards. You can draw your own animals or get some clip art from the Internet. However, we chose to go with photographs we had taken of the various farm inhabitants. For this demonstration, we chose a chicken, a cow, a sheep, a pig, and a dalek. (Yes, we know — it's a boy thing.)

Each animal is isolated on a transparent background, just like we did for the clothes in the Dress Up Doll project. This time, we kept a full-resolution version for printing on the cards, and a scaled version for the screen. The scaled versions should be between 30 to 100 pixels high, depending on the relative size of the animal. These pictures should be stored in a folder called `farmPictures` and have exactly the same name as the sung word in the `farmSounds` folder. So the chicken picture is called `chicken.png`, and the corresponding sound files are `chicken.ogg` and `chickenSound.ogg`.

Remember that `fixed` folder you created inside the `farmPictures` folder? That's where you'll want to add two files — one called `farm.png` for the backdrop of the farm and the other called `fence.png` on a transparent background, to act as the fence at the front of the screen.

Each animal appears five times on the screen on each chorus of the song, so you must enroll not only the basic animal name on the card but also the position of each of its appearances. The program to do this is in Listing 18-7.

The potential hidden file problem with the Dress Up Doll enroll program can also apply here, as can the "libpng warning."

LISTING 18-7: **Old McDonald's Farm Enroll**

```
#!/usr/bin/python3
# Old McDonald's Farm card enrolling - by Mike Cook

from rc522rfid import Rc522rfid
import pygame, time, os

pygame.init()                    # initialise graphics interface
os.environ['SDL_VIDEO_WINDOW_POS'] = 'center'
```

(continued)

LISTING 18-7: *(continued)*

```
pygame.display.set_caption("Old McDonald's Farm Enroll")
pygame.event.set_allowed(None)
pygame.event.set_allowed([pygame.KEYDOWN,pygame.QUIT])
screenWidth = 723 ; screenHeight = 369
screen = pygame.display.set_mode([screenWidth,screenHeight],  0,32)
pygame.key.set_repeat(500, 20)

rfidReader = Rc522rfid()
needsRedraw = True
enroll = False
seeAll = False

def main():
    global xPlot, yPlot, done, enroll, needsRedraw, currentA, currentType
    print ("Old McDonald's Farm - Enrolling Cards")
    print ("Return to move the next animal in the group - position using cursor ↵
      keys")
    print ("a - toggle seeAll animals together,  e - Enrol card when all five are ↵
      in position")
    init()
    drawScreen(0,0)
    while 1:
        checkForEvent()
        for i in range(0,len(animalNames)):
            currentType = i
            done = False
            needsRedraw = True
            while not done:
              checkForEvent()
              if needsRedraw:
                drawScreen(i,currentA)
            if enroll :
              enroll = False
              print("enrolling")
              enrollCard(animalNames[i])

def drawScreen(cType,Cindex):
    global needsRedraw
    pygame.draw.rect( screen,(255,255,255), (0,0,screenWidth, screenHeight),0)
    screen.blit(farmBuilding,(0,0))
    if seeAll:
        for j in range(0,len(animalNames)):
            for i in range(0,5):
                screen.blit(animalPictures[j],(xPlot[j][i], yPlot[j][i]))
    else:
        for i in range(0,5):
            screen.blit( animalPictures[cType],(xPlot[cType][i], yPlot[cType][i]))
```

```
        screen.blit(farmFence,(0,270))
        pygame.display.update()
        needsRedraw = False


def init():
    global animalNames, animalPictures, farmBuilding, farmFence
    global xPlot,yPlot,currentType,currentA, animalTypesNumber
    path = os.path.realpath(__file__)
    path = os.path.dirname(path) + "/farmPictures/"
    #get a list of files
    animalNames = [fn for fn in next(os.walk(path))[2]]
    list.sort(animalNames) # put in alphabetical order
    animalTypesNumber = len(animalNames)
    print("you have these animals\n",animalNames)
    animalPictures= [ pygame.image.load("farmPictures/" + animalNames[i]). ↵
      convert_alpha()
                    for i in range(0,len(animalNames))]
    farmBuilding = pygame.image.load( "farmPictures/fixed/farm.png").
    convert_alpha()
    farmFence = pygame.image.load("farmPictures/fixed/fence.png" ↵
      ).convert_alpha()
    xPlot = [ [144,144,144,144,144] for i in range(0,len(animalNames))]
    yPlot = [ [200,200,200,200,200] for i in range(0,len(animalNames))]
    currentType = 0 ; currentA = 0



def enrollCard(name):
    print("Enroling", name, "place a card on the reader")
    data1 = [] ; data2 = [] ; data3 = []
    for i in range(0,16): # fill with a sectors worth of zeros
        data1.append(0) ; data2.append(0) ; data3.append(0)
    for i in range(0,len(name)):
        data1[i] = ord(name[i])
    #save position of these animals
    k = 0
    for i in range(0,4):
        data2[k] = xPlot[currentType][i] >> 8
        data2[k+1] = xPlot[currentType][i] & 0xFF
        data2[k+2] = yPlot[currentType][i] >> 8
        data2[k+3] = yPlot [currentType][i] & 0xFF
        k+=4
    data3[0] = xPlot[currentType][4] >> 8
    data3[1] = xPlot[currentType][4] & 0xFF
    data3[2] = yPlot[currentType][4] >> 8
    data3[3] = yPlot [currentType][4] & 0xFF
    # This is the (default) key for authentication
    key = [0xFF,0xFF,0xFF,0xFF,0xFF,0xFF]
```

(continued)

LISTING 18-7: *(continued)*

```python
        rfidReader.RC522_WriteSector(key,12,data1)
        rfidReader.RC522_WriteSector(key,13,data2)
        rfidReader.RC522_WriteSector(key,14,data3)
        print("finished enrolling\n")

def terminate(): # close down the program
    print ("Closing down please wait")
    pygame.mixer.quit()
    pygame.quit() # close pygame
    os._exit(1)

def checkForEvent(): # see if we need to quit
    global xPlot, yPlot, needsRedraw,done, enroll, currentA
    global currentType, seeAll
    event = pygame.event.poll()
    if event.type == pygame.QUIT :
        terminate()
    if event.type == pygame.KEYDOWN :
        if event.key == pygame.K_ESCAPE :
          terminate()
        if event.key == pygame.K_RETURN :
          currentA += 1
          if currentA >= 5:
             currentA = 0
          enroll = False
        if event.key == pygame.K_SPACE :
          done = True
          enroll = False
        if event.key == pygame.K_e :
          done = True
          enroll = True
        if event.key == pygame.K_a :
          seeAll = not seeAll
          needsRedraw = True
        if event.key == pygame.K_UP :
          yPlot[currentType][currentA] -= 1
          needsRedraw = True
        if event.key == pygame.K_DOWN :
          yPlot[currentType][currentA] += 1
          needsRedraw = True
        if event.key == pygame.K_RIGHT :
          xPlot[currentType][currentA] += 1
          needsRedraw = True
        if event.key == pygame.K_LEFT :
          xPlot[currentType][currentA] -= 1
          needsRedraw = True
```

```
        if event.key == pygame.K_s :
            print (xPlot[currentType][currentA], yPlot[currentType][currentA])

# Main program logic:
if __name__ == "__main__":
    main()
```

REMEMBER

The screenWidth and screenHeight variables must be set to the size of the farm.
png image.

To start enrolling, use the cursor keys to position the first occurrence of an animal. When it's in the right position, press Return and position the next one. After
all five animals of the one type have been positioned, press the spacebar to move
on to the next animal. At any time, you can see all the animals on screen at the
same time by pressing the A key. After you have defined the position of all your
animals, you can use the spacebar to step through them, pressing the E key to
enroll each one. After you have enrolled all the cards, it's ready for your sing-
song. The final program is shown in Listing 18-8.

TIP

If you end the program by pressing the X in the corner of the window, Thonny will
give you a Backend terminated message. Simply click the red stop icon to restart
things, or use this icon to end the program in the first place.

LISTING 18-8: **Old McDonald's Farm Sing-Song**

```
#!/usr/bin/python3
# Old McDonald's Farm -- a song by Mike Cook

from rc522rfid import Rc522rfid
import pygame, time, os

pygame.init()                        # initialise graphics interface
pygame.mixer.quit()
pygame.mixer.init(frequency=22050, size=-16, channels=2, buffer=512)
os.environ['SDL_VIDEO_WINDOW_POS'] = 'center'
pygame.display.set_caption("Old McDonald's Farm")
pygame.event.set_allowed(None)
pygame.event.set_allowed([pygame.KEYDOWN,pygame.QUIT])
screenWidth = 723 ; screenHeight = 369
screen = pygame.display.set_mode( [screenWidth,screenHeight],0,32)

rfidReader = Rc522rfid()
```

(continued)

LISTING 18-8:   *(continued)*

```python
farm = [] # what animals are on the farm today
play = False
cardsRead = -1
verse = 0

def main():
    global play,xPlot,yPlot,cardsRead,farm
    print ("Old McDonald's Farm")
    print ("Present cards then Press - p = Play - n = New")
    init()
    while 1:
        drawScreen(-1,0,0)
        while not play:
            checkForEvent()
            if readCard() == 0:
                print("Farm now contains ",end="")
                for i in range(0,len(farm)):
                    print(rawNames[farm[i]]," ",end="")
                print()
                rfidReader.RC522_WaitForCardRemoved()
        singSong()
        play = False
        farm = [] ; xPlot = [] ; yPlot = []
        cardsRead = -1
        print("Place cards to populate farm")

def drawScreen(n,level,ind):
    screen.blit(farmBuilding,(0,0))
    if cardsRead != -1 and n != -1:
        if verse > 0 and level >0: # draw previous animals
            i = verse
            while i > ind:
                k = farm[i]
                for j in range(0,5):
                    screen.blit(animalPictures[k],(xPlot[i][j], yPlot[i][j]))
                i -=1
        k = farm[ind]
        for j in range(0,n): # draw latest animal
            screen.blit(animalPictures[k],(xPlot[ind][j],yPlot[ind][j]))

    screen.blit(farmFence,(0,270))
    pygame.display.update()

def readCard():
    status = rfidReader.RC522_ReadCard()
    if status == -1:
        return -1
```

```
    # This is the (default) key for authentication
    key = [0xFF,0xFF,0xFF,0xFF,0xFF,0xFF]
    block1 = rfidReader.RC522_GetSector(key,12)
    block2 = rfidReader.RC522_GetSector(key,13)
    block3 = rfidReader.RC522_GetSector(key,14)
    if block1 != -1 and block2 != -1 and block3 != -1: # if data OK
        lookAtNew(block1,block2,block3)
        return 0
    else:
        print("Error in reading card")
        rfidReader.RC522_WaitForCardRemoved()


def lookAtNew(s1,s2,s3): # look at the new element
    global xPlot,yPlot, cardsRead
    name = ""
    i = 0
    notDone = True
    while notDone and i < 16: # generate name
        c = s1[i]
        i+=1
        if c !=0:
          name = name + chr(c)
        else:
          notDone = False
    for i in range(0,len(animalNames)):
        if animalNames[i] == name: # find the name on the card
          farm.append(i)
          xPlot.append([0,0,0,0,0])
          yPlot.append([0,0,0,0,0])
          cardsRead += 1
          k = 0
          for j in range(0,4):
              xPlot[cardsRead][j] = s2[k]<< 8 | s2[k+1]
              yPlot[cardsRead][j] = s2[k+2]<< 8 | s2[k+3]
              k += 4
          xPlot[cardsRead][4] = s3[0]<< 8 | s3[1]
          yPlot[cardsRead][4] = s3[2]<< 8 | s3[3]


def init():
    global animalNames, animalPictures, farmBuilding, farmFence, animals
    global xPlot,yPlot, noises, rawNames, fixed
    path = os.path.realpath(__file__)
    path = os.path.dirname(path) + "/farmPictures/"
    #get a list of files
    animalNames = [fn for fn in next(os.walk(path))[2]]
    list.sort(animalNames) # put in alphabetical order
    rawNames = []
```

(continued)

LISTING 18-8: *(continued)*

```
    for i in range (0,len(animalNames)): # remove file extension
        rawNames.append(animalNames[i][0:len(animalNames[i])-4])
    #load in the sounds
    animals = [ pygame.mixer.Sound("farmSounds/" +rawNames[sound]+".ogg")
                    for sound in range(0,len(rawNames))]
    noises = [ pygame.mixer.Sound("farmSounds/"+ rawNames[sound]+"Sound.ogg")
                    for sound in range(0,len(rawNames))]
    files = ["start","and","ee_eye_oh","witha","here","anda", "there","herea","th ↵
      erea","everywherea","end", "well" ]
    fixed = [ pygame.mixer.Sound("farmSounds/fixed/" +files[sound]+".ogg")
                    for sound in range(0,len(files))]
    #load in the pictures
    animalPictures= [ pygame.image.load("farmPictures/" +animalNames[i]). ↵
      convert_alpha()
                    for i in range(0,len(animalNames))]
    farmBuilding = pygame.image.load("farmPictures/fixed/farm.png" ↵
      ).convert_alpha()
    farmFence = pygame.image.load("farmPictures/fixed/fence.png" ↵
      ).convert_alpha()
    xPlot = [] ; yPlot = []

def singSong():
    global verse
    verse = 0
    while verse < len(farm): # repeat for each verse
        for i in range(0,3): # start part
            fixed[i].play()
            waitFinish()
            if i == 1:
              animals[farm[verse]].play()
              waitFinish()
            if i == 2:
                farmYard(verse,verse)
                verse += 1

def farmYard(index,verse): # sing verse
    numDisplayed = 0
    level = 0
    while index != -1:
        for i in range(3,10):
            fixed[i].play()
            waitFinish()
            if i == 3 or i==5 or i==9:
                noises[farm[index]].play()
                numDisplayed +=1
                drawScreen(numDisplayed,level,index)
                waitFinish()
```

```python
                noises[farm[index]].play()
                waitFinish()
            if i == 7 or i == 8:
                noises[farm[index]].play()
                numDisplayed +=1
                drawScreen(numDisplayed,level,index)
                waitFinish()
        index -=1
        level +=1
        numDisplayed = 0
    if verse < len(farm)-1: # more verses
        fixed[0].play()
        waitFinish()
        time.sleep(0.2)
        fixed[11].play() # well
        drawScreen(-1,0,0) # clear the farm
    else:                   # end of song
        fixed[10].play()
    waitFinish()
    time.sleep(0.3)

def waitFinish():
    while pygame.mixer.get_busy():
        checkForEvent()

def terminate(): # close down the program
    print ("Closing down please wait")
    pygame.mixer.quit()
    pygame.quit() # close pygame
    os._exit(1)

def checkForEvent(): # see if we need to quit
    global play, farm, cardsRead
    event = pygame.event.poll()
    if event.type == pygame.QUIT :
        terminate()
    if event.type == pygame.KEYDOWN :
        if event.key == pygame.K_ESCAPE :
            terminate()
        if event.key == pygame.K_p :
            play = True
        if event.key == pygame.K_n :
            farm = []
            cardsRead = -1

# Main program logic:
if __name__ == "__main__":
    main()
```

To play, first present — one at a time — all the cards for animals you want on the farm. Then press the P key to play the song. Repeat this action to play the song again, maybe with different animals or in a different order. Figure 18-7 shows some of the occupants of our farm.

FIGURE 18-7:
Old McDonald's
farm.

6

The Part of Tens

Download and install ten great software packages for your Raspberry Pi.

Be inspired by ten innovative projects for the Raspberry Pi.

Find ten great add-ons for the Raspberry Pi.

Chapter **19**

Ten Great Software Packages for the Raspberry Pi

O ne of the best things about the Raspberry Pi is that you can easily download *so* many software packages over the Internet and install them. In this chapter, we give you some pointers to ten software packages to get you started.

Before you start, issue the following command in the shell to make sure your software cache is up to date:

```
sudo apt update
```

The software you run on your computer is as much a matter of taste as the music you listen to, so we hope you use this list as a starting point and then make your own software discoveries. For a full explanation of finding and installing software on your Raspberry Pi, see Chapters 4 and 5.

Penguins Puzzle

Penguins Puzzle, shown in Figure 19-1, is a 3D puzzle game in which you're tasked with safely escorting a penguin to the exit without letting it fall off the iceberg and into the freezing water. You use the cursor keys to move around, press Z to zoom out for a wider-angle view, and press R to reset the level. The game has 50 levels to test your mettle.

FIGURE 19-1:
Penguins
Puzzle is a cute
3D puzzle game.

To install the game, search for "penguinspuzzle" in the Add/Remove Software tool, located in the Preferences section of the Applications menu, or, in the shell, use the following command:

```
sudo apt install penguinspuzzle
```

Penguins Puzzle doesn't work in the desktop environment, including the Terminal window. Press Ctrl+Alt+F1 to go to a Linux virtual console. Then type **penguinspuzzle** to run the game. Press Esc to end the game and then Ctrl+Alt+F7 to return to the desktop.

The software is *charityware,* which means you're invited to make a donation to charity if you enjoy playing it.

For more information on Penguins Puzzle, go to `http://penguinspuzzle.appspot.com/.`

FocusWriter

Whether you're writing the next blockbuster from your bedroom or you just need to get your work done without distraction, FocusWriter may be the application for you. It's a word processor that's designed to be distraction-free. Most of the time when you're using it, the only thing onscreen is your writing.

When you move the mouse to the top of the screen, the menus for changing the settings and saving your files appear. To keep your motivation up, you can set a daily goal in the Preferences settings for time spent writing or (better still) words written per day. When you move the mouse to the bottom of the screen, you can see the word count and your progress toward your daily goal.

To install FocusWriter, search for "focuswriter" in the Add/Remove Software tool or, in the shell, use the following command:

```
sudo apt install focuswriter
```

To start FocusWriter, go into your desktop environment and click the program's entry in the Office category of your Applications menu.

You can find out more about the application at `https://gottcode.org/focuswriter.`

Mathematica

Mathematica is what's known as a *symbolic* package, or a computer algebra system (CAS), and it's one of the recommended applications in Raspberry Pi OS. Mathematica is one of the best systems for exploring anything to do with numbers, from mathematics to complex multidimensional graphics and music.

To get started, click the Applications menu, choose the Programming category, and click the Mathematica icon. You see a splash screen, and then two windows open: a blank notebook and, in front of it, an invitation to visit three websites.

Click the notebook to bring it to the front. Type **2^8** and press Return. This expression says "two to the power of eight," but you don't see an answer. You've entered the expression into Mathematica, but in order to tell the program to evaluate it (and give you the answer), you have to press Shift+Return.

Mathematica can expand equations for you:

```
Expand[(1+x)^6]
1 + 6x + 15x² + 20x³ + 15x⁴ + 6x⁵ + x⁶
```

It can plot graphs, such as these parametric plots:

```
For[n=1, n<4, n++,
ParametricPlot[ {Sin[n t], Sin[(n+1) t]}, {t, 0, 2Pi}] //↵
  Print]
```

Graphical output might take a moment to render, so be patient, if necessary. Mathematica even plots 3D graphics:

```
SphericalPlot3D[Sin[t] Cos[t] Sin[f], {t, 0, Pi}, {f, 0, 2 Pi}]
```

The bottom of Figure 19-2 shows what Mathematica comes up with given this input.

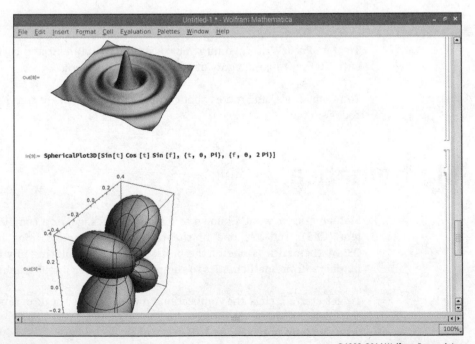

FIGURE 19-2:
Two plot examples from Mathematica.

One of Mike's favorite shapes is generated by the following code:

```
Plot3D[Sin[Sqrt[x^2 + y^2]]/Sqrt[x^2 + y^2],
{x, -6 Pi, 6 Pi}, {y, -6 Pi, 6 Pi},
Boxed -> False, Mesh -> False, PlotPoints -> 60,
PlotRange -> All, Axes -> False]
```

Try it out to see what it looks like!

Fraqtive

Fractals are patterns generated using mathematical formulas that are self-similar. That means if you zoom in on the Mandlebrot set (shown on the left in Figure 19-3), for example, you'll find that the same shape repeats in its nooks and crannies, and you can zoom in again and again and again. Fraqtive is a program for exploring fractals and generating images. You can save the images and use them as wallpaper on your Raspberry Pi (see Chapter 4). The software has a tutorial to get you started.

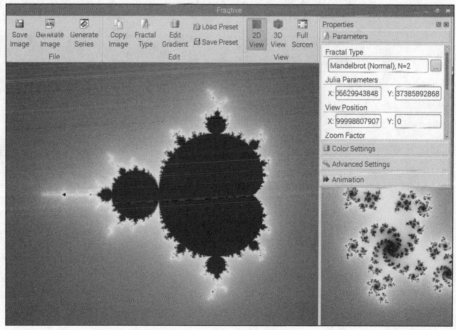

FIGURE 19-3:
Generate colorful fractal images easily using Fraqtive.

Michał Męciński

To install Fraqtive, search for "fraqtive" in the Add/Remove Software tool or, in the shell, use the following command:

```
sudo apt install fraqtive
```

After installation, you can find Fraqtive in the Education category of your Applications menu.

For more information on Fraqtive, visit the creator's website at `https://fraqtive.mimec.org`.

Tux Paint

Tux Paint, shown in Figure 19-4, is a simple drawing program for children, with tools that help them to quickly create art on the Raspberry Pi. In addition to enabling freehand drawing and the placement of shapes and lines in common with most art packages, Tux Paint has a Magic tool. This tool can be used to create effects such as brick walls, flowers, snowballs, rainbows, waves, and various creative image distortions. The Stamp tool is used to stamp clip art onto the screen, including animals, penguins, hats, food, and musical instruments.

FIGURE 19-4:
Tux Paint turns every child into an artist. And us, too.

The Tuxpaint Project (www.tuxpaint.org)

Tux Paint is named in tribute to Tux, the penguin who is the official mascot of the Linux kernel. The application has been created with the help of more than 300 contributors worldwide and has been downloaded tens of millions of times.

To install Tux Paint, search for "tuxpaint" in the Add/Remove Software tool or, in the shell, use the following command:

```
sudo apt install tuxpaint
```

After you've installed Tux Paint, you can start it from the Education category of your Applications menu.

The official website for Tux Paint can be found at www.tuxpaint.org.

Grisbi

If you want to manage your financial accounts on your Raspberry Pi, Grisbi is a free application you can use to keep track of your regular and one-off payments. Although other programs are also available, Grisbi is the easiest one we've tried, both to set up and keep updated. Many banks enable you to download your bank statements in a format that can be used in Grisbi, so you may be able to analyze your financial situation without too much rekeying.

To install Grisbi, search for "grisbi" in the Add/Remove Software tool or, in the shell, use the following command:

```
sudo apt install grisbi
```

You can find it in the Office category of your Applications menu.

Beneath a Steel Sky

The Beneath a Steel Sky game, shown in Figure 19-5, tells a science-fiction story about Robert Foster, a boy who survived a helicopter crash and was raised by indigenous Australians in a wasteland called The Gap. Many years later, after Robert has grown up, armed forces arrive in another helicopter, kidnap him, and fly him back to the city. He escapes, and you pick up the controls to guide him on his journey of discovery. Why is he here? Who is in charge?

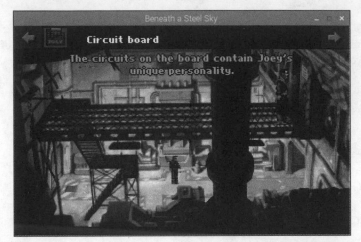

FIGURE 19-5:
Beneath a Steel Sky, an interactive science-fiction story.

It's a point-and-click adventure game, which means you solve puzzles and interact with the environment using the mouse cursor and clicking objects and people. You use the left mouse button to examine things and the right mouse button to take an action (such as opening or closing a door, picking up an object, or looking through a window). You can talk to characters in the game by clicking them and choosing from the provided phrases. When you move the cursor to the top of the screen, the inventory of items you're carrying appears so that you can use items you're carrying. To walk through an exit, click it.

The game's fantastic opening sequence and witty dialog draw you in, and the solution is available online if you'd like to experience the full story but you've gotten stuck on one of the puzzles.

This hit game from 1994 was officially released as freeware in 2003 and is now available for you to install on your Raspberry Pi. Search for "steel-sky" in the Add/Remove Software tool or, in the shell, enter the following command:

```
sudo apt install beneath-a-steel-sky
```

The game is installed into the Games category of your Applications menu.

Brain Party

If you fancy a few minutes of gaming to tune up your brain between programming sessions, Brain Party is here for you. It's a series of fun minigames, designed to stretch the gray matter. You complete five randomly selected tests to get your "brain weight" score, and you can practice each minigame when you've unlocked it in a test. Puzzles will challenge your memory, observation skills, logic skills, and reactions. It's family friendly, too.

You can install it by searching for "brainparty" in the Add/Remove Software tool (the game is listed as "36 puzzle games for all the family") or, in the shell, enter the following command:

```
sudo apt install brainparty
```

You can run Brain Party from the Applications menu (find it under Games) or from the command line by entering **brainparty**.

Pure Data

Pure Data (Pd) is a programming language that allows you to create and manipulate sound using a visual interface. You create programs, or *patches* as they're known, by creating boxes that "do something" and connecting them together with "signal" wires.

Pure Data was created in the '90s by Miller Puckette but is now an open-source project. A similar commercial product also created by Miller Puckette is called Max/MSP. Although the interface and graphics look a lot prettier in Max/MSP, Pd is essentially the same thing.

You can use Pd to create sound synthesizers of many types (for example, FM, wave table, sample players, or granular synths). You can make sequencers for drums or notes and connect them to the sound generators. You can input sounds from a microphone or MP3 player and manipulate the sound, adding distortion, echo, reverberation, and pitch shifting, to name just a few effects.

You can take digital inputs from a MIDI keyboard or even from the Raspberry Pi's general-purpose input/output (GPIO) pins. Pd can output MIDI signals for both control and notes. All that is just scratching the surface of what Pd can do!

You can install it by searching for "puredata" in the Add/Remove Software tool or, in the shell, enter the following command:

```
sudo apt install puredata
```

After it's installed, you'll find it in the Applications menu under Sound & Video. When it asks whether you want Pd to create a documents directory, select Yes.

From the menu (shown in Figure 19-6), choose Media and then Test Audio and MIDI. Click the Test Tones 60 box, and you should hear a pure sine wave tone. Click and drag on the pitch box, and you'll change the frequency of the tone.

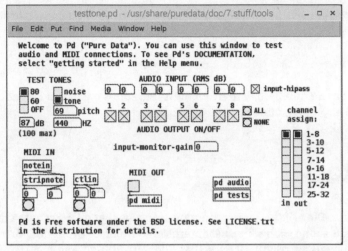

FIGURE 19-6: Testing Pure Data (Pd).

To test the audio input, you need a USB sound card. Select it by choosing Media ➪ Audio Settings. You see changing numbers, indicating the volume, in the first two boxes of the audio input row. When you plug in a MIDI keyboard, it should be recognized. In the Media menu, make sure that OSS-MIDI is selected. Go to the MIDI Settings in the Media menu to select your MIDI device — it'll look something like /dev/midi3.

For a keyboard, pressing a note key will show the note number and velocity, and turning a control knob will show the value and control number.

We made a simple theremin with Pd, shown in Figure 19-7. You just drag the bars for changing the volume and frequency with the mouse. We then extended this to use sensors connected to the Raspberry Pi's GPIO pins.

For more information on Pd, go to https://puredata.info. There's a tutorial at www.pdpatchrepo.info/puredata. For advice on adding GPIO access, go to http://nyu-waverlylabs.org/rpi-gpio.

Although we've only used it for sound, you can also use Pd to manipulate video. Go to https://youtu.be/XyS2M0mM5iA for a demonstration.

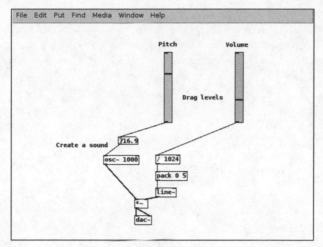

File   Edit   Put   Find   Media   Window   Help

Pitch   Volume

Drag levels

Create a sound   716.9

osc~ 1000   / 1024

pack 0 5

line~

*~

dac~

Inkscape

Inkscape is a vector drawing package like Adobe Illustrator. Unlike Illustrator, it's open-source and free to download and use. In contrast to photo-editing packages like Gimp, Inkscape doesn't manipulate the actual pixels of a drawing. Those sorts of systems are known as *bitmapped* systems. With a *vector* system, what's stored and manipulated are instructions for creating a drawing. At its simplest level, you have objects like squares, circles, and lines, and these combine to form a drawing when you play the drawing through a rendering engine.

This sounds complex — why bother? Well, when dealing with a bitmap, there is only so much you can enlarge a picture without people being able to see the pixels. With a vector image, you can scale it to however big or small you want and you'll *never* see any jagged lines (known as *pixelation*). This is just what you need when it comes to rendering an image on a device with a much greater resolution than your computer screen (for example, a printer).

But more than this, a vector image is what you need if you're going to pass a computer drawing onto some computer numerical controlled (CNC) device like a milling machine, laser cutter/engraver, vinyl cutter, or 3D printer. Using those machines, the individual lines and curves can be made into a real object. Figure 19-8 shows the design for beer mat for a singer/songwriter Mike knows. This mat was cut out of ⅛-inch plywood and engraved on a laser cutter; then it was coated in matte-finish varnish and sold at her gigs.

You can install it by searching for "inkscape" in the Add/Remove Software tool or, in the shell, enter the following command:

```
sudo apt install inkscape
```

Sean McManus

Chapter **20**

Ten Inspiring Projects for the Raspberry Pi

I f you've read the rest of this book and worked through the projects, you now know how to program and how to create your own electronics projects on the Raspberry Pi. What you learn next, and what you create with that knowledge, is up to you.

It's amazing to see what people of all ages are doing with their Raspberry Pis. In this chapter, we've collected some of the most interesting and inspiring projects we've come across. Each one has a link so that you can find out more and perhaps follow instructions to replicate the project — or get some advice for similar projects of your own.

One-Button Audiobook Player

```
https://github.com/exitnode/theonebuttonaudiobookplayer
```

Michael Clemens has used the Raspberry Pi to create an audiobook player for his wife's grandmother, who is visually impaired and finds digital audio players difficult to use.

This project requires some electronics work — you add transistors, an LED, a pair of speakers, and a large button to a plastic case and link the button and LED to the Raspberry Pi's GPIO pins.

A Python script enables the button to control the media player software: Pressing the button pauses or plays the audiobook, and holding it down for 4 seconds sends it back one track.

To change the audiobook, you just plug in a USB drive with the new audiobook on it. It's automatically copied across to the Raspberry Pi, replacing the old audiobook.

Heartbeat Monitor

https://magpi.raspberrypi.org/articles/heartbeat-monitor

Daniel Fernandez connected a heart monitor to a Raspberry Pi so that he could have more flexibility in capturing and using its data. The aim was to be able to see a graph showing the results while he ran on the treadmill, and to be able to export the results in a standard format for in-depth analysis, such as CSV, which can be read by Excel or LibreOffice.

He used a Polar H7 heartbeat sensor and a Raspberry Pi 3 with a 3½-inch screen, which displays an easy-to-read graph as he runs. Everything was packaged in a protective case.

Smart Fridge

https://github.com/InitialState/smartbeerfridge/wiki

Jamie Bailey created a beer fridge that keeps count of how many bottles there are and sends a text message to raise the alarm if houseguests try to pilfer your beer. It uses a Wii Balance Board underneath the fridge to weigh its contents and work out how many bottles are left.

The Raspberry Pi talks to the Wii Balance Board using Bluetooth, and an online dashboard shows how many beers have been taken and how many remain.

The dashboard uses the Initial State platform for visualizing Internet of Things (IoT) projects. Initial State has published detailed instructions for making your own smart beer fridge at the link earlier.

The Next Verse

www.stewarteaston.net/archive-1#/the-next-verse

Stewart Easton and Gawain Hewitt created The Next Verse, an interactive embroidered artwork that depicts the cycle of life. Each family scene includes a triangle of conductive thread that can be touched to play relevant sound effects. The childhood scene, for example, includes sounds recorded in a school playground.

The project uses the Touch Board, made by Bare Conductive. It enables sounds to be triggered when the board (or conductive material connected to it) is touched. By adding a Raspberry Pi, it was possible to use higher-quality sounds and to have more than one sound playing at the same time. The piece includes a 23-minute piece of music by Michael Tanner, which plays in the background.

The software to trigger the sound effects and play the music was written using Pure Data (Pd), a visual programming language for multimedia (see Chapter 19).

The Next Verse has been exhibited at London's Victoria and Albert (V&A) Museum.

For more information, see the article in Issue 67 of *The MagPi*. It's available online as a free PDF at https://magpi.raspberrypi.org/issues/67.

Electric Skateboard

https://youtu.be/2WLEur3M8Yk

You can use a Raspberry Pi as the guidance system on a motorized skateboard, as TheRaspberryPiGuy demonstrated on YouTube. He attached an Alien Power Systems motor to the skateboard and used a Pi Zero to control it.

Acceleration is controlled using a remote control from Nintendo's Wii console, sending signals to the Pi over Bluetooth. In his video, you can see him speeding through the streets of Cambridge, UK hometown of the Raspberry Pi Foundation. His top speed is 30 kilometers an hour, and he estimates that the battery power is enough for at least 10 kilometers.

As an upgrade, you could add a Pi-powered speedometer, as shown in this tutorial, which also shows you how to build the board itself: www.instructables.com/The-Longboard-Speedometer.

T-Shirt Cannon

http://drstrangelove.net/2014/01/raspberry-pi-powered-t-shirt-cannon

David Bryan and Lucas Saugen turned to the Raspberry Pi when they were asked to repair the T-shirt cannon used at the Minnesota RollerGirls' roller derby matches. The cannon is used to fire T-shirts into the audience during the time-outs, which last for up to about a minute-and-a-half. They wanted a design that would enable more than one T-shirt to be fired in that period, and that would also enable the cannon to tweet a photo when a T-shirt was fired.

The resulting design uses four clear PVC tubes for the barrels, and compressed air to fire the T-shirts. The cannon has three buttons: one to choose which barrel to fire from, and the other two that are pressed together to fire a shirt. As an additional safety measure, a key is required to arm the cannon. The Raspberry Pi is used to control the device, with software written in Python that uses the GPIO Python libraries.

To find out more about how the cannon was built, watch videos, and download the code, visit David's website.

Magic Mirror

www.raspberrypi.org/blog/twitter-triggered-photobooth

For London Fashion Week, Photobot.Co created a photo booth for The Body Shop that's guaranteed to capture your best side. They built three vertical panels, each of which has five Raspberry Pis and Raspberry Pi cameras in it. Together, they make a Magic Mirror (see Figure 20-1) that compiles full-body portraits of subjects, angled from the left, the right, and face-on.

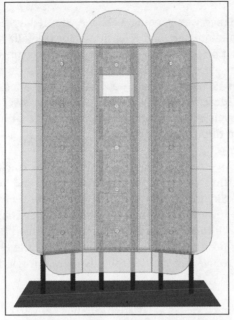

FIGURE 20-1:
The design for
the Magic Mirror,
showing where
the Raspberry Pi
cameras (the
small circles) are
mounted in the
mirror surfaces.

Sean McManus

The photo booth was triggered when somebody tweeted the booth to say what color she was wearing. The composite photo, a single image that showed all three angles, was tweeted back to her. The tweet also included a response that was appropriate for the color she said she was wearing if it was one of the hundred recognized color names. The system used a Mac Mini as the central controller (although the developers say they later found that a Raspberry Pi could have done this job, too). The controller ran software in Python that listened for tweets, told the Raspberry Pis to capture images and send them to it, and then created the composite image. The controller connected to the Raspberry Pis using Secure Shell (SSH), so it could control them remotely and access their image data.

Pi in the Sky

www.pi-in-the-sky.com

Dave Akerman uses the Raspberry Pi for high-altitude ballooning, using hydrogen-filled balloons. The balloons go high enough to photograph a slight curvature in the earth, with the darkness of space visible above the atmosphere. After two or three hours of ascent, the balloons burst and release their payload to fall to Earth on a parachute.

Dave's favorite experience was sending Babbage, the teddy bear mascot of the Raspberry Pi, up to 39 kilometers and releasing it for freefall. "The first attempt failed due to a combination of factors — a windy launch meant Babbage's support line wasn't in the right place around the heated resistor that was to cut him free, and for weight reasons I'd been overcautious in using a smaller and less powerful battery pack than was ideal," says Dave. "Success is always sweeter after an initial failure, and it was a lovely moment when I knew for sure that he'd been released properly on the second flight."

Dave's early projects involved hand-building systems for tracking flights using GPS and for transmitting positioning signals, photos, and other data to the ground. The Pi in the Sky kit is now available to provide these features off the shelf. You can also use the kit in combination with the Sense HAT (see Chapter 15) to send sensor data back to the ground. Among other things, Dave's used the Pi to predict the landing position on the way down. Part of the challenge of high-altitude ballooning is recovering the payload, given that predictions can be off by 5 miles or more.

High-altitude ballooning can be a satisfying hobby, but it's one that requires planning and commitment. You will most likely need permission for each flight so that you don't endanger aircraft, and you'll need to budget several hundred dollars or pounds for the equipment, and a couple of hundred for the balloon, gas, and fuel for the chase vehicle on each flight. Plenty of help is available, though, to help you make a success of it: Start by reading Dave's tutorial at `www.daveakerman.com/?p=1732`, join the #highaltitude channel at `https://webchat.freenode.net`, and visit the UK High Altitude Society website (`https://ukhas.org.uk`).

Raspberry Turk

`www.raspberryturk.com`

You might have heard of the Mechanical Turk, an elaborate illusion that was claimed to be a chess-playing robot from 1770 until 1854. In fact, a chess master was hidden inside the table, making the moves and controlling the dummy of a man in Turkish dress who appeared to be playing the game. The Mechanical Turk went on to give its name to a service from Amazon where people complete tasks over the Internet, and now to Raspberry Turk, a genuine chess-playing robot.

Raspberry Turk is built around the open source Stockfish chess engine, which plays the games. A Raspberry Pi is used to control a robot arm that moves the chess pieces. A Raspberry Pi Camera Module is used with computer vision software to see when it's the computer's turn to play, by checking for the player's move on the board. The whole thing is built into a table, which has the chess board painted onto its surface.

You can find out more about this project, which combines elements of artificial intelligence, robotics and computer vision, at the project's website.

Sound Fighter

www.foobarflies.io/pianette

To mark the reopening of the Maison de la Radio in Paris as a cultural center, Jean Weessa and Mélanie Pennec had the idea of creating an installation that uses two upright pianos as game controllers for Street Fighter on the PlayStation.

Eric Redon and Cyril Chapellier brought the idea to life using a Raspberry Pi Model B+ as the brains of the operation, receiving all the piano keypresses, and piezo sensors attached to the piano hammers that detect when a key is struck. The project also required them to create custom-printed circuit boards, and use an Arduino Uno to feed the keypresses into the PlayStation. Their code was written in Python 3.

To create a gameplay experience that made good pianists good gamers, and also made it sound as musical as possible, player movements were assigned to the left-hand keys, actions were assigned to the right-hand keys and both pianos played in the same musical scale. Later modifications of the project include using the foot pedals for long button presses and expansion to the additional games Tekken5 (another beat-'em-up) and Crash NitroKart (a racing game).

You can read about the challenges involved in building this project, and how they were overcome, at the website. You can also find the code and designs if you want to replicate the project.

IN THIS CHAPTER

» **Building an arcade machine**

» **Making a robot**

» **Setting your Raspberry Pi to turn on or off on schedule**

» **Adding new inputs: a piano keyboard, temperature sensor, and gesture sensor**

» **Adding new outputs: LEDs, LCDs, electronic paper, and a speaker**

Chapter **21**

Ten Great Add-Ons for the Raspberry Pi

I n addition to being able to connect your own electronics projects to the Raspberry Pi's general-purpose input/output (GPIO) pins, you can use the pins to mount add-on boards. Extension boards that sit on top of the Pi boards are often called HATs (*HAT* is short for "hardware attached on top"). HATs comply with a standard and are sized to sit on a full-size Raspberry Pi without overhanging it. Add-ons that are the same size as the Pi Zero are often called pHATs, short for *partial HATs*. You can use HATs or pHATs with any Raspberry Pi computer.

If you're using a Raspberry Pi 400, the GPIO pins are parallel to the desk, pointing away from you. On other devices, they point up to the ceiling. To use HATs as designed with the Raspberry Pi 400, you'll need an adapter to point the pins in the right direction. You can use a simple extension cable, but there are plug-in converters that turn the GPIO pins the right way up for you, too. Pimoroni (www.pimoroni.com) makes the Flat Hat Hacker, and SB Components (https://shop.sb-components.co.uk/) makes the Raspberry Pi 400 GPIO Adapter (see Figure 21-1).

Sean McManus

Although HATs cover all the GPIO pins, they often only use a couple of them. The Mini Black Hat Hack3r by Pimoroni enables you to use a HAT and access an additional set of GPIO pins at the same time. The company's pHAT Stack enables you to connect three HATs or five pHATs at the same time. At `https://pinout.xyz`, you can look up HATs and pHATs to see which GPIO pins they use, so you can work out which add-ons would clash and which would work well together.

Picade

If you've ever wanted to have your own arcade cabinet, Pimoroni's Picade is here. At its simplest, it's a HAT that you can use to connect a joystick and arcade buttons to your Raspberry Pi. Then you can attach a screen and build your own cabinet to house it all. There's also the Picade Console, which includes the buttons, joystick, and speaker. Everything is housed in a unit you can connect to your TV.

For the ultimate Raspberry Pi–based games machine, Pimoroni make a complete kit with either a 10-inch or an 8-inch display. It takes about half a day to build it using the online instructions. Because it uses the Retropie software (see Chapter 2), you can play a wide range of games on it from classic computers and consoles. We found that classic arcade games offered the best experience, because they were designed to work with a joystick and a few buttons. Old computer games run, but they often require a keyboard. You can plug one in, no problem, but it feels more like a real arcade machine when you don't need a keyboard. The sound quality through the built-in speaker is excellent. You'll need to provide your own Raspberry Pi, power supply, and microSD card.

The 10-inch Picade (shown in Figure 21-2) also includes a license for PICO-8, a so-called "fantasy console." PICO-8 is a programming environment that has similar limitations to the computers of the '80s, but it isn't based on any real machine. You can use it to develop your own arcade games, based on the Lua programming language.

FIGURE 21-2:
The Picade console with 10-inch screen.

CamJam EduKit 3

Let's build a robot! The CamJam EduKit 3 contains all the components you need. It has two chunky wheels with motors, a board to connect them to your Raspberry Pi, a distance sensor, and a line-following sensor. You can use the box it comes in as a chassis or create your own case out of LEGO blocks or anything else you have lying around. The instructions show you how to control the motors and sensors in Python. This kit is a solid, affordable introduction to robotics. Find the documentation at `https://camjam.me/?page_id=1035` and order at `www.thepihut.com`.

Piano HAT

This is one of Sean's favorite add-ons for the Raspberry Pi: a tiny touch-sensitive piano, with lights in the keys (see Figure 21-3). It covers an octave (from C to C, including the black keys in between). You can program it using Python to create musical instruments. Sean wrote a simple game that teaches you how to read sheet music using the Piano HAT. You can download it at `https://news.sean.co.uk/2016/03/learn-to-read-sheet-music-on-raspberry.html`.

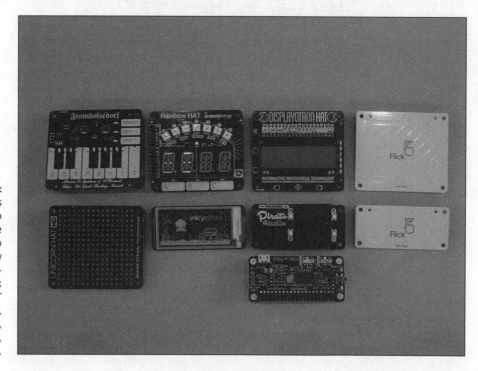

FIGURE 21-3: The HATs in this chapter, left to right from the upper left: Piano HAT, Rainbow HAT, Display-O-Tron HAT, Flick HAT, Unicorn HAT HD, Inky pHAT, Pirate Audio, Flick pHAT, Witty Pi Mini.

Pimoroni also makes a Drum HAT, which gives you eight touch pads with LEDs that you can tap to beat out a rhythm. Drum HAT and Piano HAT can be used together using a device such as the pHAT Stack.

Rainbow HAT

The Rainbow HAT from Pimoroni (refer to Figure 21-3) packs a lot of inputs and outputs into a regular, HAT-size board:

>> An arc of seven multicolored LEDs and four green alphanumeric displays. These displays have more segments than a digital clock, to help you represent letters as well as numbers.

>> A red, a green, and a blue LED above three buttons marked A, B, and C.

>> A buzzer that you can use to make beepy music.

>> A temperature and pressure sensor, making it a great base for a weather station.

>> Pins you can connect to if you want to experiment with the I2C, SPI, or UART interfaces.

Finally, using the Python library, you can program your own projects based on the HAT.

Display-O-Tron HAT

If you want to display a bit more information within a compact HAT, the Display-O-Tron (refer to Figure 21-3) is a good choice. It gives you an LCD screen that can show three rows of 16 characters. The screen has multicolored backlighting, split into six areas from left to right. Up the side of the screen, there are six white LEDs. Around the left and bottom edges are six buttons (labeled as left, right, up, down, confirm, and back — although you can program them to do anything). If you want to add simple circuits on top, you can solder breakout pins onto the board for access to five GPIO pins and several other interfaces.

Flick

The Flick HAT from Pi Supply (https://uk.pi-supply.com/) is a flat board that not only detects swipes and taps, but also can detect hand movements above the board. It's available as a pHAT, too (refer to Figure 21-3), as well as in a large version that measures 5¾ inches by 4 inches. The software can detect taps on the edge and in the center of the board, a hand's height above the board, and a spinning finger gesture above the board. One of the best features is that the non-contact gestures can be detected through some surfaces. You may be able to mount a Flick Large under your desk to turn it into a sci-fi-style control panel. For more information, see Sean's article at `https://news.sean.co.uk/2017/11/adding-gesture-control-to-raspberry-pi.html`.

Pimoroni makes a similar gesture detection board called Skywriter, in the standard HAT size.

Unicorn HAT HD

Everything looks better with twinkly LEDs on it. Pimoroni's Unicorn HAT provides an 8 x 8 grid of multicolored LEDs that you can program from Python. The Unicorn HAT HD (refer to Figure 21-3) increases the resolution to 16 x 16 lights, so you can create more sophisticated displays. The Unicorn HATs can be used to display scrolling text, make simple games, or create dazzling light shows.

TIP

Use a diffuser layer to soften the light. They make the light look nicer and help to shield your eyes from the bright LEDs. You can buy diffuser layers for Raspberry Pi cases from Pimoroni, too.

Inky pHAT

The Inky family of add-ons from Pimoroni adds electronic paper displays to your Raspberry Pi. When you power off, the image remains. The pHAT is available in black and white with red or yellow as a third color (refer to Figure 21-3). The screen takes 15 seconds to refresh and flashes during the refresh, so the Inky pHAT is best used for information that doesn't change frequently. Pimoroni also makes the Inky Impression (a large display at 5¾ inches and with seven colors) and Inky wHAT (with a 4¼-inch display).

Pirate Audio

Pirate Audio, also by Pimoroni, is a range of pHATs you can use to add audio capabilities to your Raspberry Pi. The Pirate Audio Speaker (refer to Figure 21-3) includes a small, high-res display for artwork and a built-in 1W speaker. Four buttons are mounted around the display, too, so you can use the pHAT to create your own audio player. It can also be used to play sound effects when making games in Scratch or Python or music played from VLC.

Witty Pi

The Witty Pi and Witty Pi Mini (refer to Figure 21-3) add real-time clocks to your Raspberry Pi. They can be used to wake up and switch off your Raspberry Pi at set times, so you can run a program on a regular schedule and power down when the Pi isn't needed. The devices are made by UUGear (www.uugear.com).

IN THIS CHAPTER

» Troubleshooting and fixing common problems

» Adjusting the settings on your Raspberry Pi

» Fixing audio problems

» Mounting external storage devices in the Linux shell

» Fixing software installation issues

» Troubleshooting your network connection

» Connecting using SSH

Appendix

Troubleshooting and Configuring the Raspberry Pi

M any people find that they can just connect up their Raspberry Pi, and everything works fine the first time. Fingers crossed that this will apply to you!

Sometimes people experience problems, however, or want to make more advanced changes to their computer's settings (also known as *configuring* it).

In this appendix, we show you how to resolve some common complaints and how to change some of the settings. Hopefully, you won't need to consult this appendix much, but it might prove valuable if you experience undesirable behavior when you first set up the Pi or if you have an unusual setup.

TIP

Whatever you're doing on the Raspberry Pi (or any computer, come to that), save your work regularly. If it crashes, you'll be able to pick things up from your last saved version, which will hopefully prevent you from losing too much work.

Troubleshooting the Raspberry Pi

The Raspberry Pi is reliable and has a strong ecosystem of compatible products you can depend on. If you do have a problem, we recommend you work through this checklist to try to identify the cause.

These steps are listed in a rough order of priority, with the quickest tests and simplest solutions first. You can try any of these solutions at any time, but if you respect this order (more or less), you can minimize any expense and hassle.

1. **Be patient.**

   With the huge increase in performance over the years, this is less of an issue than it was when the Pi first came out, but it's worth saying nonetheless: When your Raspberry Pi is busy, it can appear to be unresponsive, so you might think it's crashed. Often, if you wait, it recovers when it finishes its tasks. If it's not doing anything you particularly care about, you can always just restart the machine, but that loses any data in memory, and it's not a good idea to reset during operations like software installations (if you can avoid it), because it leaves them half-finished. There is also a risk of corrupting the microSD card if you don't shut down properly. Note that the Raspberry Pi has a screensaver built in, so you can recover the Pi from a blank screen by wiggling the mouse (when in the desktop environment) or pressing any key (in the command line). You can use the Shift key so that nothing appears onscreen.

2. **Restart your Raspberry Pi.**

   Very occasionally, the machine has crashed in a way that we haven't been able to replicate, so a simple reset can sometimes do the trick. If you're using a Raspberry Pi 400, try using the Fn+F10 key combination to shut down, pause a moment and then restart. On other models, remove the power, pause a moment, and then reconnect it. If you can, it's better to shut down safely (see Chapters 4 and 5).

3. **Check your connections.**

   Switch off your Raspberry Pi and make sure that all its cables are firmly fixed in the right sockets. Start with the source of the problem: For example, if the screen is blank, check the video cable; if the keyboard is unresponsive, check its connection. Connect everything before turning on your Raspberry Pi.

Chapter 3 is a guide to setting up your Raspberry Pi, including connecting its peripherals and cables.

4. Ensure that your microSD card is inserted correctly.

If your Raspberry Pi's red PWR light comes on but the green OK light does not flicker or light, the Raspberry Pi is having difficulty using the microSD card. In the first instance, check to see that the card is correctly inserted (see Chapter 3).

5. Try a new SD or microSD card.

If the red light comes on but the green one still won't, try a new card. We've occasionally had problems with SD cards or microSD cards, and with adapters that convert a microSD card to fit an SD card slot. You can find a list of SD cards that have been reported as compatible with the Raspberry Pi at https://elinux.org/RPi_SD_cards. Compatible microSD cards are also available from Raspberry Pi resellers, often with Raspberry Pi OS preinstalled.

6. Disconnect peripherals.

Try disconnecting the keyboard and mouse and then restart. Obviously, this won't help much if the problem you're experiencing requires input devices for you to replicate it, but it can help to identify any device incompatibilities that might stop the Pi from starting up correctly. If the Pi works fine without anything connected, use the process of elimination (connecting devices one at a time and restarting) to identify which one is causing problems. Try connecting the device directly, instead of using a USB hub if you have one.

7. Try different peripherals.

If possible, try a different keyboard and mouse. Official Raspberry Pi products are available and there is a list of devices at https://elinux.org/RPi_VerifiedPeripherals that are known to work. Many of the problems people experience are the result of using incompatible devices with the Raspberry Pi, so replacing the keyboard, mouse, and USB hub can resolve a wide range of apparently different problems (including a strange experience Sean had with his Internet connection not working in the desktop environment, even though it worked in the command line — the problem was an incompatible keyboard). The previous step can help you to identify which peripherals might be causing problems.

8. Try different cables.

Especially if you're having problems with the network connection and audio or visual output, try using different cables to rule out faulty cables as the cause of the problem.

9. **Try a different screen.**

   If you can't see anything on the screen but the Raspberry Pi appears to be powering up (the red light comes on and the green light flickers), try connecting to a different monitor or TV. (See Chapter 3 for advice on this.)

10. **Update your software.**

    Assuming your Internet connection is working, you can update the operating system and other software on your Raspberry Pi (without overwriting any of your work files) using this Linux command (see Chapter 5):

    ```
    sudo apt-get update && sudo apt-get upgrade
    ```

    When new Raspberry Pi models launch, a new version of Raspberry Pi OS is often required to work with them. If you're using an old microSD card with a new device, it may be incompatible. (See Chapter 2 for advice on creating a new microSD card.)

11. **Try a different power supply.**

    We've put this near the end of our steps list because it's probably hardest to do, although dodgy power has been reported to cause a wide range of different problems. If you have a friend with a Raspberry Pi and theirs works fine, try using their power supply to see whether it fixes the issues you're seeing on yours. Alternatively, you might need to buy a new power supply. Note that the Raspberry Pi 3 needs more power than earlier models even though it has a compatible power socket, so if you've upgraded your Pi but not your power supply, you might experience problems. For best results, use an official Raspberry Pi power supply, or buy one from a Raspberry Pi reseller.

12. **Check online for a solution.**

    It's not possible to cover every eventuality here, so if you're still experiencing difficulties, check the rest of this appendix and then see the troubleshooting guide at https://elinux.org/R-Pi_Troubleshooting, search the forums at www.raspberrypi.org/forums, or search the web for a solution. You're highly likely to find that someone else has already overcome any difficulties you encounter.

Troubleshooting Your Network Connection

In the desktop environment, you can easily test whether your network is working by using the web browser. In the Linux shell, you can test whether it's working with the ping command:

```
ping -c 5 www.google.com
```

This command makes five attempts to connect with Google and reports on its success. If the network is working perfectly, you should see that five packets were transmitted and five were received. Firewalls can sometimes interfere with the ping command, but this is rare. If the command works, it's a guarantee that the Pi is connected to the Internet.

You can query the network devices on your Raspberry Pi using ifconfig. This command shows you the information for eth0 (your Ethernet connection), wlan0 (your Wi-Fi connection, if available), and the local loopback, which is how the Raspberry Pi refers to itself and which you can safely ignore. If there is an inet entry (this is not the same as the inet6 entry) for eth0 or wlan0, it means your Raspberry Pi has connected to the router and been assigned an IP address successfully.

The Ethernet connection should be automatically activated, but if it isn't, you can manually activate it with this command:

```
sudo ifup eth0
```

You can deactivate the Ethernet connection using this command:

```
sudo ifdown eth0
```

Your Raspberry Pi should automatically connect to home routers using Dynamic Host Configuration Protocol (DHCP), but these tips can help you to identify where the problem lies if you experience difficulties.

TIP

If you experience network problems, try a different cable to rule out problems with the physical connection, and make sure your power supply is strong enough for the Raspberry Pi. (See Chapter 1 for more on power supplies.)

See Chapter 3 for advice on configuring your Wi-Fi connection.

Adjusting the Settings on Your Raspberry Pi

The best way to change the settings on your Raspberry Pi is using the configuration programs in the desktop or the shell. If that doesn't work. Try editing the settings file directly. In this section, we talk you through those options.

Changing the screen resolution in the settings file does not override the settings in the desktop, so try using the desktop first.

Changing settings in the desktop and Raspi-config

Try using the menus on the desktop (see Chapter 3) or running the Raspi-config program, which gives you a menu for changing some of the most frequently used options, including some that are not included in the desktop tool. You can run the program at any time using the following command in the shell:

```
sudo raspi-config
```

Raspi-config can help with

>> **Keyboard configuration:** Under Localization Options, you can select your keyboard type. You can also use the Localization options in the Raspberry Pi Configuration tool, as described in Chapter 3.

>> **Camera problems:** Ensure the camera is enabled in Raspi-config. You'll find the camera settings in the Interface Options. You can also enable the camera from the Interfaces section of the Raspberry Pi Configuration tool (see Chapter 3).

>> **Audio problems:** In System Options, you can force the audio to use the headphone jack or HDMI output.

>> **Missing space on the card:** You might need to expand the file system to use all the space available. Under Advanced Options, choose Expand Filesystem.

>> **Boot options:** Under System Options, you can choose whether to boot into the shell or the desktop and whether to require a password. You can use the System tab in the Raspberry Pi Configuration tool to manage this, too. Booting to the command line interface (CLI) goes straight to the shell.

The shell is covered in Chapter 5, but, in brief, it is the way of giving text instructions to your Raspberry Pi. You can open the shell by clicking the Terminal icon at the top of the screen in the desktop environment.

Raspi-config and the Raspberry Pi Configuration tool can make changes for you without your having to edit any configuration files, so it's more convenient than editing config.txt yourself, and there is less risk of error too. If the option you need isn't covered on the Raspi-config menu, you need to edit the configuration file manually.

Using Nano to edit config.txt

The settings that your Raspberry Pi uses are stored in files on the microSD card, and many of them are in a file called `config.txt` that's in the `/boot` directory. You can edit this file directly to change your computer's settings using a simple text editor called Nano that is preinstalled on your Raspberry Pi.

WARNING

Before you start tampering with the `config.txt` file, make sure you've backed up any important data on your Raspberry Pi. (See "Mounting External Storage Devices" later in this appendix and the section about backing up your data in Chapter 4.) There is a risk that you could, for example, render the screen display unreadable, which would make it difficult to use the Raspberry Pi to access your files.

To open the `config.txt` file in the Nano editor, enter the following command in the shell, all in lowercase:

```
sudo nano /boot/config.txt
```

The Nano text editor, with `config.txt` open, looks like Figure A-1.

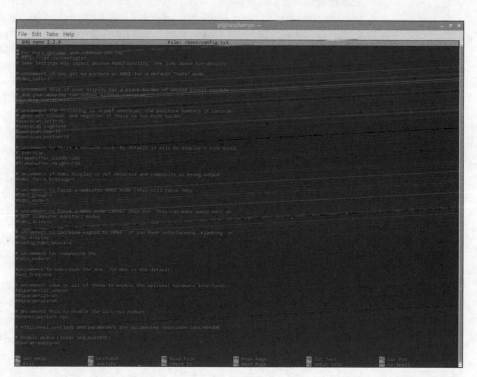

FIGURE A-1:
The Nano text editor with the config.txt file open.

Use the cursor keys to move around the document. At the bottom of the window is a menu explaining Nano's controls, where the upward arrow represents the Control key. The shortcuts here are different to what you might be used to, but the main ones you should know about are

» **Ctrl+W:** Search for a word or phrase. This option (short for Where Is?) enables you to jump straight to the configuration option you want to edit.

» **Ctrl+V:** Next page.

» **Ctrl+Y:** Previous page.

» **Ctrl+K:** Cut the current row of text.

» **Ctrl+U:** Uncut text, which means paste the text you previously cut at the cursor's location.

» **Ctrl+G:** Get help, which provides more detailed instructions.

» **Ctrl+O:** Write out, or save, the current file.

» **Ctrl+X:** Exit Nano and return to the shell.

The first thing you'll notice about `config.txt` is that the # (hash mark) symbol is used at the start of most lines. This symbol has a special meaning to the computer, which is "ignore the rest of this line." You might wonder why anyone would enter information into a computer that they want it to ignore, but this concept is often used (not often enough, some would say) to help the human users of a particular program or file. Any line with a # symbol at the start of it isn't actually doing anything, but it's there to guide you as you edit `config.txt`. Lines like this are called *comments*. (They are also used in Python, as you see in Chapter 11.)

The first two lines of settings in `config.txt` say

```
# uncomment if you get no picture on HDMI for a default "safe" mode
#hdmi_safe=1
```

The first line is obviously intended for you to read, but the second line shows the settings you need to use to turn the HDMI Safe mode on. This takes the form that all settings in `config.txt` do — namely:

```
setting_name=value
```

Each setting needs a line of its own. If you wanted to turn the HDMI Safe mode on, you would remove the comment symbol (the hash mark) before the second line, or "uncomment" that line, so that the first two lines now read

```
# uncomment if you get no picture on HDMI for a default "safe" mode
hdmi_safe=1
```

WARNING

Don't remove the # symbol from the line of instructions. It remains a comment that's intelligible only to human readers. You should remove the # symbol only from lines you want the computer to do something with.

Just taking out that single hash mark makes all the difference! Save the file (Ctrl+O) and reboot the computer, and Safe mode is activated. You can reboot the Raspberry Pi with the following command:

```
sudo reboot
```

If you need to disable a setting, you can just put a # symbol in front of it again to turn its line into a comment that the computer will ignore.

TIP

You can add your own comments too. It's a good idea to add a line starting with a # symbol to remind yourself what you changed and when, in case you need to change the settings back later.

REMEMBER

You can change multiple settings at the same time, but each setting must be on its own line.

The config.txt file has lots of comments to tell you what to change for various settings. We don't have space to document them all here. You can find a detailed list at https://elinux.org/RPiconfig.

Fixing Audio Problems

In the desktop environment, you can use the Speaker icon in the top right to adjust the volume of sound and mute or unmute it. In the command line, you can call up a utility to adjust the sound using this instruction:

```
sudo alsamixer
```

Use the cursor keys or the mouse scroll wheel to adjust the volume level.

The sound output device is automatically detected. Note that if you're using HDMI to connect to a screen, the audio might be directed there by default, even if the monitor does not have speakers. You can use the audio options in Raspi-config (found under System Options) to direct the audio to the headphone jack or the

HDMI cable. In LibreELEC (see Chapter 8), you can click the cogwheel at the top of the menu to go into Settings, choose System Settings, and click Audio to find the option to change the output device.

Fixing Software Installation Issues

The apt package manager should enable you to cleanly install and remove software. If software isn't working, try removing it and then reinstalling it as described in Chapter 5.

Packages often require other packages (called *dependencies*) to work. The package manager looks after these dependencies for you, but if they get broken, you can fix dependencies using the following command:

```
sudo apt -f install
```

Mounting External Storage Devices

When you plug in an external storage device such as a USB key or flash drive, the desktop environment recognizes it automatically and opens it in File Manager for you. Not so when using the shell. You need to mount the device yourself, which means you need to connect the device to a folder in the directory tree where you want to browse its contents.

TIP

If your only goal is to back up your data to an external storage device, it's probably easier to use File Manager or the SD Card Copier application in the desktop environment (see Chapter 4).

To use external storage in the shell, you first need to create a directory that will be the mount point for the USB key, which means when you look in that directory, you are actually looking at the contents of the external storage device. You can reuse this directory, but the first time you mount a device, you need to create the directory. You can create this directory anywhere (including inside your home directory), but it's conventional to mount temporary devices in the /mnt directory:

```
sudo mkdir /mnt/usbdrive
```

Next, you need to investigate the device you're connecting. To do that, connect your storage device and then enter this command:

```
sudo fdisk -l
```

The last character of this command is a letter *l* (lowercase *L*), and not a number 1. At the end, after a number of entries for RAM devices, the output should look something like this:

```
Disk /dev/mmcblk0: 14.9 GiB, 15931539456 bytes, 31116288 sectors
Units: sectors of 1 * 512 = 512 bytes
Sector size (logical/physical): 512 bytes / 512 bytes
I/O size (minimum/optimal): 512 bytes / 512 bytes
Disklabel type: dos
Disk identifier: 0xcbd582ec


Device         Boot  Start      End  Sectors   Size Id Type
/dev/mmcblk0p1        8192   532479   524288   256M  c W95 FAT32 (LBA)
/dev/mmcblk0p2      532480 31116287 30583808  14.6G 83 Linux


Disk /dev/sda: 59 GiB, 63333990400 bytes, 123699200 sectors
Disk model: Rainbow Line
Units: sectors of 1 * 512 = 512 bytes
Sector size (logical/physical): 512 bytes / 512 bytes
I/O size (minimum/optimal): 512 bytes / 512 bytes
Disklabel type: dos
Disk identifier: 0x59e52127


Device    Boot Start       End    Sectors Size Id Type
/dev/sda1 *       64 123699199 123699136  59G  7 HPFS/NTFS/exFAT
```

This lists the different storage devices that are connected to the Pi. In the preceding example, you can see the first disk (Disk /dev/mmcblk0) is 14.9GiB MB, which is a 16GB SD card, and the second one (Disk /dev/sda) is 59GiB, which is a 60GB USB key we've connected. The important information we need from this is the device name and the partition number, which is shown at the bottom of the output and is sda1.

To mount the drive for the user pi (uid=pi) and the group pi (gid=pi), we then use

```
sudo mount -o uid=pi,gid=pi /dev/sda1 /mnt/usbdrive
```

To view the contents of the USB key, you can then use

```
ls /mnt/usbdrive
```

To back up your home directory to the USB key, use

```
cp -R ~/* /mnt/usbdrive
```

Index

About the Authors

Sean McManus: Sean is an expert technology and business author. His other books include *Mission Python* (No Starch Press), *Coder Academy* (Kane Miller Books), *Cool Scratch Projects in Easy Steps* (In Easy Steps Limited), *Scratch Programming in Easy Steps* (In Easy Steps Limited), and *Web Design in Easy Steps* (In Easy Steps Limited). His novel for adults, *Earworm*, goes undercover in the music industry, exposing a conspiracy to replace bands with computer-generated music. His tutorials and articles have appeared in magazines including *The MagPi*, *Internet Magazine*, *Internet Works*, *Business 2.0*, *Making Music*, and *Personal Computer World*. He has been a Code Club volunteer, helping children at a local school to learn computer programming. Visit his website at www.sean.co.uk for bonus content from his books.

Mike Cook: Mike has been making electronic things since he was in school. A former senior lecturer in physics at Manchester Metropolitan University, he wrote more than 300 computing and electronics articles in the pages of computer magazines for 20 years starting in the 1980s. After leaving the university after 21 years when the physics department closed down, he got a series of proper jobs where he designed digital TV set-top boxes and access control systems. In 2015, he started writing a monthly column in *MagPi* magazine; he has covered 73 projects to date. His other books include *Raspberry Pi Projects* (Wiley), *Raspberry Pi Projects For Dummies* (Wiley), and *Arduino Music and Audio Projects* (Apress). He also works with Drake Music Labs North, a charity for disabled musicians, developing accessible music equipment. Now retired and freelancing, he spends his days surrounded by wires, patrolling the forums as Grumpy Mike.

Dedication

To my wife, Karen, with thanks for all her support throughout this project and always. And to Leo, our wonderful son.

—Sean

To my wife, Wendy, who always acts delighted whenever I show her yet another blinking LED. And also to the late Leicester Taylor, World War II radar researcher and inspirational supervisor of my post-graduate research at the University of Salford.

—Mike

Authors' Acknowledgments

Sean McManus: Thank you to my coauthor, Mike, for bringing his electronics expertise and fantastic project ideas. Thank you to Kelsey Baird, our acquisitions editor on this edition, and previous acquisitions editors Craig Smith and Katie Mohr. Thanks to Elizabeth Kuball, our editor on this edition, and previous editors Linda Morris, Paul Levesque, and Becky Whitney.

Our technical editors, Guy Hart-Davis (this edition), Jason E Geistweidt (3rd edition), Ryan Walmsley (2nd edition), and Paul Hallett (1st edition) cast a careful eye over the text and code and made much appreciated suggestions. Olivier Engler, who translated the first edition into French, provided helpful feedback, too. Thanks also to Lorna Mein and Natasha Lee in marketing, and to the *For Dummies* team for making it all happen.

Many people helped with research or permissions requests, including Karen McManus, Sam Aaron, Eben Upton, Liz Upton, Leo McHugh, Mark Turner, Peter Sayer, John Hartnup, Bill Kendrick, Simon Cox, Jon Williamson, Paul Beech, Peter de Rivaz, Michał Męciński, Ruairi Glynn, Stephen Revill, Lawrence James, Bram Stolk, Adam Kemeny, Will Jessop, David Bryan, Pimoroni, and Pi Supply. We wouldn't have a book to write if it weren't for the wonderful work of the Raspberry Pi Foundation, the manufacturers who took a gamble on it, and the many thousands of people who have contributed to the Raspberry Pi's software.

Mike Cook: I would like to thank Sean McManus for inviting me to contribute to this book and the staff at Wiley for making the process of producing this book as painless as possible.

Publisher's Acknowledgments

Acquisitions Editor: Kelsey Baird
Project Editor: Elizabeth Kuball
Copy Editor: Elizabeth Kuball
Technical Editor: Guy Hart-Davis

Production Editor: Vivek Lakshmikanth
Cover Image: Courtesy of Mike Cook; Raspberry Pi logo courtesy of Raspberry Pi Foundation

Leverage the power

Dummies is the global leader in the reference category and one of the most trusted and highly regarded brands in the world. No longer just focused on books, customers now have access to the dummies content they need in the format they want. Together we'll craft a solution that engages your customers, stands out from the competition, and helps you meet your goals.

Advertising & Sponsorships

Connect with an engaged audience on a powerful multimedia site, and position your message alongside expert how-to content. Dummies.com is a one-stop shop for free, online information and know-how curated by a team of experts.

- Targeted ads
- Video
- Email Marketing

- Microsites
- Sweepstakes sponsorship

20 **MILLION**
PAGE VIEWS
EVERY SINGLE MONTH

15
MILLION
UNIQUE
VISITORS PER MONTH

43%
OF ALL VISITORS
ACCESS THE SITE
VIA THEIR MOBILE DEVICES

700,000 NEWSLETTER
SUBSCRIPTIONS
TO THE INBOXES OF
300,000 UNIQUE **INDIVIDUALS EVERY WEEK**

# of dummies

Custom Publishing

Reach a global audience in any language by creating a solution that will differentiate you from competitors, amplify your message, and encourage customers to make a buying decision.

- Apps
- Books
- eBooks
- Video
- Audio
- Webinars

Brand Licensing & Content

Leverage the strength of the world's most popular reference brand to reach new audiences and channels of distribution.

For more information, visit dummies.com/biz

PERSONAL ENRICHMENT

Staying Sharp
9781119187790
USA $26.00
CAN $31.99
UK £19.99

Facebook
9781119179030
USA $21.99
CAN $25.99
UK £16.99

Guitar
9781119293354
USA $24.99
CAN $29.99
UK £17.99

Investing
9781119293347
USA $22.99
CAN $27.99
UK £16.99

Beekeeping
9781119310068
USA $22.99
CAN $27.99
UK £16.99

Digital Photography
9781119235606
USA $24.99
CAN $29.99
UK £17.99

Meditation
9781119251163
USA $24.99
CAN $29.99
UK £17.99

Pregnancy
9781119235491
USA $26.99
CAN $31.99
UK £19.99

Samsung Galaxy S7
9781119279952
USA $24.99
CAN $29.99
UK £17.99

iPhone
9781119283133
USA $24.99
CAN $29.99
UK £17.99

Crocheting
9781119287117
USA $24.99
CAN $29.99
UK £16.99

Nutrition
9781119130246
USA $22.99
CAN $27.99
UK £16.99

PROFESSIONAL DEVELOPMENT

Windows 10
9781119311041
USA $24.99
CAN $29.99
UK £17.99

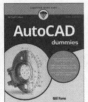
AutoCAD
9781119255796
USA $39.99
CAN $47.99
UK £27.99

Excel 2016
9781119293439
USA $26.99
CAN $31.99
UK £19.99

QuickBooks 2017
9781119281467
USA $26.99
CAN $31.99
UK £19.99

macOS Sierra
9781119280651
USA $29.99
CAN $35.99
UK £21.99

LinkedIn
9781119251132
USA $24.99
CAN $29.99
UK £17.99

Windows 10 All-in-One
9781119310563
USA $34.00
CAN $41.99
UK £24.99

SharePoint 2016
9781119181705
USA $29.99
CAN $35.99
UK £21.99

Fundamental Analysis
9781119263593
USA $26.99
CAN $31.99
UK £19.99

Networking
9781119257769
USA $29.99
CAN $35.99
UK £21.99

Office 2016
9781119293477
USA $26.99
CAN $31.99
UK £19.99

Office 365
9781119265313
USA $24.99
CAN $29.99
UK £17.99

Salesforce.com
9781119239314
USA $29.99
CAN $35.99
UK £21.99

Coding
9781119293323
USA $29.99
CAN $35.99
UK £21.99

dummies.com

dummies
A Wiley Brand

Learning Made Easy

ACADEMIC

9781119293576
USA $19.99
CAN $23.99
UK £15.99

9781119293637
USA $19.99
CAN $23.99
UK £15.99

9781119293491
USA $19.99
CAN $23.99
UK £15.99

9781119293460
USA $19.99
CAN $23.99
UK £15.99

9781119293590
USA $19.99
CAN $23.99
UK £15.99

9781119215844
USA $26.99
CAN $31.99
UK £19.99

9781119293378
USA $22.99
CAN $27.99
UK £16.99

9781119293521
USA $19.99
CAN $23.99
UK £15.99

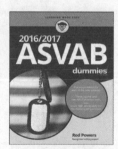

9781119239178
USA $18.99
CAN $22.99
UK £14.99

9781119263883
USA $26.99
CAN $31.99
UK £19.99

Available Everywhere Books Are Sold

dummies.com